AF361988

Application of Power Electronics Converters in Smart Grids and Renewable Energy Systems

Application of Power Electronics Converters in Smart Grids and Renewable Energy Systems

Editors

Irfan Ahmad Khan
S. M. Muyeen

MDPI • Basel • Beijing • Wuhan • Barcelona • Belgrade • Manchester • Tokyo • Cluj • Tianjin

Editors
Irfan Ahmad Khan
Texas A&M University
USA

S. M. Muyeen
Qatar University
Qatar

Editorial Office
MDPI
St. Alban-Anlage 66
4052 Basel, Switzerland

This is a reprint of articles from the Special Issue published online in the open access journal *Energies* (ISSN 1996-1073) (available at: https://www.mdpi.com/journal/energies/special_issues/Application_Power_Electronic_Converter).

For citation purposes, cite each article independently as indicated on the article page online and as indicated below:

LastName, A.A.; LastName, B.B.; LastName, C.C. Article Title. *Journal Name* **Year**, *Volume Number*, Page Range.

ISBN 978-3-0365-4373-4 (Hbk)
ISBN 978-3-0365-4374-1 (PDF)

Contents

Article

Simulation Study of Allied In-Situ Injection and Production for Enhancing Shale Oil Recovery and CO_2 Emission Control

Haiyang Yu [1],*, Songchao Qi [1], Zhewei Chen [1], Shiqing Cheng [1], Qichao Xie [2] and Xuefeng Qu [2]

[1]   State Key Laboratory of Petroleum Resources and Prospecting, China University of Petroleum, Beijing 102249, China; qisongchao95@163.com (S.Q.); zhewei.chen@outlook.com (Z.C.); chengsq973@163.com (S.C.)

[2]   Research Institute of Exploration and Development, Petro China Changqing Oilfield Company, Xi'an 710018, China; xqc_cq@petrochina.com.cn (Q.X.); qxf_cq@petrochina.com.cn (X.Q.)

*   Correspondence: haiyangyu.cup@139.com

Received: 20 September 2019; Accepted: 15 October 2019; Published: 18 October 2019

Abstract: The global greenhouse effect makes carbon dioxide (CO_2) emission reduction an important task for the world, however, CO_2 can be used as injected fluid to develop shale oil reservoirs. Conventional water injection and gas injection methods cannot achieve desired development results for shale oil reservoirs. Poor injection capacity exists in water injection development, while the time of gas breakthrough is early and gas channeling is serious for gas injection development. These problems will lead to insufficient formation energy supplement, rapid energy depletion, and low ultimate recovery. Gas injection huff and puff (huff-n-puff), as another improved method, is applied to develop shale oil reservoirs. However, the shortcomings of huff-n-puff are the low sweep efficiency and poor performance for the late development of oilfields. Therefore, this paper adopts firstly the method of Allied In-Situ Injection and Production (AIIP) combined with CO_2 huff-n-puff to develop shale oil reservoirs. Based on the data of Shengli Oilfield, a dual-porosity and dual-permeability model in reservoir-scale is established. Compared with traditional CO_2 huff-n-puff and depletion method, the cumulative oil production of AIIP combined with CO_2 huff-n-puff increases by 13,077 and 17,450 m^3 respectively, indicating that this method has a good application prospect. Sensitivity analyses are further conducted, including injection volume, injection rate, soaking time, fracture half-length, and fracture spacing. The results indicate that injection volume, not injection rate, is the important factor affecting the performance. With the increment of fracture half-length and the decrement of fracture spacing, the cumulative oil production of the single well increases, but the incremental rate slows down gradually. With the increment of soaking time, cumulative oil production increases first and then decreases. These parameters have a relatively suitable value, which makes the performance better. This new method can not only enhance shale oil recovery, but also can be used for CO_2 emission control.

Keywords: allied in-situ injection and production (AIIP); CO_2 huff and puff; shale oil reservoirs; enhanced oil recovery

---

1. Introduction

The global greenhouse effect makes CO_2 emission reduction an important task for the world, however, CO_2 can be applied to develop shale oil reservoirs. Shale oil reservoirs are related to high total organic carbon (TOC), with an estimated reserve that is equal to 345 billion barrels of oil worldwide [1,2]. Although shale oil is a potential unconventional energy resource with huge reserves, it is difficult to exploit efficiently due to its small pore throats and ultra-low permeability of micro to nano Darcy [3]. At present, long horizontal wells and multi-stage hydraulic fracturing technology are used to develop shale oil reservoirs [4], which makes economic development of shale oil reservoirs possible. However, the primary oil recovery factor is usually less than 8% due to the ultra-low matrix permeability [5], leading to large amounts of crude oil remaining in the shale formation. Horizontal well depletion development begins with a high production rate, then quickly declines and stabilizes at the low production rate [6]. Therefore, it is extremely important to inject fluid into the formation, supplementing formation energy. Nevertheless, in shale oil reservoirs, water is hard to inject into the formation due to extremely low porosity and permeability, leading to high injection pressure and low injection rate [7]. The existence of natural and hydraulic fractures aggravates the heterogeneity of shale reservoirs, which makes injected water tend to inflow high permeability channels, resulting in much limited sweep efficiency and incremental water cut [8].

For shale oil reservoirs, gas injection is easier to inject into the formation than water injection or other fluids. Commonly injected gases include carbon dioxide, hydrocarbon gases, nitrogen, and gas mixtures [9]. Cheng studied the effect of different gas injection on oil recovery through experiments. Experimental results show that under the same conditions, the oil recovery of CO_2 is the highest, CH_4 is the second, and N_2 is the lowest. Similar results are obtained from numerical simulation [10]. Therefore, it is generally believed that carbon dioxide is a more suitable gas type for gas injection [11]. CO_2 injection is currently used in 65% of the world's gas-injection projects [12]. The main mechanism of CO_2 flooding is that CO_2 is miscible with crude oil through multiple contacts. When carbon dioxide flows in porous media, it can cause the change of components between carbon dioxide and crude oil then generate miscible fluid. This multi-contact miscible process is a phase equilibrium process with the change of reservoir temperature, pressure, and crude oil composition [13]. Immiscible CO_2 flooding also produces oil based on the principle of phase equilibrium. The mechanism of CO_2 injection for enhancing recovery chiefly includes the following aspects [14]: (1) increasing formation pressure; (2) reducing oil viscosity; (3) inflating crude oil; (4) extracting the light component of crude oil; and (5) reducing interfacial tension.

The low sweep efficiency caused by gas channeling leads to the insufficient effect of CO_2 injection [15], hence some experts put forward the method of gas injection huff-n-puff to improve the effect of gas injection. Song used experimental techniques to estimate the performance of the CO_2 huff-n-puff process in Bakken tight oil reservoir [16]. A string of displacement experiments was conducted with Bakken tight core samples to compare water flooding, immiscible CO_2 huff-n-puff, near-miscible CO_2 huff-n-puff, and miscible CO_2 huff-n-puff. Compared with water flooding, immiscible CO_2 huff-n-puff has a higher recovery of 51.5%, while near-miscible and miscible CO_2 huff-n-puff have a higher recovery of 61% and 63%, respectively. Pu investigated CO_2 huff-n-puff process through core experiments, indicating that CO_2 huff-n-puff is a viable technique to enhance oil recovery (EOR) [17]. Janiga Damian conducted laboratory core experiments to obtain valuable information. At the same time, experiments were combined with the numerical representation of core samples to generate a reliable model for process optimization [18]. By matching experimental data, Song further evaluated the performance of oilfield-scale CO_2 huff-n-puff through numerical simulation [16]. Based on the historical matching model, the development parameters such as gas injection volume, soaking time, and production pressure are optimized in the CO_2 huff-n-puff process. Pavel Zuloaga simulated and analyzed the effect of the CO_2 huff-n-puff process on enhanced oil recovery by using a numerical model with oilfield-scale [19]. The sensitivity analyses of four parameters, such as well number, well pattern, matrix permeability, and half-length of fracture, were studied through simulation design. The results showed that the influence of matrix permeability is the largest, followed by the influence of well pattern, half-length of fracture and well number. Zhang Yuan studied the effects of matrix permeability, CO_2 injection rate, and stress-dependent deformation mechanism on the CO_2 huff-n-puff process by numerical simulation [20]. It is concluded that the effects of CO_2 diffusion and nano-pore constraints on enhanced oil recovery should be considered into the reservoir simulation model. The CO_2 huff-n-puff process is notably affected by CO_2 injection rate, matrix permeability, and stress-related deformation mechanism. Besides, Daniel Sanchez-Rivera also optimized the huff-and-puff process in the Bakken shale through numerical reservoir simulation [21]. It is found that the final recovery factor will be reduced if the huff-n-puff is carried out too early, while shorter soaking time is preferable to longer soaking time. Natural fracture networks make huff-n-puff better, allowing the injection fluid to penetrate the formation and come into contact with more fluids.

Although gas injection huff-n-puff improves development efficiency, its coverage is limited, only around horizontal wells [22]. Chen et al. put forward the method of multi-stage injection-production between fractures in the same fractured horizontal well [23]. The injection-production separation device and injection-production valve are used to achieve the injection-production between fractures. Odd hydraulic fractures inject fluid and even hydraulic fractures produce fluid, which turns inter-well displacement into inter-fracture displacement. The new technology has a higher oil production than the CO_2 huff-n-puff, with longer stable production period, smaller decline rate, and better performance. Based on the principle of inter-fracture injection and production in the same well, Yu et al. proposed the technology of asynchronous injection alternating production for different wells [24]. Two horizontal wells, as an injection-production unit, can be designed as symmetrical distribution of injection fracture and dislocation distribution of injection fracture according to the location of injection fracture. At the same time, the working system of synchronous or asynchronous injection-production can be adopted. This method further expands the sweep range.

Although other experts have proposed some innovative exploit methods with great theoretical value, it is difficult to implement them in oil fields because of the complex technology and high production cost [25]. This paper firstly proposes to use Allied In-Situ Injection and Production (AIIP) technology combined with CO_2 huff-n-puff to develop shale oil reservoirs, which can not only improve shale oil recovery, but also be used for CO_2 emission control.

2. Allied In-Situ Injection and Production Technology

2.1. Principles of AIIP

In previous work, our research team proposed a method for improving oil recovery in tight oil reservoirs, Cumulative In-situ Injection and Production (CIIP) or inter-fracture injection and production (IFIP) [23,25]. The CIIP method divides a multi-stage hydraulic fracturing horizontal well into several parts. As shown in Figure 1, some of the fractures inject fluids (water or gas) into the formation and the neighboring fractures produce oil. There are three periods for CIIP process: (1) injection period: production fractures are closed and gas is injected through injection fractures; (2) soaking period: both production and injection fractures are closed and soak for several days; (3) production period: production segments are opened to produce oil while injection segments remain closed. In previous simulation work, we studied the feasibility of CIIP for improving oil recovery in tight oil reservoirs. CIIP has better EOR performance than traditional CO_2 huff-n-puff due to higher displacement efficiency among hydraulic fractures. Although CIIP can significantly improve the performance of gas injection, there are several disadvantages: first, CIIP can improve oil recovery in the stimulated reservoir volume (SRV) area, but CIIP has limited displacement efficiency for the area far from the SRV (NSRV) zone; second, the implementation for CIIP is complicated and expensive, it requires high level quality for modification; third, with the existence of complex natural fractures and micro hydraulic fractures in SRV area, gas breakthrough might easily occur during CIIP process, resulting in limited sweep efficiency [26].

In order to improve the limitation of CIIP and make an easier way to modify the horizontal well, Allied In-Situ Injection and Production Technology (AIIP) is proposed. AIIP is a novel approach, conducting fluid injection and oil production in different segments of the same well. Figure 2 illustrates the design of AIIP graphically. Instead of dividing horizontal wells into several parts, AIIP divides hydraulic fractures in a horizontal well into three parts: the adjacent wellhead part as the production section (Red segment), the far wellhead part as the fluid injection section (Blue segment), and the middle part as the packer section between the production section and the injection section (Grey segment). AIIP needs to design special strings and devices for injection and production in the same well. The packer section needs to inject gelling agent to seal the formation and prevent mutual interference of injection and production in the same well. Figure 3 illustrates the well pattern and schedule of AIIP. The production section accounts for about two-thirds of the total length of the horizontal section and is suitable for the use of long hydraulic fractures. The injection section accounts for about one-third of the total length of the horizontal section and is suitable for the use of short hydraulic fractures. There are three periods for AIIP process: (1) injection period: the production segments are closed and gas is injected through injection segment; (2) soaking period: both of the production and injection segments are closed and soak for several days, like huff-n-puff; (3) production period: the production segments are opened to produce oil while the injection segments remain closed. Compared with the CIIP method, AIIP method has an easier implementation procedure and cheaper cost than CIIP. Additionally, AIIP can not only enhance oil recovery in the NSRV area, but ignore the potential possibility of gas breakthrough among hydraulic fractures. These advantages remedy the shortcoming of CIIP. Therefore, AIIP is a promising and more practical method to improve oil recovery in tight oil reservoirs than CIIP.

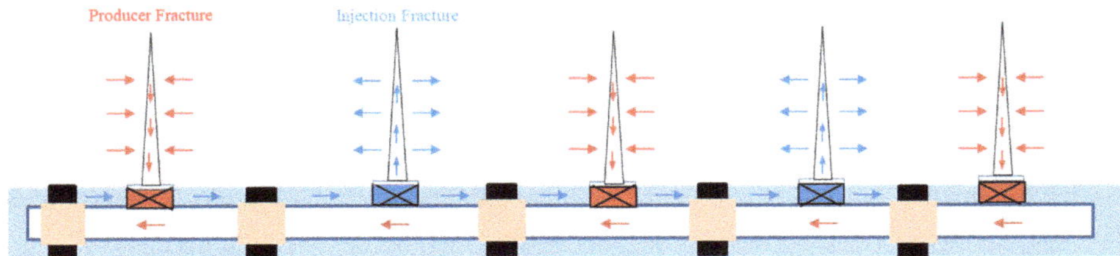

Figure 1. Sketch of Cumulative In-situ Injection and Production (CIIP).

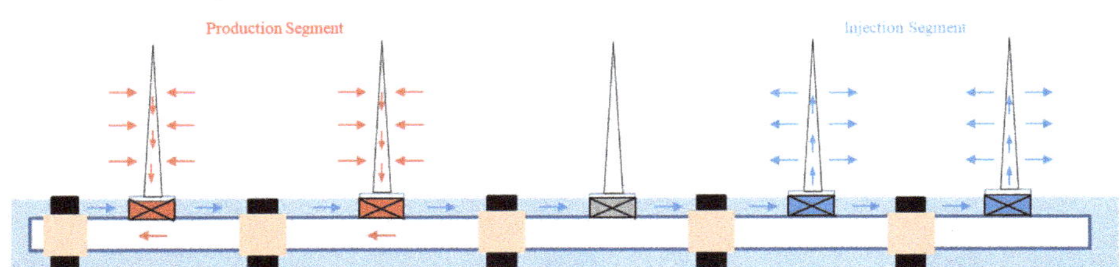

Figure 2. Sketch of Allied In-Situ Injection and Production Technology (AIIP).

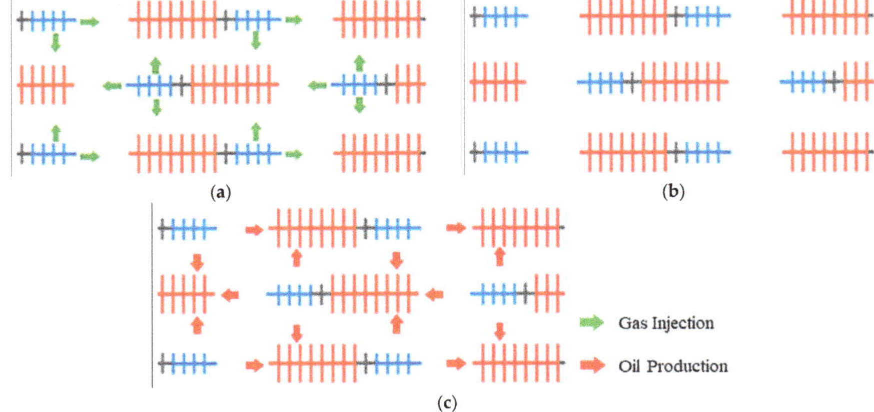

Figure 3. Design schematic diagram of AIIP. (**a**) Injection period; (**b**) soaking period; (**c**) production period.

2.2. Implementation Procedure

AIIP can be used for displacement or huff-n-puff, as well as for different injection fluids. Therefore, it is feasible to develop shale oil reservoirs by combining AIIP with CO_2 huff-n-puff. In order to meet the requirements of injection and production in the same well, the packer device enables the tubing to be used for production and the tubing-casing annulus to be used for fluid injection. The electric valves control well injection volume and production volume to meet the production requirements of tight and shale oil reservoirs. Besides, the flowmeter can monitor the injection rate in real time, and then technicians can optimize the injection and production on the ground. For example, when gas channeling occurs, the injection rate can be reduced or the valve can be closed for some time. The explicit implementation process of AIIP can follow the steps below.

(1) Select an appropriate block for drilling a series of horizontal wells where no fault exists nearby.
(2) Horizontal wells use the multi-stage hydraulic fracturing technique, and well logging or micro-seismic monitoring is used for obtaining fracture conditions.
(3) The reservoir undergoes depletion until the pressure drops bubble point pressure.
(4) Insert special pipe string and device, then select appropriate fractures as production section and injection section respectively.
(5) Select a section between the production section and the injection section as the packer section and inject the gelling agent into it to isolate the formation.
(6) First, open the electric injection valve to inject a certain amount of carbon dioxide. Then close the injection valve and soak the well for some time. In the end, open the production valve and begin oil production.

For ensuring the performance of AIIP, it is necessary to identify the distribution of faults and fractures through micro-seismic monitoring, so as to prevent horizontal wells from being arranged in areas with faults or dense fractures. As a novel technology, AIIP has higher well completion requirements than conventional horizontal wells. First of all, it is necessary to put down tubing and packers to ensure the injection and production of fluid in the same well. At the same time, the electric injection valve is an important component, which must ensure that the valve can work durably and steadily in the underground.

3. Numerical Simulation

3.1. Reservoir Characteristics

In order to evaluate the performance of AIIP combined with CO_2 huff-n-puff in shale oil reservoirs, this paper chooses the shale oil reservoir in Z Sag of Shengli Oilfield as the target block and establishes a numerical simulation model. Z Sag belongs to a third-level tectonic unit in Bohai Bay Basin and is located in the northeast of J depression with an area of about 2800 km^2 [27]. The whole is distributed in a trumpet-like shape from southwest to northeast, and the depression can be subdivided into several secondary tectonic units. The reservoir depth is about 3025–3075 m and the pay thickness is 20–30 m. The lithology is mainly argillaceous limestone with a high brittle mineral content of 81.6% on average. The average pore volume of the reservoir is 4.1 μL/g, and the pore connectivity is good. The average proportion of centrifugal movable shale oil is as high as 38.80%. Other reservoir parameters are listed in Table 1, including effective reservoir thickness, porosity, permeability, oil saturation, formation pressure, formation temperature, etc. In summary, this reservoir is a favorable section for shale oil exploration in this area [28].

The shale oil density of 0.79–0.954 g/cm^3, the gas-oil ratio of 5–350.1 m^3/m^3, is medium crude oil. The viscosity of crude oil is 0.4–40 mPa·s (at the temperature of 20 °C and the pressure of 0.1 MPa). The bubble point pressure of the crude oil is 17 MPa. Formation fluids are difficult to flow due to the poor reservoir physical properties, but formation overpressure can provide energy for flowing. Initial oil saturation averages 60%. The immovable oil mainly exists in the small pore, while the movable oil mainly exists in the large pore [29]. The salinity of formation water is 10,000–20,000 mg/L and the formation water is a $CaCl_2$ type.

Table 1. Reservoir properties.

Variable Name	Value
Reservoir depth (m)	3025–3075
Pay thickness (m)	20–30
Average porosity (%)	5.56
Average permeability (mD)	0.002
Initial oil saturation (%)	60
Formation pressure (MPa)	53
Formation temperature (°C)	124
Average content of TOC (%)	4.8
Grade of maturity (%)	0.7–0.9
Crude oil density (g/cm^3)	0.79–0.954
Viscosity of crude oil (mPa·s)	0.4–40
Bubble point pressure (MPa)	17

3.2. Model Description

This work uses CMG simulation software. The compositional model was built by CMG GEM model. The CMG-WINPROP module was used for oil component lumping, according to the fluid properties of the target reservoir. In order to speed up the numerical simulation and ensure the accuracy of the calculation results, we divided the crude oil components into seven pseudo-components by analyzing the data of crude oil components in the study area. The data of seven pseudo-components are listed in Table 2 and Figure 4 shows the pressure-temperature (P-T) phase diagram of crude oil. Based on the above reservoir characteristics and properties, a model using the number of $258 \times 49 \times 1$ grid blocks with dimensions of $15 \times 15 \times 20$ m was built (Table 3). Considering the characteristics of natural fractures reservoir, according to reservoir parameters in the study area and referring to the commonly used methods for numerical simulation of shale oil reservoirs in the United States [30,31], a dual-porosity and dual-permeability model with reservoir-scale was established to simulate and evaluate the performance of AIIP combined with CO_2 huff-n-puff in shale oil reservoirs. According to the development experience of shale reservoirs at home and abroad, the study area adopted a long horizontal well row for development (Figure 5a) [32]. For simulating AIIP's process, two additional horizontal wells were arranged in the same location of one horizontal well. Producer with whole horizontal length (e.g., Well-5 in Figure 5b) was used for simulating oil production during primary stage. Producer with around 2/3 of horizontal length (e.g., Producer-5 in Figure 5b) was used for simulating oil production through production segment during AIIP stage. Injector with around 1/3 of horizontal length (e.g., Injector-5 in Figure 5b) was used for simulating gas injection through the injection segment during AIIP stage. Hydraulic fractures were simulated by using the Log Grid Refinement (LGR) method (Table 4). At the initial stage, the method of depletion development was adopted. After the production declines, the AIIP development was adopted.

Table 2. Data of seven pseudo-components.

Component	N$_2$	CO$_2$	C$_1$	C$_2$–C$_3$	C$_4$–C$_6$	C$_7$–C$_{10}$	C$_{11}{}^+$
Content (%)	3.22	0.37	26.83	18.53	13.20	18.15	19.70

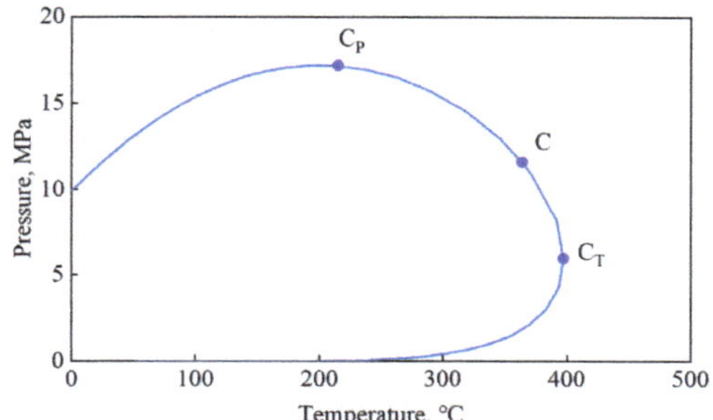

Figure 4. P-T phase diagram of crude oil.

Table 3. Numerical simulation model parameters.

Model Parameters	Value
Number of grid blocks, x y z	$258 \times 49 \times 1$
Dimensions, x y z (m)	$15 \times 15 \times 20$
Length of horizontal well (m)	1500
Horizontal well spacing (m)	300
Row spacing of horizontal wells (m)	360
Long/Short hydraulic fracture half-length (m)	130/70
Hydraulic fracture spacing (m)	105
Hydraulic fracture conductivity (mD·m)	50
Matrix porosity (%)	5.56
Matrix permeability (mD)	0.002
Fracture porosity (%)	0.2
Fracture permeability (mD)	0.02
Natural fracture spacing (m)	30
Water saturation (%)	40
Initial reservoir pressure (MPa)	53
Minimum bottom-hole pressure (MPa)	25
Maximum injection pressure (MPa)	60

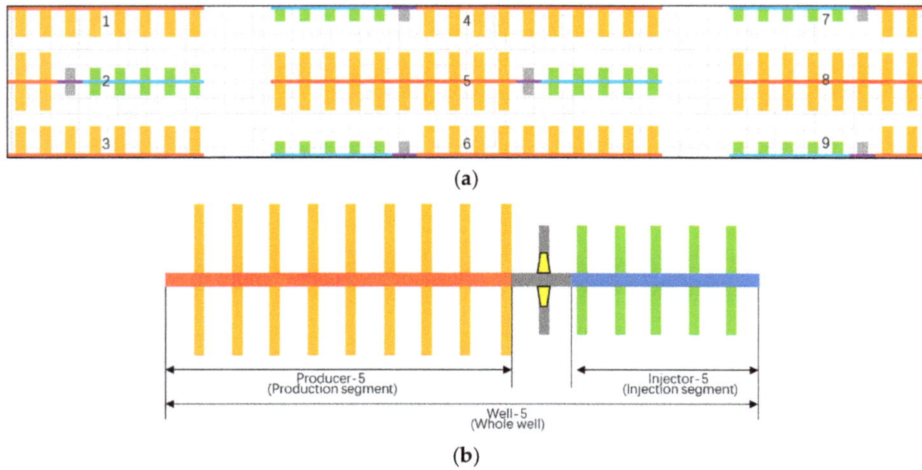

Figure 5. Diagram of simulation model. (**a**) Injection-production well pattern in the study area; (**b**) sketch for three horizontal wells in the same location in the model.

Table 4. Parameters for the Log Grid Refinement (LGR) method.

Parameters	Value
Fracture width (m)	0.003
Intrinsic permeability (mD)	10,000
Effective permeability (mD)	15
Grid cell width (m)	2
Number of refinements in the I/J/K direction	$5 \times 5 \times 1$

3.3. Case Comparison

In order to compare the performance of depletion, CO_2 huff-n-puff, and AIIP, we set up three cases for comparison.

(1)　Case A: Depletion for 12 years.
(2)　Case B: Depletion for 3 years, CO2 huff-n-puff development for 4 years, depletion development for 5 years.
(3)　Case C: Depletion for 3 years, AIIP development for 4 years, depletion for 5 years.

Depletion produces, with the minimum bottom-hole flowing, a pressure of 25 MPa and the maximum liquid production of 30 m^3/d. The difference of schedule is that AIIP and CO_2 huff-n-puff require cyclic gas injection and huff-n-puff. The schedule of CO_2 huff-n-puff and AIIP development is shown in Table 5. Finally, the injection and production parameters are optimized through sensitivity analysis, including injection rate, injection volume, soaking time, hydraulic fracture half-length, and hydraulic fracture spacing.

Table 5. Schedule of CO_2 huff-n-puff and AIIP development.

Parameters	Value
Minimum bottom-hole pressure (MPa)	25
Maximum liquid production (m^3/d)	60
Gas injection volume per well per cycle (10^4 m^3)	150
Daily gas injection volume of the single well (10^4 m^3)	5
Gas injection period duration (month)	1
Soaking period duration (month)	1
Production period duration (month)	4

4. Results and Discussion

4.1. EOR Performance of AIIP

Figure 6 compares the performances of AIIP, CO_2 huff-n-puff, and depletion development. The cumulative oil production of depleted development in the single well is 30,510 m^3, that of CO_2 huff-n-puff development in the single well is 34,883 m^3, and that of AIIP development in the single well is 47,960 m^3. The cumulative oil production of AIIP in the single well is the highest. Under the same CO_2 injection volume in the single well, the AIIP method has a better formation energy supplement effect and higher cumulative oil production in the single well because of short injection horizontal section length and high injection pressure (Figure 7). With the increment of huff-n-puff cycles, the oil production of the traditional horizontal well using gas injection huff-n-puff method decreases in each cycle, but the oil production of the AIIP method increases first and then decreases, and the oil production of each cycle is greater than that of the traditional huff-n-puff method (Figure 8). Through the above comprehensive comparison, the performance of AIIP method is the best.

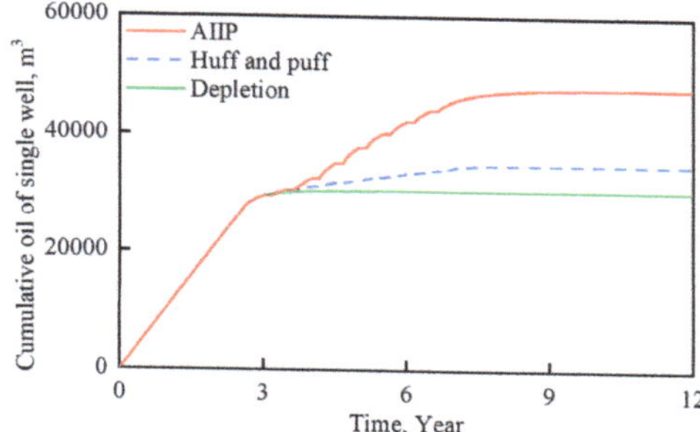

Figure 6. Comparisons of cumulative oil production in different development modes.

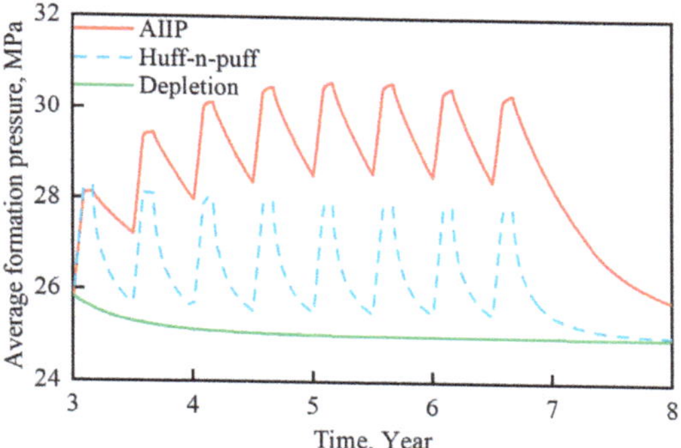

Figure 7. Average formation pressure variation in the production process.

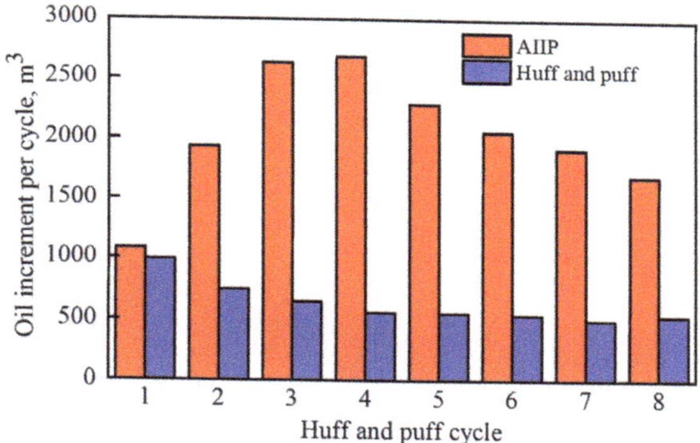

Figure 8. Comparisons of oil increment in different huff-n-puff cycles.

4.2. Sensitivity Analysis

The feasibility analysis shows that AIIP combined with CO_2 huff-n-puff has a good application prospect for the development of shale oil reservoirs. In this part, the influence of different parameters on the performance of AIIP will be studied through sensitivity analyses. The major influential parameters include gas injection rate, gas injection volume, soaking time, hydraulic fracture length, and hydraulic fracture spacing.

4.2.1. Effect of CO_2 Injection Rate

With the same CO_2 injection volume, the difference of injection rate will also exert certain influence on the performance of AIIP theoretically. A higher injection rate can inject the set CO_2 volume within a shorter time with higher injection pressure. A lower injection rate can extend the duration of the injection stage, allowing more CO_2 to dissolve into the oil for a longer time. Hence, it is necessary to study the effect of injection rate on AIIP.

For sensitivity analyses of this parameter, a total of four groups of different injection rate are designed, as shown in Table 6. In the four groups of schemes, injection rate and duration of the injection stage are different in each cycle. In order to ensure comparability of each scheme, the injection volume, duration of soaking stage, and duration of production stage were set to be the same within each cycle of each scheme, so as to compare the influence of gas injection rate on the performance of AIIP. As shown in Figure 9 and Table 7, with the increment of gas injection rate, the cumulative oil production of the single well increases in turn within 5 years of development. In a certain period of time, the higher injection rate can shorten the time required for one cycle, and the total production period is longer in a certain period, but the cumulative oil production in one cycle is slightly lower than that in the lower injection rate. However, the influence of injection rate is relatively small for the cumulative oil production of the single well over a given period time, so the injection rate is not the main factor affecting the performance of AIIP (Figure 10).

Table 6. CO_2 injection rate in different schemes.

Scheme Number	1	2	3	4
CO_2 Injection Rate of the Single Well (10^4 m³/d)	2.5	3.3	5.0	7.5
Duration of the Injection Stage (day)	60	45	30	20
CO_2 Injection Volume of The Single Well Per Cycle (10^4 m³)	150	150	150	150

Table 7. Calculation results of different injection rates.

Injection Rate of the Single Well (10^4 m³/d)	2.5	3.3	5.0	7.5
Cumulative Oil Production of the Single Well (m³)	49,267	49,304	49,729	50,015

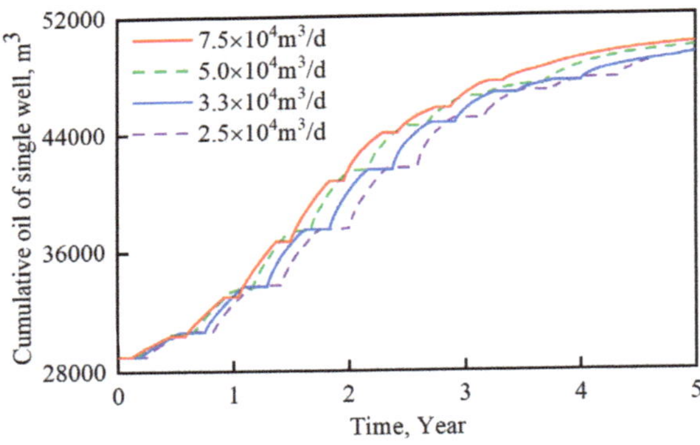

Figure 9. Comparison of cumulative oil production of the single well in each scheme under different CO_2 injection rates.

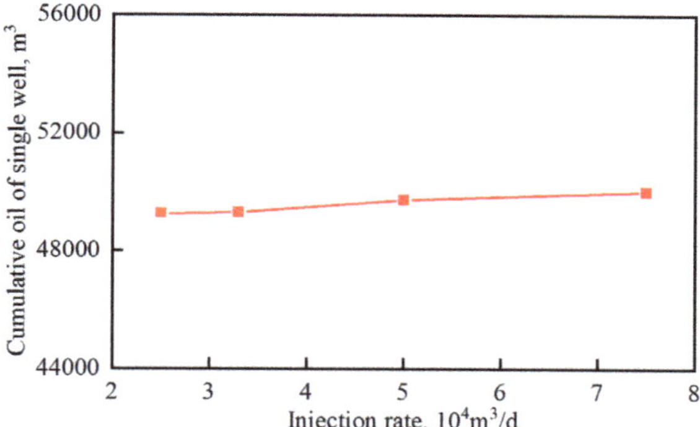

Figure 10. The relationship between cumulative oil production and CO_2 injection rate in the single well.

4.2.2. Effect of CO_2 Injection Volume

The CO_2 injection volume is one of the important factors that affect the performance of AIIP. With the increment of CO_2 injection volume, formation energy can be fully supplemented, and better development results can be achieved. On the other hand, excessive CO_2 injection volume will lead to the high bottom-hole pressure rise in the injection section. Limited by the capacity of gas injection equipment, there is a maximum CO_2 injection pressure, beyond which, it is impossible to continuously inject CO_2 into the formation. Therefore, it is an important task to determine a reasonable CO_2 injection volume for the target block.

For sensitivity analyses of this parameter, a total of five groups of different injection volume were designed, as shown in Table 8. In the five groups of schemes, the volume of CO_2 injected into the single well is different during each cycle. To ensure the comparability between schemes, different CO_2 injection rates and the same injection stage time were set for each scheme, so as to compare the influence of CO_2 injection volume on the performance of AIIP. As shown in Figure 11 and Table 9, with the increment of CO_2 injection volume, the injection pressure at the end of the injection stage increases in turn, and the cumulative oil production of AIIP also increases in turn, but the incremental rate gradually slows down (Figure 12). In addition, when the injection volume is 1.5×10^6 m^3, the highest bottom-hole pressure in the injection section can reach 43.4 MPa, which is close to the maximum injection pressure of the injection equipment. Therefore, CO_2 injection volume has a great impact on the performance of the AIIP method.

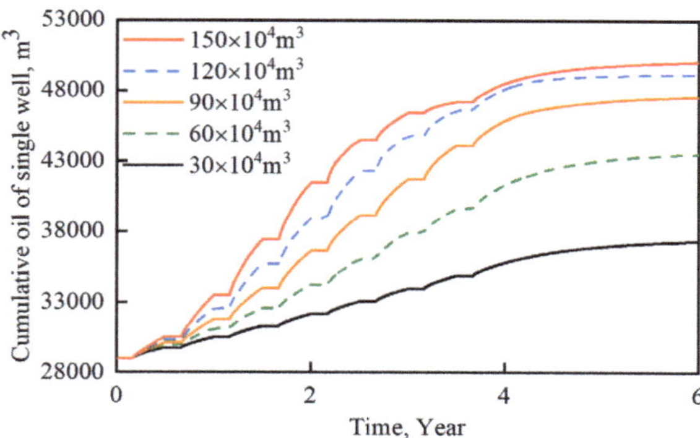

Figure 11. Comparison of cumulative oil production of the single well in each scheme under different CO_2 injection volumes.

Table 8. CO_2 injection volume in different schemes.

Scheme Number	1	2	3	4	5
CO_2 **Injection Volume of the Single Well Per Pycle (10^4 m^3)**	30	60	90	120	150
CO_2 **Injection Rate of the Single Well Per Cycle (10^4 m^3/d)**	1	2	3	4	5

Table 9. Calculation results of different injection volumes.

CO_2 **Injection Volume of tde Single Well Per Cycle (10^4 m^3)**	30	60	90	120	150
Cumulative Oil Production of the Single Well (m^3)	36,931	43,148	47,150	48,705	49,643
Maximum Injection Pressure (MPa)	30	33.9	37.2	40.4	43.4

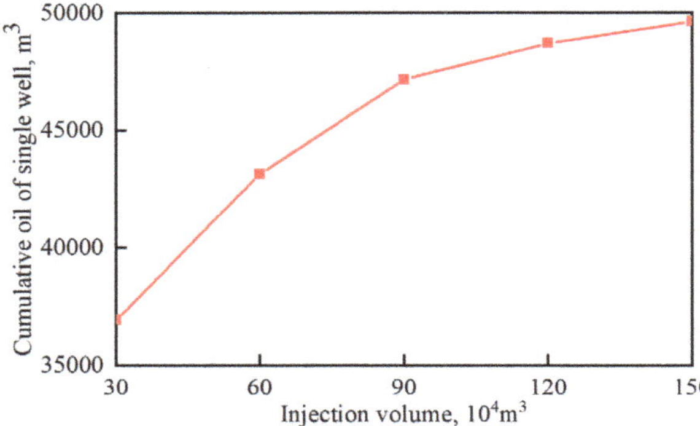

Figure 12. The relationship between cumulative oil production and CO_2 injection volume in the single well.

4.2.3. Effect of CO_2 Soaking Time

For the CO_2 huff-n-puff process, the soaking stage is of great significance to the performance of CO_2 huff-n-puff. During the soaking stage, injected CO_2 will gradually dissolve into the crude oil and bring about oil expansion, viscosity reduction, and other oil-increment effects, thereby promoting pressure propagation. Similarly, soaking time also has a certain impact on the performance of the AIIP method. If the soaking time is too short, CO_2 cannot be fully dissolved, and pressure cannot be fully propagated. Excessive soaking time will lead to a large pressure propagation range and poor formation energy supplement effect. At the same time, a longer shut-in period will lead to lower cumulative oil production in a certain amount of time. Hence, the reasonable soaking time of AIIP method should be determined.

For sensitivity analyses of this parameter, a total of five groups of different soaking time were designed, as shown in Table 10. In the five groups of schemes, the soaking time is different during each cycle. In order to ensure the comparability of each scheme, the duration of injection stage, duration of production stage and gas injection volume were set to be the same within each cycle of each scheme, so as to compare the impact of soaking time on the performance of AIIP. As shown in Figure 13 and Table 11, with the increment of soaking time, the cumulative oil production of the single well within 5 years of development increases first and then decreases. In a certain period, shorter soaking time can shorten the time required for one cycle, and thus the AIIP with more cycles can be carried out. Although longer soaking time can improve AIIP in one cycle, excessive soaking time leads to fewer cycles of AIIP, shorter total production period, and lower oil production. Compared with different schemes, the 30-day soaking time is shorter, and the cumulative oil production of the single well is the highest (Figure 14).

Table 10. Soaking time in different schemes.

Scheme Number	1	2	3	4	5
Soaking Time (day)	15	30	45	60	90

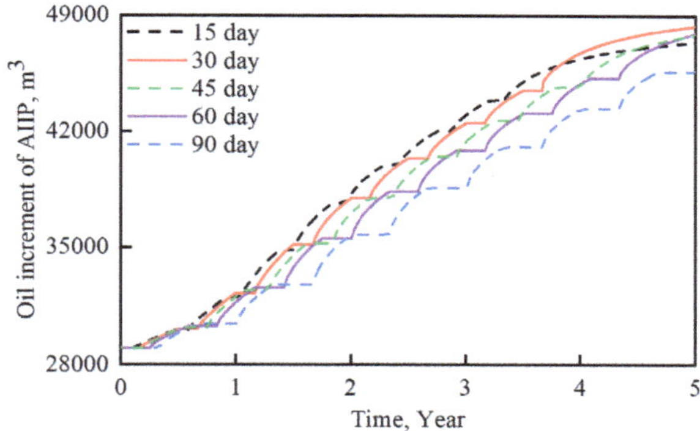

Figure 13. Comparison of oil increment of AIIP in each scheme under different soaking times.

Table 11. Calculation results of different soaking times.

Soaking Time (day)	15	30	45	60	90
Cumulative Oil Production of the Single Well (m^3)	47,385	48,382	47,948	47,872	45,659

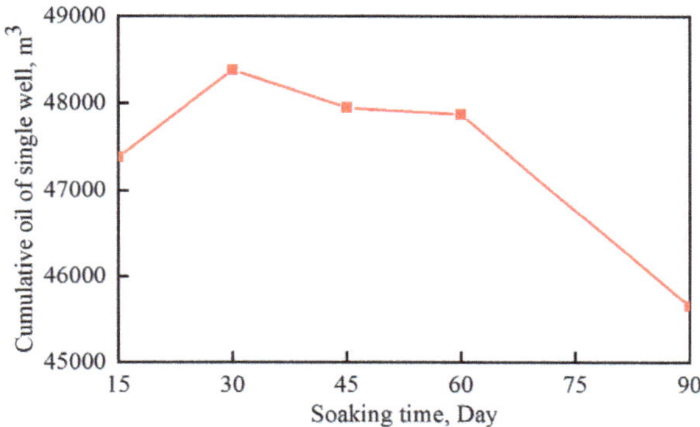

Figure 14. The relationship between cumulative oil production and soaking time in the single well.

4.2.4. Effect of Fracture Half-Length

Fracture length has a great influence on shale oil reservoir development. Effective fracturing can facilitate fluid flow in the shale reservoir and increase oil production of the horizontal well. For CO_2 huff-n-puff or AIIP, on the one hand, the increment of fracture length can facilitate the flow of injected gas, expand the contact area between injected gas and matrix, and promote the effect of oil increment. On the other hand, the excessively long length of some fractures will lead to the rapid flow of injected gas along with the fractures, forming gas channeling, and pressure leakage during the soaking stage, resulting in poor pressure level. Therefore, it is necessary to determine the effect of fracture length on the performance of the AIIP method.

For sensitivity analyses of this parameter, a total of four groups of different fracture lengths were designed, as shown in Table 12. Among the four groups of schemes, three groups of horizontal wells have different hydraulic fracture half-lengths, comparing the influence of different fracture half-length on the effect of AIIP. In another scheme, a small number of hydraulic fracture lengths were set to be relatively long to simulate the pressure leakage caused by excessively long hydraulic fractures. As shown in Figure 15 and Table 13, the cumulative oil production of the single well increases in turn with the increment of fracture half-length, and the long fracture scheme produces more oil in the first three cycles of AIIP. With the increment of huff-n-puff cycles, the oil production of the long-fracture scheme in a single cycle is gradually lower than that of the short-fracture scheme after three cycles. Figures 16 and 17 compare the cumulative oil production and daily oil production between the ideal model and the model with fracture connectivity. When there is fracture connectivity between wells, the existence of long fractures can achieve a better effect in the first two cycles, but the oil production gradually decreases in the subsequent cycles. The connection of fractures leads to gas channeling, resulting in a poor pressure maintenance effect in the soaking period (Figure 18). Therefore, if the hydraulic fracture is connected, the performance of AIIP will become worse.

Table 12. Fracture half-length in different schemes.

Scheme Number	1	2	3	4
Fracture Half-length (m)	70	100	130	130 (normal fracture) 150 (long fracture)

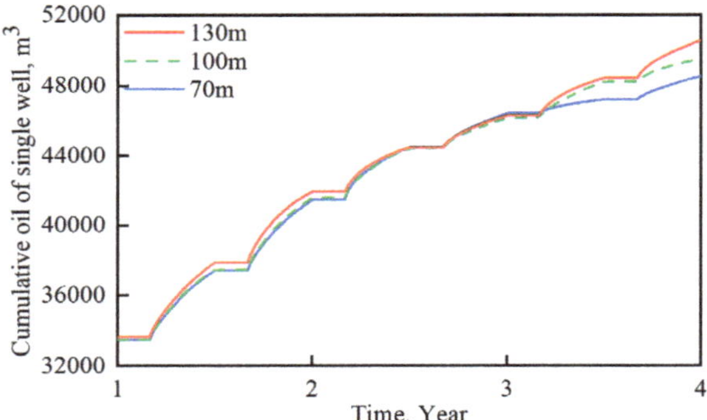

Figure 15. Comparison of cumulative oil production of the single well in each scheme under different fracture lengths.

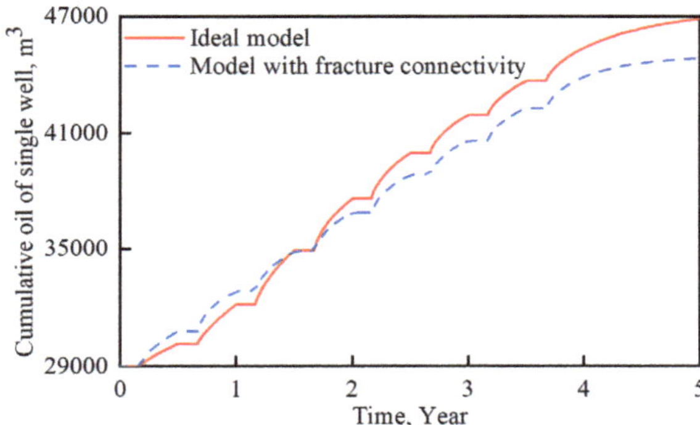

Figure 16. Comparison of cumulative oil production of the single well between the ideal model and the model with fracture connectivity.

Table 13. Calculation results of different fracture half-lengths.

Fracture Half-Lengtd (m)	70	100	130
Cumulative Oil Production of the Single Well (m^3)	48,535	49,487	50,570

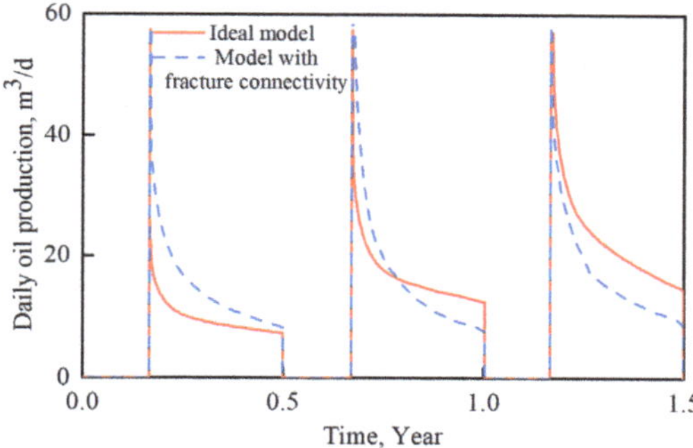

Figure 17. Comparison of daily oil production of the single well between the ideal model and the model with fracture connectivity.

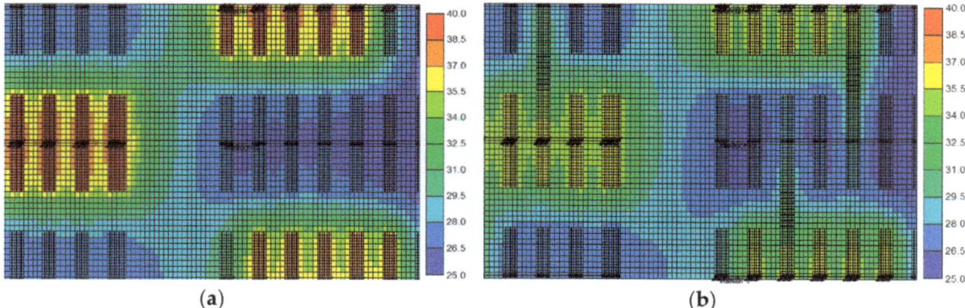

Figure 18. Comparison of formation pressure at the end of gas injection period between the ideal model and the model with fracture connectivity (unit: MPa). (**a**) Ideal model; (**b**) model with fracture connectivity.

4.2.5. Effect of Fracture Spacing

The hydraulic fracture spacing is also one of the important factors affecting the oil production of the shale oil reservoir. The smaller the spacing between hydraulic fractures is, the larger the contact area between fracture and matrix is, which is beneficial to the flow of fluid in the shale oil reservoir and increases oil production of horizontal wells. For CO_2 huff-n-puff or AIIP, the decrease of fracture spacing can also facilitate the flow of injected gas, expand the contact area between injected gas and matrix, and promote the effect of oil increment. However, too small fracture spacing will greatly increase the difficulty and cost of hydraulic fracturing, and when the fracture spacing reaches a certain value, the increment of oil production is also limited. Therefore, it is necessary to determine the effect of fracture spacing on the performance of AIIP method.

For sensitivity analyses of this parameter, four groups of different fracture spacing were designed to compare the influence of fracture spacing on the effect of AIIP, as shown in Table 14. As shown in Figure 19 and Table 15, the cumulative oil production of the single well increases in turn with the shortening of fracture spacing, but when the fracture spacing is less than 30 m, the cumulative oil production growth of the single well slows down gradually (Figure 20). Considering the fracturing technology and economy, there may be a suitable fracture spacing, which can give full play to the performance of AIIP and have a better economy at the same time.

Table 14. Fracture spacing in different schemes.

Scheme Number	1	2	3	4
Fracture Spacing (m)	7.5	15	30	45

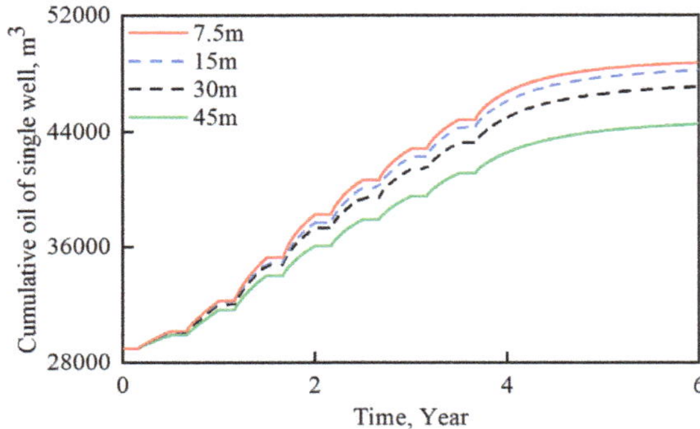

Figure 19. Comparison of cumulative oil production of the single well in each scheme under different fracture spacing.

Table 15. Calculation results of different fracture spacing.

Fracture Spacing (m)	7.5	15	30	45
Cumulative Oil Production of the Single Well (m^3)	48,772	48,229	47,093	44,540

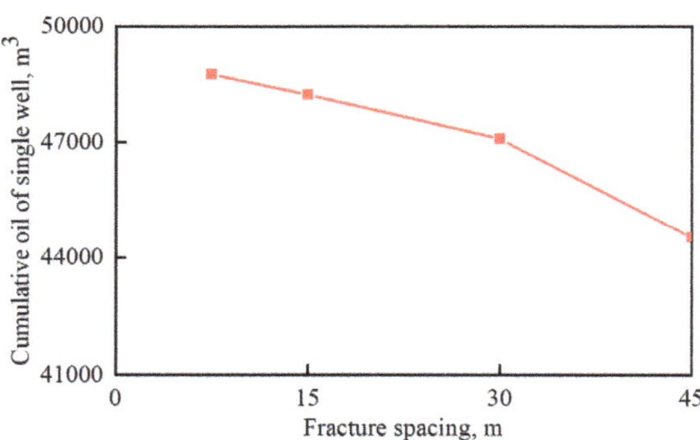

Figure 20. The relationship between cumulative oil production and fracture spacing in the single well.

5. Conclusions

This paper first proposes to use AIIP combined with CO_2 huff-n-puff to develop shale oil reservoirs and reduce CO_2 emissions. Compared with depletion and conventional CO_2 huff-n-puff, the AIIP method has higher cumulative oil production and maintains higher formation pressure. Therefore, AIIP combined with CO_2 huff-n-puff has a good prospect for the development of shale oil reservoirs. In the end, the effect of several parameters on AIIP development was studied through sensitivity analyses. The following conclusions can be obtained from this study:

(1) Compared with horizontal well depletion and CO_2 huff-n-puff, cumulative oil production of AIIP increases by 17,450 and 13,077 m^3 respectively in the 20 years of development prediction. The oil production of AIIP first increases, and then decreases, and is higher than that of conventional huff-n-puff in each cycle. It shows that AIIP combined with CO_2 huff-n-puff is an effective technique to improve shale oil recovery.

(2)   The performance improves with the increment of gas injection volume. When the injection volume is 1.5 million cubic meters, the oil increment slows down significantly, and the injection pressure is 43.4 MPa, which is close to the maximum injection pressure of the gas injection equipment. Injection rate has little impact on performance, but the injection volume exerts a great influence.

(3)   Shorter soaking time can increase more oil production in the process of huff-n-puff, but the oil production improves with the increment of soaking time after huff-n-puff. Increasing the soaking time can make the pressure wave spread more widely, and make more gas dissolve into oil, leading to a better performance. However, oil cannot be produced during the soaking period, and oil production is low in the short term.

(4)   The longer the fractures are, the higher the oil production is in each cycle of AIIP. With the increment of huff-n-puff cycles, the difference of oil production gradually decreases in different fracture length schemes. If some hydraulic fractures are too long and connected, the effect of AIIP will be affected. With the shortening of fracture spacing, the cumulative oil production of the single well increases in turn, but when the fracture spacing is less than 30 m, the cumulative oil production of the single well gradually slows down. Considering the fracturing technology and economy, there is a suitable fracture length and fracture spacing.

Author Contributions: Conceptualization, H.Y. and S.C.; Formal analysis, Q.X. and X.Q.; Methodology, H.Y. and S.C.; Software, S.Q. and Z.C.; Supervision, H.Y.; Visualization, S.Q; Writing—original draft, S.Q.; Writing—review and editing, H.Y. and Z.C.

Funding: This research was funded by National Natural Science Foundation of China (U1762101) and National Science and Technology Major Projects (2017ZX05009-004).

Acknowledgments: The authors appreciate the Computer Modeling Group for offering CMG software. Special thanks for the help from Shengli Oilfield.

Conflicts of Interest: The authors declare no conflict of interest.

References

1.   Wang, Q.; Ma, Y.; Li, S.; Qian, J. Global oil shale research, development and utilization today—Two international oil shale symposiums held in 2016. *Sino-Global Energy* **2017**, *22*, 23–29.

2.   Zhou, Q.; Jin, Z.; Yang, G.; Dong, N.; Shang, Z. Shale oil exploration and production in the US: Status and outlook. *Oil Gas Geol.* **2019**, *40*, 469–477.

3.   Wang, X.; Sheng, J. Effect of low-velocity non-Darcy flow on well production performance in shale and tight oil reservoirs. *Fuel* **2017**, *190*, 41–46. [CrossRef]

4.   Lv, X.; Wu, W.; Li, G.; Ben, C.; Lin, L.; Yang, Q. Key technologies for shale gas development in North America. *Pet. Geol. Oilfield Dev. Daqing* **2015**, *34*, 158–162.

5.   Sorensen, J.; Braunberger, J.; Liu, G.; Smith, S.; Hawthorne, S.; Steadman, E. Characterization and evaluation of the Bakken petroleum system for CO_2 enhanced oil recovery. In Proceedings of the Unconventional Resources Technology Conference, San Antonio, TX, USA, 20–22 July 2015.

6.   Zeng, F.; Peng, F.; Guo, J.; Xiang, J.; Wang, Q.; Zhen, J. A transient productivity model of fractured wells in shale reservoirs based on the succession pseudo-steady state method. *Energies* **2018**, *11*, 2335. [CrossRef]

7.   Yang, L.; Kang, H.; Lu, X. Experimental investigation of CO_2 flooding in tight oil reservoirs using CT scanning method. In Proceedings of the 4th ISRM Young Scholars Symposium on Rock Mechanics, Jeju, Korea, 10–13 May 2017.

8.   Wang, Y.; Cheng, S.; Zhang, K. Case studies: Pressure-transient analysis for water injector with the influence of waterflood-induced fractures in tight reservoir. In Proceedings of the SPE Improved Oil Recovery Conference, Tulsa, OK, USA, 14–18 April 2018.

9.   Sheng, J.J. Enhanced oil recovery in shale reservoirs by gas injection. *J. Nat. Gas Sci. Eng.* **2015**, *22*, 252–259. [CrossRef]

10.   Cheng, J. Laboratory Experimental Study on Oil Displacement Effects of Different Gases. Ph.D. Thesis, Northeast Petroleum University, Daqing, China, 2014.

11. Yu, Q. Study on Gas Injection Development Mode of Low Permeability Tight Reservoir in Block L3. Ph.D. Thesis, Southwest Petroleum University, Chengdu, China, 2015.
12. Shyeh-Yung, J.G. Mechanisms of miscible oil recovery effects of pressure on miscible and displacements of oil by carbon dioxide. In Proceedings of the SPE Annual Technical Conference and Exhibition, Dallas, TX, USA, 6–9 October 1991.
13. Wahaibi, A.; Yahya, M. First-contact-miscible and multicontact-miscible gas injection within a channeling heterogeneity system. *Energy Fuels* **2010**, *24*, 1813–1821. [CrossRef]
14. Ampomah, W.; Balch, R.; Cather, M. Evaluation of CO_2 storage mechanisms in CO_2 enhanced oil recovery sites: Application tomorrow sandstone reservoir. *Energy Fuels* **2016**, *30*, 8545–8555. [CrossRef]
15. Gao, Y.; Zhao, M.; Wang, J. Performance and gas breakthrough during CO_2 immiscible flooding in ultra-low permeability reservoirs. *Pet. Explor. Dev.* **2014**, *41*, 88–95. [CrossRef]
16. Song, C.; Yang, D. Experimental and numerical evaluation of CO_2 huff-n-puff processes in Bakken formation. *Fuel* **2017**, *190*, 145–162. [CrossRef]
17. Pu, W.; Wei, B.; Jin, F. Experimental investigation of CO2 huff-n-puff process for enhancing oil recovery in tight reservoirs. *Chem. Eng. Res. Des.* **2016**, *111*, 269–276. [CrossRef]
18. Janiga, D.; Czarnota, R.; Stopa, J. Huff and puff process optimization in micro scale by coupling laboratory experiment and numerical simulation. *Fuel* **2018**, *224*, 289–301. [CrossRef]
19. Zuloaga, P.; Yu, W.; Xu, Y. Simulation study of CO_2-EOR in tight oil reservoirs with complex fracture geometries. *Sci. Rep.* **2016**, *6*, 33445. [CrossRef] [PubMed]
20. Zhang, Y.; Yu, W.; Li, Z.; Sepehrnoori, K. Simulation study of factors affecting CO_2 Huff-n-Puff process in tight oil reservoirs. *J. Pet. Sci. Eng.* **2018**, *163*, 264–269. [CrossRef]
21. Sanchez-Rivera, D.; Mohanty, K.; Balhoff, M. Reservoir simulation and optimization of huff-and-puff operations in the Bakken Shale. *Fuel* **2015**, *147*, 82–94. [CrossRef]
22. Sun, J.; Zou, A.; Sotelo, E.; Schechter, D. Numerical simulation of CO_2 huff-n-puff in complex fracture networks of unconventional liquid reservoirs. *J. Nat. Gas Sci. Eng.* **2016**, *31*, 481–492. [CrossRef]
23. Cheng, S.; Wang, Y.; Lang, H.; Yu, H. Feasibility of inter-fracture injection and production in the same well for multistage fractured horizontal wells in tight reservoir. *Acta Pet. Sin.* **2017**, *38*, 1411–1419.
24. Yu, H.; Yang, Z.; Ma, T.; Lei, Z.; Cheng, S.; Chen, H. The feasibility of asynchronous injection alternating production for multistage fractured horizontal wells in a tight oil reservoir. *Pet. Sci. Bull.* **2018**, *1*, 32–44.
25. Yu, H.; Yang, Z.; Luo, L.; Liu, J.; Cheng, S. Application of cumulative-in-situ-injection-production technology to supplement hydrocarbon recovery among fractured tight oil reservoirs: A case study in Changqing Oilfield, China. *Fuel* **2019**, *242*, 804–818. [CrossRef]
26. Yu, H.; Rui, Z.; Chen, Z.; Lu, X.; Yang, Z.; Liu, J.; Qu, X.; Patil, S.; Ling, K.; Lu, J. Feasibility study of improved unconventional reservoir performance with carbonated water and surfactant. *Energy* **2019**, *182*, 135–147. [CrossRef]
27. Dong, D.; Qiu, L.; Ma, Y.; Yang, Y.; Zou, Y.; Dai, L.; Teng, B.; Zhang, Z. Control of tectonics on sedimentation of sandstone and process of sediment filling in multi-fault lacustrine basins: A case study on the Eocene in eastern Zhanhua Sag, Jiyang Depression in Bohai Bay Basin. *Oil Gas Geol.* **2008**, *39*, 653–663.
28. Chen, G.; Lu, S.; Zhang, J.; Wang, M. Estimation of enriched shale oil resource potential in E2s4L of Damintun Sag in Bohai Bay Basin, China. *Energy Fuels* **2017**, *31*, 3635–3642. [CrossRef]
29. Zhu, G.; Yao, J.; Li, A.; Sun, H.; Lei, Z. Pore-scale investigation of carbon dioxide-enhanced oil recovery. *Energy Fuels* **2017**, *31*, 5324–5332. [CrossRef]
30. Sun, R.; Yu, W.; Xu, F.; Pu, H.; Miao, J. Compositional simulation of CO_2 huff-n-puff process in middle bakken tight oil reservoirs with hydraulic fractures. *Fuel* **2019**, *236*, 1446–1457. [CrossRef]
31. Shen, Z.; Sheng, J.J. Experimental study of permeability reduction and pore size distribution change due to asphaltene deposition during CO_2 huff and puff injection in Eagle Ford shale. *Asia Pac. J. Chem. Eng.* **2017**, *12*, 381–390. [CrossRef]
32. Chen, P. Field application of the horizontal well sectional fracturing technology in shale oi of well Jinye-1HF. *Reserv. Eval. Dev.* **2012**, *2*, 73–77.

Review

Recent Approaches of Forecasting and Optimal Economic Dispatch to Overcome Intermittency of Wind and Photovoltaic (PV) Systems: A Review

Manzoor Ellahi [1], Ghulam Abbas [1,*], Irfan Khan [2], Paul Mario Koola [3], Mashood Nasir [4], Ali Raza [1] and Umar Farooq [5]

[1]   Electrical Engineering Department, The University of Lahore, Lahore 54000, Pakistan; engr.manzoorellahi@hotmail.com (M.E.); ali.raza@ee.uol.edu.pk (A.R.)

[2]   Marine Engineering Technology Department in a joint appointment with Electrical and Computer Engineering Department, Texas A&M University, Galveston, TX 77554, USA; irfankhan@tamu.edu

[3]   Ocean Engineering Department, Texas A&M University, Galveston, TX 77554, USA; paulmkoola@tamu.edu

[4]   Department of Energy Technology, Aalborg University, Aalborg 9220, Denmark; mnas@et.aau.dk

[5]   Department of Electrical Engineering, University of the Punjab, Lahore 54590, Pakistan; engr.umarfarooq@yahoo.com

*   Correspondence: ghulam.abbas@ee.uol.edu.pk; Tel.: +92-304-285-4035

Received: 1 October 2019; Accepted: 6 November 2019; Published: 19 November 2019

Abstract: Renewable energy sources (RESs) are the replacement of fast depleting, environment polluting, costly, and unsustainable fossil fuels. RESs themselves have various issues such as variable supply towards the load during different periods, and mostly they are available at distant locations from load centers. This paper inspects forecasting techniques, employed to predict the RESs availability during different periods and considers the dispatch mechanisms for the supply, extracted from these resources. Firstly, we analyze the application of stochastic distributions especially the Weibull distribution (WD), for forecasting both wind and PV power potential, with and without incorporating neural networks (NN). Secondly, a review of the optimal economic dispatch (OED) of RES using particle swarm optimization (PSO) is presented. The reviewed techniques will be of great significance for system operators that require to gauge and pre-plan flexibility competence for their power systems to ensure practical and economical operation under high penetration of RESs.

Keywords: renewable energy sources; forecasting; Weibull distribution; neural networks; optimal economic dispatch; particle swarm optimization

1. Introduction

Renewable energy sources (RESs) are the primary solution to the growing environmental concerns, which include carbon and nitrogen emissions and power shortages around the world. The climate change, the variable cost, continuously increasing environmental issues, and fast depletion of fossil fuels have urged the electric power suppliers to incorporate RESs more strappingly into the power system [1]. Nuclear energy was considered to be a source of cheap electricity production. Accidents in the Nuclear reactors during the last two decades, issues-oriented with disposing-off the nuclear waste and increasing awareness regarding global warming [2] have led to immense emphasis on integrating renewable RESs such as wind and solar energy into the power system.

RESs such as wind and solar-based photovoltaic (PV) with all their benefits have the issue of being unpredictable as weather conditions keep on changing throughout the year; besides, they require high initial cost for installation. These sources have maximum and minimum generation limits that vary over time, unlike conventional power plants, where we know the maximum generation possible [3,4]. Wind power has more probability as well as variability. Solar power is less uncertain and less variable, as compared to wind power. The intermittent nature of wind and solar power is a significant issue in employing them on a priority basis as long term planning gets tough. The feasible solution to this problem is accurate resource forecasting. Due to the sporadic nature of these resources, there has been widespread interest in the optimal integration of wind and solar power over various times windows [5]. Considering all these facts, the first half of the paper proposes a review of the forecasting techniques to predict wind and PV power potential.

Electric power extracted from RESs can be classified in three possible ways: (a) Distributed Generation (DG), in which RESs have been installed individually by the consumers. DG is mostly employed by domestic and small commercial consumers [6–9]; (b) Micro-Grid (MG), in which a small number of PV panels or wind turbines have been installed and integrated to supply power to a small community. These MGs are mostly near to the load centers and they are employed by small towns, villages, very small industrial units and shopping malls [10–13]; (c) Large-Scale Generation (LG), in which a large number of power generation units have been installed to form large scale wind farms or solar panel farms. These farms have the potential to produce power from a few hundred to thousands of MWs. This category requires a properly planned ED scheme to supply the major load centers.

Stochastic distribution, especially Weibull distribution (WD) is used to predict the life of products or outputs of a system according to the statistical distribution of the sample measurement [14]. This stochastic technique can provide forecasted wind and solar power data at a particular location but, accuracy remains an issue, as there may be a difference in the predicted and actual values. Neural networks are more robust to minimize this error between the forecasted and actual values [15]. The accuracy in the forecasted values is critical since the system requires predictive planning to meet the varying demand from the supply and load side to make the system more reliable [16].

Neural networks (NNs) or Artificial neural networks (ANNs) or connectionist systems are computing mechanisms inspired by the biological neural networks mimic brains [17]. The effectiveness of these NN systems is because they can learn and improve performance like self-repair, be fault-tolerant, and handle nonlinear data processing [15,18]. The incorporation of NNs or their variants into the forecasting system helps in reducing error thereby ensuring the predicted value falls in the permissible range to better balance the supply and demand in real-time [19]. Although the author in [20] performed an extensive review of the forecasting techniques, it lacked the survey of techniques used for the removal of error emergence in the predicted values.

The low energy density of RES requires a large area to generate an adequate amount of power to become a significant load bearer. Hence, to be sustainable over a more extended time, power sources should be economically dispatchable [21]. The requirement of large areas forces locations to be hundreds or even thousands of miles away from load centers. Hence, a comprehensive review of the Optimal Economic Dispatch (OED) techniques is also required [22–24]. Of the many optimization techniques, particle swarm optimization (PSO) stands out due to its fast convergence, flexibility in application, simplified approach and good adaptability to variation [25–27]. The second half of the paper, thus, proposes a review on the solution of OED of RES using PSO.

The researchers around the world have reviewed and analyzed PSO for the solution of the economic dispatch problem, and WD for the forecasting of RESs. Authors in [28,29] investigated PSO and its variants in a very comprehensive manner. Their analysis, however, was confined to the application of PSO for the solution of the Economic Dispatch Problem (EDP) in thermal systems only. Reference [30] reviewed the impact PSO had on improving the performance of flexible alternating current transmission system (FACTS) devices by providing a solution to the FACTS allocation problem. However, the author in [30] focused only on the FACTS allocation problem through standard PSO and did not discuss the dispatch constraints that can be solved by the variants of PSO. The authors in [31] reviewed independent and hybridized optimization techniques employed for the optimization of wind-PV hybrid power systems. In [31] the authors studied the optimization techniques for generation but ignored the source variability issue of RESs in the modern power system. Power fluctuation control is essential for a sustainable power system as regulatory bodies are compelled to provide a check on moment-to-moment variations in system load and inconsistent power generation [32]. Reference [20] described the uncertainties related to the modern power system and reviewed the techniques used for the solution of this problem. The paper comprehensively gives a review of PSO-based solution to OED incorporating RESs and considers the realistic constraints associated with OED problems.

To be precise, the first half of the paper provides a review of forecasting mechanisms employed using WD, both with and without the incorporation of NNs for the removal of error in forecasted value, whereas the second half of the paper emphasizes on the solution of EDP using PSO and its variants while considering all constraints that emerge for the Economic Dispatch (ED) of RESs. This paper will provide researchers, power system engineers, planners, and developers a comprehensive survey on forecasting through WD for understating the working and possible issues in the regular, stable, reliable and efficient operation of power systems, and on solving ED problem incorporating RESs through PSO.

More descriptively, the paper is organized as follows. Section 2 reviews wind power generation forecasting through WD with and without the incorporation of NNs. Section 3 provides forecasting of solar (PV) generation using WD with and without incorporation of NN. Comprehensive tables are also designed to provide the objective function(s) and mathematical relations for forecasting power generation from the said resources. Section 4 discusses a detailed survey of the application of PSO in OED of RESs without and with the incorporation of resource forecasting, followed by the conclusion and references at the end of the paper.

2. Forecasting of Wind Power Generation

2.1. Wind Power Generation Fundamentals

Amongst the various solutions of RESs, wind energy is a popular source that works by converting the kinetic energy of the wind using a turbo-generator into electricity. The power output of a wind turbine is given by Equation (1).

$$P = \frac{\rho A}{2} v^3 c_p(\lambda) \tag{1}$$

where P, C_p, ρ, A, v and λ are mechanical output power, turbine's performance coefficient, air density, turbine swept area, wind speed, tip speed ratio, respectively [33].

Wind power is more certain to occur, although its speed is varied during 24 h [34,35]. Forecasting and optimization techniques must be worked upon to make them more dependable. Forecasting also requires the consideration of ramp event and potentially high-risk scenarios of wind power. Amongst numerous wind power forecasting techniques, many authors proposed to employ the WD function due to its reliability, accuracy, stability, and sophisticated computations and accurate results as compared to other techniques like the Rayleigh distribution [36].

2.2. Weibull Distribution (WD) for Wind Power Forecasting

Wind power forecasting through stochastic techniques (especially WD) has been a very important topic for researchers, power producers, and planning engineers. It enhances the share of wind power in meeting the load demand [37]. The basic formula for probability density function (PDF) $f(v, v_0, \beta)$ of WD in its most basic form is given in Equation (2).

$$f(v, v_0, \beta) = \begin{cases} \frac{\beta}{v_0} \left(\frac{v}{v_0} \right)^{\beta-1} e^{-\left(\frac{v}{v_0} \right)^{\beta}}; & v \geq 0 \\ 0; & v < 0 \end{cases} \tag{2}$$

where $\beta > 0$ is the shape parameter, and $v_0 > 0$ is the scale parameter of the distribution. Depending on the complexity of the problem, WD can have multiple variables that are defined according to the requirement like the translated distribution containing three variables has the Equation (3) shown below [38].

$$f(t) = \frac{\beta}{\eta}\left(\frac{t-\gamma}{\eta}\right)^{\beta-1} e^{-(\frac{t-\gamma}{\eta})\beta} \tag{3}$$

where β, η and γ are WD shape, scale parameters, and wind speed, respectively.

PDF and multi-variable forms of WD have been mentioned in Equations (2) and (3), respectively. Cumulative density function (CDF) of WD is given by Equation (4).

$$F(v, v_0, \beta) = \begin{cases} 1 - e^{-(\frac{v}{v_0})^{\beta}}, & v \geq 0 \\ 0, & v < 0 \end{cases} \tag{4}$$

Reliability function (RF) in WD finds major significance in forecasting, as it computes the amount of time for a particular item can operate without failure. Mathematically, it is given by Equation (5) [39].

$$F(t) = \int_{0,\gamma}^{t} f(s)ds \tag{5}$$

RF is a function of time and is important for life data analysis. It can be computed using the CDF of WD as given in Equation (4).

One of the major apprehensions in implementing wind power at a large scale is the impact of ramps, and it requires proper handling. An event took place at the Electric Reliability Council of Texas (ERCOT) system on 26 February 2008, which caused a system emergency due to the occurrence of a large down ramp [40]. Therefore, proper forecasting of the ramp emergence in wind power is essential for the sustainability of the power system.

A brief comparison of the stochastic techniques like Weibull Distribution (WD), Rayleigh Distribution (RD), and Normal Distribution (ND) is presented in Table 1.

Table 1. Comparison of stochastic forecasting techniques.

Characteristics	Weibull Distribution (WD)	Rayleigh Distribution (RD)	Gaussian/Normal Distribution (ND)	Ref. No.
Mathematical representation & parameters	$f(v) = \begin{cases} \frac{\beta}{v_0}\left(\frac{v}{v_0}\right)^{\beta-1} e^{-(\frac{v}{v_0})^{\beta}}; & v \geq 0 \\ 0; & v < 0 \end{cases}$ where $\beta > 0$ is the shape parameter, and $v_0 > 0$ is the scale parameter of the distribution.	$f(v) = \begin{cases} \frac{v}{\sigma^2}e^{-v^2/2\sigma^2}, & v > 0 \\ 0, & v \leq 0 \end{cases}$ where σ is the scale parameter of the distribution.	$f(v) = \frac{1}{\sqrt{2\pi\sigma^2}}e^{-\frac{(v-\mu)^2}{2\sigma^2}}$ where μ is the mean, whereas σ is the standard deviation.	[41–44]
Flexibility	WD is very flexible as a small sample size; the estimated shape of the distribution may be altered considerably.	Not flexible as a response to the out of range parameters are strict.	Not flexible as the shape doesn't vary.	
Accuracy	Fatigue test results follow WD, showing it to be more accurate. It is effective for both values above and below the sample size N.	Close to WD.	Effective only for values below the sample size N.	[45–49]
Reliability	WD is more reliable even in situations where distribution parameters (shape and scale) tend to vary.	RD loses its effectiveness in situations where variables undergo variation.	Reliability in ND suffers severely at the hands of variation in variables.	

2.3. Review of Wind Power Forecasting without NN

Wind power has captured the biggest share amongst all RESs due to its certainty, but the variability of wind speed still poses a hurdle in its implementation at large scale discussed. This issue can be addressed by predicting the wind speed for a particular time period and planning the power dispatch mechanism accordingly.

In [50], the authors discussed that the probability of wind speed at a particular site has to be modeled for calculation of the energy production by a wind farm. Methodical computation of the generation capacity factor of a wind turbine at the planning stage is of vital importance. The authors performed the comparison of WD computation using graphical, empirical, modified maximum likelihood, and energy pattern factor methods used monthly for the estimation of parameters at the hub height of 65 m. They concluded that parameters of WD assessed through the proposed modification of the maximum likelihood method, complemented the measured values accurately and the graphical method provided the most erroneous results.

The authors in [51] presented an analysis of wind speed data based on fitting curve methods applied for wind farms in Galicia, Spain. The results of the fitting methods applied for determining the Weibull parameters were calculated using a set of pointers defined by wind speeds and wind power density distributions. The authors concluded when the energy produced by the wind turbine generators (WTGs) is considered, the proposed part density energy method (PDEM) demonstrated the best results when the energy produced by the wind turbine generators (WTGs) is considered.

Reference [52] proposed an approach that used three procedures (maximum likelihood, least squares, and method of moments) for the estimation of the Weibull parameters. The approach was based upon the shape parameter k and scale parameter λ. They concluded that the presented methodology indicated good agreement between the data obtained from actual measurements, and enabled the investigators with the knowledge of calculating the wind potential of a region for future installations. The authors in [52] modeled wind power forecasting mechanisms for effective management and balancing of the power grid. The designed system followed a data-driven approach to overcome uncertainties attached to the wind source. The designed model was implemented at a location in the state of Karnataka, India for analysis over a period of three months. The authors used the graphical method, maximum likelihood estimation and Monte Carlos methods for estimation and prediction of wind power. The authors concluded that the wind power followed WD as compared to wind velocity that followed Rayleigh distribution. The plots derived from the computed Weibull parameters also showed significant accuracy. They also projected that the forecasting systems for longer durations, possessing required accuracy can also be designed as discussed in [53].

The optimal power dispatch of multi-microgrid (MMG) has gained its significance in the replacement of fossil fuel, as discussed in reference [53]. The authors presented a stochastic and probabilistic model for both small-scale energy resources (SSERs) and load demand at each micro-grid (MG) and emphasized the significance of forecasting through the WD based model. They emphasized that the stochastic approach must be considered for load-supply modeling. The normal distribution function for modeling of the load on MG was computed using Equation (6).

$$f(P_l) = \frac{1}{\sqrt{2\pi} \times \sigma} \exp\left(-\frac{(P_l - \mu)^2}{2 \times \sigma^2}\right) \tag{6}$$

where μ is the mean value and σ is the standard deviation. A predefined number of load samples were considered. The authors concluded that the power-sharing between MGs and the main grid reduced the total operational cost of the future distribution network. They also proved that by probabilistic modeling of the input variables, the output variables could be represented as random variables. The proposed WD model provided the distribution of wind speed and wind power estimation for long term planning as compared to [52] where WD provided only wind power forecasting. This WD model enabled them to calculate the wind power available during a specified time duration.

In [54], the authors presented a mechanism of wind power forecasting and load estimation for a wind-thermal system. They developed energy and spinning reserve market clearing (ESRMC) system. The authors targeted overall cost minimization and reduction in system-risk levels through the designed system. The solution provided the best comparison in the resource availability and decision-makers' judgment while meeting the system requirements and avoiding uncertainties more accurately than other research models. The system security level is described by a linear fuzzy membership function λ for wind penetration, which is given by Equation (7).

$$\mu(F_i) = \begin{cases} 1, & F_i \leq F_i^{\min} \\ \frac{F_i^{\max} - F}{F_i^{\max} - F_i^{\min}}, & F_i^{\min} < F_i < F_i^{\max} \\ 0, & F_i^{\min} \geq F_i^{\max} \end{cases} \tag{7}$$

where $\mu(F_i)$ was assumed to be strictly monotonic and decreasing.

The authors emphasized after analyzing the results that the variability in the supply may cause the system cost to rise, which was minimized by the implementation of the designed system. They also concluded that the reserve contribution in calculations improved the efficiency of the system. However, the authors ignored the significance of the initialization method of WD.

The authors in [55] presented an analysis of the wind energy conversion system using WD. Four methods were used to calculate the shape factor "k" and the Weibull scale factor "λ". Six statistical tools were employed for analyzing the goodness of curve fittings and precisely ranked the methods. The designed methodology was tested with the available data of the Hatiya island, Bangladesh. The authors concluded that the method of moments (MOM) was most efficient amongst the four tested methods because it endured much lower error percentage and better forecasted the wind power and energy density for a particular site. However, dispatching of the generated power to the load was not considered.

Reference [56] presented model predictive control (MPC) based control of DED using WD. The cost related to lifted states was worked-out at every step during the MPC formulation of DED. An optimal input sequence was determined by resolving the optimization problem on-line in the MPC optimization procedure. The formulation used for the linear relationship between wind speed and power by the authors is given by Equation (8).

$$P(v) = \begin{cases} 0, & v < v_{ci} \, or \, v > v_{co} \\ P_r\left(\frac{v - v_i}{v_r - v_i}\right), & v_{ci} \leq v \leq v_r \\ P_r, & v_r \leq v \leq v_{co} \end{cases} \tag{8}$$

The authors concluded that the proposed method provided a better prediction of the source availability and attained more ED. The presented technique managed a good efficiency over dispatch duration, both in the long-term plan and occurrence of an abrupt event.

The authors in [57] performed a statistical analysis of the wind energy available at the Iskenderun region in Turkey using WD and RD. The data taken was of one year and was analyzed on an hourly basis. The mathematical equations adopted by the authors are mentioned in Table 2. Based on the analyzed data, the authors concluded that the WD provided better-fitting when tested for monthly PDF as compared to the RD. They also concluded that the WD also provided better power density for the whole year under analysis.

Table 2 summarizes the discussed approaches, proposed by different researchers to use a stochastic distribution like WD for wind speed and power forecasting while considering different constraints and OED of the produced energy.

Table 2. Wind speed and power prediction with Weibull distribution (WD) and without incorporating neural networks (NN).

Wind Speed Probability Distribution	Wind Power Distribution	Explanation
$f(v) = \left(\frac{\beta}{v_0}\right)\left(\frac{v}{v_0}\right)^{(\beta-1)} e^{-\left(\frac{v}{v_0}\right)^{\beta}}$	$P_e(v) = P_r \times \begin{cases} 0, & v < v_{ci}\,or\,v > v_{co} \\ P_{cinr} & (v), v_{ci} \leq v \leq v_r \\ 1, & v_r \leq v \leq v_{co} \end{cases}$	where $P_e(v)$ is electric output power of WT; v_{ci}, v_{co} and v_r represent cut-in, cut-out, and the rated wind speed, respectively [50,51].
$f(v) = \begin{cases} \frac{v}{\sigma^2} e^{-v^2/2\sigma^2}, & v > 0 \\ 0, & v \leq 0 \end{cases}$	$P(v) = \frac{\beta}{v_0}\left(\frac{v}{v_0}\right)^{\beta-1} e^{-\left(\frac{v}{v_0}\right)\beta}$ with $\beta = 1/3$; $v_0 = b$;	The authors suggest that wind velocity follows Rayleigh distribution, whereas the power follows Weibull distribution [52].
$f(v) = \left(\frac{\beta}{v_0}\right)\left(\frac{v}{v_0}\right)^{(\beta-1)} e^{-\left(\frac{v}{v_0}\right)^{\beta}}$	$P(v) = \begin{cases} P_r \cdot \left(\frac{v_{co}^n - v^n}{v_{ci}^n - v_r^n}\right), & v_{ci} \leq v \leq v_r \\ P_r, & v_r \leq v \leq v_{co} \\ 0, & \text{otherwise} \end{cases}$	where $P(v)$ is generated power at speed v, and v_{ci}, v_{co} and v_r are wind turbine parameters [53].
$f(v) = \left(\frac{\beta}{v_0}\right)\left(\frac{v}{v_0}\right)^{(\beta-1)} e^{-\left(\frac{v}{v_0}\right)^{\beta}}$	$P(v) = \begin{cases} 0, & v < v_{ci}\,or\,v > v_{co} \\ P_r\left(\frac{v-v_i}{v_r-v_i}\right), & v_{ci} \leq v \leq v_r \\ P_r, & v_r \leq v \leq v_{co} \end{cases}$	Here β and v_0 are the shape and scale parameters; P is power output against wind speed [54,56].
$f(v) = \left(\frac{\beta}{v_0}\right)\left(\frac{v}{v_0}\right)^{(\beta-1)} (e)^{-\left(\frac{v}{v_0}\right)^{\beta}}$	$P = \frac{\rho A}{2} v^3 c_p(\lambda)$	[55]
$f(v) = \left(\frac{\beta}{v_0}\right)\left(\frac{v}{v_0}\right)^{(\beta-1)} e^{-\left(\frac{v}{v_0}\right)^{\beta}}$	$P_{m,R} = \sum_{j=1}^{n}\left[\frac{1}{2}\rho v_{m,j}^3 f(v_j)\right]$	[57]
$f_R(v) = \left(\frac{\pi}{2}\right)\left(\frac{v}{v_m^2}\right)\exp\left[-\left(\frac{\pi}{4}\right)\left(\frac{v}{v_m^2}\right)^{\beta}\right]$	$P_R = \frac{3}{\pi}\rho v_m^3$	

2.4. Review of Wind Power Forecasting with the Incorporation of NN

WD has emerged as the most accurate and effective technique for wind power forecasting; however, errors may occur in the calculated value. This requires continuous monitoring and correction of the forecasting to make the RESs based system more sustainable. The appearance of error in the forecasted value can be handled through its computation using NNs and incorporating it further in the calculations. For the purpose, researchers have incorporated NN or Artificial Neural Networks (ANN) with WD to make forecasted values more precise [58,59].

The authors in [60] proposed a model that utilized the ANN to foresee the wind speed data, which had similar sequential and seasonal features to the actual wind data. The model was tested on wind speed databases from Mersing, Kudat, and Kuala Terengganu in Malaysia for its authentication. The results indicated that the presented hybrid artificial neural network (HANN) model had the capability of illustrating the fluctuations in the wind speed during different seasons of the year at different locations. However, the model did not consider a range satisfying lower and upper bound of estimations that are important in forecasting mechanisms. The authors presented a wind speed prediction model designed by the integration of WD and ANN. They addressed the crucial need for wind speed forecasting and seasonal variations. The proposed model utilized ANN to predict the wind speed data, which had similar sequential and seasonal characteristics as of the actual wind data. The proposed mechanism was authenticated by applying it to the wind speed data collected from different locations in Malaysia. The inverse transform of the PDF, designed by combination of WD and random variables that were implemented for modeling of wind speed is given in Equations (9) and (10).

$$U = F(v) = 1 - \exp\left[-\left(\frac{v}{\lambda}\right)^k\right] \tag{9}$$

$$v = c\left[-\ln(1 - U)^{\frac{1}{k}}\right] \tag{10}$$

where U is the uniformly distributed random variable between [0, 1]. The authors concluded that the proposed Hybrid ANN (HANN) improved the effectiveness and efficiency of WD by considering the variations in the seasonal characteristics. The authors did not consider the duration of the forecast short, medium, or long term.

Reference [61] proposed a lower-upper bound estimation (LUBE) method and extended it to develop prediction intervals (PIs) using NN models sorting out the problem in [60]. The presented technique translated the primary multi-objective problem into a constrained single-objective problem [62]. In comparison to the cost function, the presented mechanism was nearer to the principal problem and had lesser parameters. PSO was combined with the mutation operator to solve the problem. Comparative analysis of the obtained results showed that the proposed method could construct higher-quality PIs for load and wind power generation forecasts in a much shorter time.

The authors in [63] investigated two different methods for wind power forecasting. A comprehensive comparison was performed using ANN, and a hybrid prediction method was used for wind power perdition, and a comprehensive comparison was performed. The authors performed short-term wind power prediction for a wind farm having 40 wind generators. The computations concluded that the individual ANN prediction method yielded the estimated results swiftly but, precision in forecasted data was low and the root mean squared error (RMSE) was 10.67%. On the other hand, the hybrid prediction method operated slowly but, the prediction accuracy was high and the RMSE was 2.01%. Also, in contrast to [62], the authors in [63] considered the impact of wind speed on error emergence. They concluded that the prediction errors were small when the wind speeds were lower than 5 m/s or higher than 15 m/s.

References [64,65] designed an algorithm based on an extreme learning machine (ELM) for computation of shape and scale parameters of WD. The authors in [65] tested the algorithm developed in [64] and compared the results obtained with those obtained from support vector machine (SVM) and genetic programming (GP) for estimation of the same Weibull parameters. The wind density calculated using the wind speeds was computed through Equation (1). The coefficient of determination used by the authors in [64] was calculated using Equation (11).

$$\mathcal{R}^2 = \frac{\sum_{i=1}^{n}\left(X_{i,act} - X_{act,avg}\right)^2 - \sum_{i=1}^{n}\left(X_{i,est} - X_{i,act}\right)^2}{\sum_{i=1}^{n}\left(X_{i,act} - X_{act,avg}\right)^2} \tag{11}$$

where the value of co-efficient of determination provided a linear relationship between the actual and estimated values. The authors concluded that the developed algorithm improved the precision level in the estimation of Weibull parameters and also performed the calculation for the available wind power that was not done in [63].

In [66], the authors emphasized mechanisms to integrate multiple wind farms with the aim of enhancement in the wind power capacity and decrease in the wind power curtailment. The authors used WD approximation and maximum-likelihood estimation methods for forecasting. They also presented an algorithm QRCNN, designed by the combination of the convolutional neural network (CNN) and quantile regression technique to achieve detailed quantiles of corresponding predicted wind power output from the system. The forecasted wind power was calculated using Equation (12).

$$P_t = F\left(P_{t-k}, W^h_{t-k}, W^h_t, W^h_{t+m}\right) \tag{12}$$

where P_t denotes the forecasted wind power at time t; P_{t-k} denotes all historical wind power data between $t - k$ and t; W denotes the wind speed data from numerical weather prediction (NWP); h represents all wind speed components, which contains horizontal and vertical information, in general. They concluded that based on the forecasted multidimensional random variables and joint distribution function, the output generation scenario for multiple wind farm power could be achieved.

Reference [67] presented a wind speed forecasting mechanism called WindNet that was based on Convolutional Neural Networks (CNN). The forecasting mechanism was designed to provide the predicted data for the following three days. Wind speed data was accumulated on an hourly basis for the previous seven days, and the data set was developed as $24 \times 7 = 168$ sets. The WindNet performed 1D convolution on the collected data and authors used 16 filters that developed 168×16 1D convolved map shapes. The authors tested WindNet for a wind site in Taiwan to examine its efficiency and compared the results with following four techniques; Support vector machine (SVM), Random Forest (RF), Decision tree (DT), and Multi-layer perception (MLP). The authors used MAE and RMSE as indicators to estimate the performance of the presented architecture. Based on the comparative results, the authors concluded that the designed architecture presented better and more efficient results than MLP and DT while SVM showed the worst performance.

The authors in [68] combined Principal Component Analysis (PCA) and Independent Component Analysis (ICA) to improve the accuracy of the wind forecasting, and it was then forwarded to Radial Based Function (RBF) for improving the prediction accuracy of wind speed. The developed mechanism worked by processing the sample that can weaken the mutual interference among multiple factors to obtain precise independent components, resulting in improved accuracy of the predicted wind speed. The authors compared the results with the traditional wind speed forecasting models like backpropagation (BP) and Elman neural network (ENN). They concluded on the basis of extracted results that their developed architecture performed better, as it made proper and complete use of the available information in contrast to the other NN based wind prediction schemes.

Table 3 summarizes the discussed approaches, proposed by different researchers to use WD and NN for wind power forecasting and prediction error.

Table 3. Wind resource and power prediction with WD and NN and prediction error.

Resource/Power Forecasting Model	Prediction Error	Description
$P = \frac{\rho A}{2} v^3 c_p$	$MAPE = \frac{1}{n} \sum_{t=1}^{n} \frac{y_{i(ANN)} - y_{k(measured)}}{y_{k(measured)}}$	Here $\bar{v}$, σ and Γ are mean wind speed, standard deviation, and gamma function, respectively. Also, n, $y_{i(ANN)}$ and $y_{k(measured)}$ are total input and output pairs, forecasted wind speed, and actual wind speed for one hour, respectively [61].
$\bar{P} = \frac{1}{2n} \rho \sum_{i=1}^{n} \bar{v^3}$	$MAPE = \frac{1}{n} \sum_{i=1}^{n} \left\| \frac{P_{i,pred} - P_{i,means}}{P_{i,means}} \right\| \times 100$ $MABE = \frac{1}{n} \sum_{i=1}^{n} \left\| P_{i,pred} - P_{i,means} \right\|$ $R^2 = \frac{\sum_{i=1}^{X} \left(P_{i,means} - P_{means,avg} \right)^2 - \sum_{i=1}^{n} \left(P_{i,pred} - P_{i,means} \right)^2}{\sum_{i=1}^{X} \left(P_{i,means} - P_{means,avg} \right)^2}$	Here n denotes the specified time period $P_{i,pred}$ and $P_{i,means}$ are predicted and calculated wind powers [64].
$\bar{v} = \frac{1}{n} \sum_{i=1}^{h} v_i$ $\sigma = \left[\left(\frac{1}{n-1} \sum_{i=1}^{n} (v_i - \bar{v})2 \right) \right]^{0.5}$ $P = \frac{\rho A}{2} v^3 c_p$	$R^2 = \frac{\sum_{i=1}^{n} \left(X_{i,act} - X_{act,avg} \right)^2 - \sum_{i=1}^{n} (X_{i,est} - X_{i,act})^2}{\sum_{i=1}^{n} \left(X_{i,act} - X_{act,avg} \right)^2} \frac{\partial^2 \Omega}{\partial u^2}$ $MAPE = \frac{1}{n} \sum_{i=1}^{n} \left\| \frac{X_{i,est} - X_{i,act}}{X_{i,act}} \right\| \times 100$ $MABE = \frac{1}{n} \sum_{i=1}^{n} \left\| X_{i,est} - X_{i,act} \right\|$ $RMSE = \sqrt{\frac{1}{n} \sum_{i=1}^{n} \left(X_{i,est} - X_{i,act} \right)^2}$	[65]
$P_t = F\left(P_{t-k}, W_{t-k}^h, W_t^h, W_{t+m}^h \right)$	$\underset{\theta \ = \ weights}{min} \sqrt{\sum_{t=1}^{n} \left[y_t - F(X_{t,i}, \theta) \right]^2}$	Here n denotes the specified time period, and $P_{i,pred}$ and $P_{i,means}$ are predicted and calculated wind powers, respectively [66].

In addition to the above-reviewed techniques for wind forecasting, the authors in this paper also give consideration to machine learning for wind prediction and have presented a brief review of the research work based on machine learning-based wind forecasting.

The authors in [69] presented a wind forecasting model as an application of machine learning. They developed a neuro evolutionary technique of Cartesian genetic programming to evolve ANN for the development of the resource prediction model. The proposed model was developed using three different forecasting models, and each model predicted the generation of wind power for next one hour. The authors calculated percentage error such as MAPE, NRMSE of the calculated values in time series. The NRMSE was calculated using Equation (13).

$$NRMSE = \sqrt{\frac{1}{N}\sum_{i=1}^{N}\left(\frac{P_{ia} - P_{if}}{P_{ia}}\right)^2} \tag{13}$$

where P_{ia} is the observed power, P_{if} is the forecasted power at the instant i, and N is the number of hours. The authors concluded the MAPE improved the accuracy of the forecasted value by the models making the system more reliable and consistent. They also concluded that the proposed model could be further improved by introducing parameters like wind flow direction at the site, instantaneous humidity, atmospheric temperature, and pressures.

3. Forecasting Solar PV Power Generation

3.1. Solar PV Power Generation Fundamentals

Solar energy is the most abundantly available source for electric power generation. Amongst RESs, Solar PV has been widely implemented throughout the world, and countries are continuously shifting towards such sources from conventional fossil fuels. According to a report published by World Energy Council (WEC) in 2016, the installed potential of solar PV generation reached 227 GW till the end of 2015, and future projects around the globe have a target to at least double this generation around the globe by 2022 [70]. PV power generation depends upon the solar irradiance and the power in relation to it is calculated using Equation (14).

$$P = \gamma S\eta(1 - n\Delta t) \tag{14}$$

where P, γ, η, S, Δt and n stand for solar active power, amount of solar irradiance, efficiency, a total area of PV modules, PV cell temperature's forecast error, and co-efficient of the temperature, respectively [71].

Although the variability of solar power is less than the wind, and it also has one factor affecting its predictability (i.e., cloud cover) besides, the consistency of PV generation is also highly region-dependent. Still, it requires substantial consideration of its forecasting [4,72,73]. Initially, WD was used for forecasting of wind power; however, now, many researchers have effectively used it for the forecasting of solar power as well.

PV systems suffer a major issue of being highly dependent on the direct impact of sunlight; this results in a 10–25% loss of efficiency if a properly designed tracking system is not employed [74]. Besides, cloud cover, dust accumulation, and impediments present in the atmosphere also reduce the power output [75]. The mentioned issues make it essential for power system planners and utility companies to have an accurately designed forecasting mechanism for sun irradiance. WD after making its mark in the wind power forecasting has started finding its significance in the PV power forecasting as well. Like in wind power forecasting, solar power forecasting also may require the inclusion of NN to bound prediction of error. The following sub-sections consider both calculations with and without inclusion of NN.

3.2. Weibull Distribution (WD) for Solar (PV) Forecasting

WD, for its reliability and accuracy, has started to find its mark in the forecasting of solar radiation as well. PV cells generally follow a bathtub curve, where they have three stages for their working mechanism. The output of PV modules directly depends on the irradiance, so its forecasting has a major part in the planning of the working mechanism of the power system. The basic formulas of PDF and CDF of WD have already been discussed in Equations (2) and (4), respectively. The reliability function $R(t)$ and the failure rate $\lambda(t)$ for irradiance calculation are given in Equations (15) and (16), respectively.

$$R(t) = e^{-(\frac{t-\gamma}{\eta})\beta} \tag{15}$$

where η, β and γ are the parameters for scale, shape, and location, respectively.

$$\lambda(t) = \frac{f(t)}{R(t)} = \frac{\beta}{\eta}\left(\frac{t-\gamma}{\eta}\right)^{\beta-1} \tag{16}$$

where $\lambda(t)$ is the failure rate function over time, and it requires the values of η and β to be computed for finding the failure rate [76]. The failure rate is important as it provides information about the accuracy of the system and makes the computations more reliable.

The preceding sub-sections explain the mechanisms that have already been developed for the forecasting using stochastic techniques (such as WD) for PV power without and with the incorporation of the NN or ANN.

3.3. Review of PV Power Forecasting without NN

PV power has been widely used but, the problem with these panels is of their low efficiency and impact of environmental conditions on their performance. These efficiency and performance issues require an adequately designed forecasting mechanism of solar irradiance as the PV power output is directly proportional to the amount of irradiance available.

Reference [77] presented a very comprehensive solar irradiation forecasting analysis by computing Global Horizontal Solar Irradiation (GHI) and annual Direct Normal Solar Irradiation (DNI) probability density functions. Annual DNI and GHI distributions were defined through WD and normal distribution functions, respectively. They concluded that Weibull fitting of annual DNI distributions provided very appreciable results as the yielded uncertainties in scale and shape parameters were ~1% and ~15%, respectively. In [78] the author presented a suite of largely applicable and value-based metrics for solar forecasting to accommodate a comprehensive set of scenarios like different time horizons, geographic locations, and applications to improve the accuracy of solar forecasting. The results showed that the proposed metrics proficiently calculated the quality of solar forecasts and also assessed impacts on economics and reliability due to the improved solar forecasting. Sensitivity analysis resulted in achieving the suitability of the proposed scheme to enhance precision in solar irradiance forecasting with uniform forecasting improvements.

The authors in [79] proposed a scheme, designed by using the Pearson system based on the calculation of probability distribution by matching theoretical moments with empirical moments. The authors developed a data processing system to perform distribution fitting and future potential analysis. The equation used by the authors for computation is given in Equation (17).

$$\frac{f'(x)}{x(x)} = \frac{A(x)}{B(x)} = \frac{x-a}{c_0 + c_1 x + c_2 x^2} \tag{17}$$

where f' is the density function, a, c_0, c_1 and c_2 are the distribution parameters, and x is the variable. The authors compared the results through plots with the computed values of WD and the actual value. The results ignored several seasonal variations but confirmed that WD maintained its efficiency, in comparison to Pearson parameter irradiance forecasting [79].

Reference [80] proposed a model by combining different distribution schemes for different purposes understudy in their research. They also considered a clear-sky index for PV power production that was ignored in [79], and the mathematical formulas used for calculation of mean and variance of WD are mentioned in Equations (18) and (19), respectively.

$$\mu = \lambda\Gamma(1 + 1/k) \tag{18}$$

$$\sigma^2 = \lambda^2\Gamma(1 + 2/k) - \mu^2 \tag{19}$$

The authors used the data obtained from the above equations in their simulation. They concluded that besides forecasting, seasonal variations in PV power production needs to be considered as the application of forecasting model could be helpful in grid designing and future power system planning. However, the possible emergence of error was not considered by the authors.

In [81], the author determined the most efficient distribution for global radiation modeling and measured it for the Iadan site in Nigeria. The WD function used by the author for solar irradiance forecasting is given in Equation (20).

$$f(x) = \frac{\beta}{x_0}\left(\frac{x}{x_0}\right)^{\beta-1}\exp\left[-\left(\frac{x}{x_0}\right)^{\beta}\right] \tag{20}$$

where $f(x)$ is the WD for solar radiation x. The author concluded that the logistic distribution along-with WD appeared as the most appropriate distribution function for global solar radiation modeling as the percentage error calculated was the lowest.

The authors in [82] developed and tested the solar radiation models for the city of Tirana, Albania. The models were used for the estimation of the monthly average total solar radiation on the horizontal surface based on the measured data for solar radiation intensity and the time duration. The monthly average daylight hours were calculated using Equation (21).

$$N = \frac{2}{15}\cdot\omega_s \tag{21}$$

Along-with WD, other models were also considered but, the WD stood-out because of its accuracy and reliability and hence provided better results. After statistical analysis, this model presented the most appropriate results for the solar radiation prediction model. However, this designed system lacked tractability, and the authors in [83] tackled this issue.

Reference [83] proposed a new mechanism for enhancing the overall efficiency by using the tracking system designed by the V-trough technique. The formula for the calculation is given in Equation (22).

$$f_{(x)} = \frac{k\Gamma(1 + 1/k)}{\mu}\left(\frac{x\Gamma(1 + 1/k)}{\mu}\right)^{k-1}\times e^{-\left(\frac{k\Gamma(1+1/k)}{\mu}\right)^{k}} \tag{22}$$

where Γ is the gamma function. The authors concluded that the presented techniques permitted adjustment of the range, shape, and the bias function towards the desired mean of WD. The proposed mechanism also provided better control, more flexibility.

Table 4 summarizes the discussed mechanisms for solar irradiance distribution functions and PV based power production.

Table 4. Solar prediction with WD and without incorporation of NN.

Solar Distribution Functions for Prediction	PV Power Production	Reference
$f(t) = \frac{\beta}{\eta}\left(\frac{t-\gamma}{\eta}\right)^{(\beta-1)} \times e^{-\left(\frac{t-\gamma}{\eta}\right)^{\beta}}$ $R(t) = e^{-\left(\frac{t-\gamma}{\eta}\right)^{\beta}}$ $f(t) = \frac{\Gamma\left(\frac{v+1}{2}\right)}{\sqrt{v\pi}\Gamma\left(\frac{v}{2}\right)}\left(1+\frac{t^2}{v}\right)^{\left(-\frac{v+1}{2}\right)}$	$P = \gamma S\eta(1-n\Delta t)$	Here $R(t)$, β, γ and η are the reliability function, slope, location and scale parameters, respectively [78]; P, γ, η, S, Δt and n stand for solar active power, amount of solar irradiance, efficiency, the total area of PV modules, PV cell temperature's forecast error, and co-efficient of the temperature, respectively.
$f\left(\overline{X}; x, k\right) = \begin{cases} \frac{\beta}{x}\left(\frac{\overline{X}}{x}\right)^{(k-1)} e^{-\left(\frac{\overline{X}}{x}\right)^{\beta}}, & x \geq 0 \\ 0, & x < 0 \end{cases}$	$P = \max\limits_{P \leq P_{SET}} \{\eta P_{PV}(V)\}$ $P = \gamma S\eta(1-n\Delta t)$	Here P, $P_{PV}(V)$ and η are active power, active power-voltage relationship, and converter efficiency, respectively [80].
$f(x) = \frac{\beta}{x_0}\left(\frac{x}{x_0}\right)^{\beta-1} \exp\left[-\left(\frac{x}{x_0}\right)^{\beta}\right]$ $\beta = \left[\frac{\sum\limits_{i=1}^{n} T_I^k \ln(x)}{\sum\limits_{i=1}^{n} x^k} - \frac{\sum\limits_{i=1}^{n} \ln(x)}{n}\right]^{-1}$ & $x_0 = \left[\frac{1}{n}\sum\limits_{i=1}^{n} x^k\right]^{\frac{1}{k}}$	$P = \gamma S\eta(1-n\Delta t)$	[81]
$f_{(x)} = \frac{k\Gamma(1+1/k)}{\mu}\left(\frac{x\Gamma(1+1/k)}{\mu}\right)^{k-1} \times e^{-\left(\frac{k\Gamma(1+1/k)}{\mu}\right)^{k}}$		Here $f_{(x)}$ is the solar irradiance function [83].

3.4. Review of PV Power Forecasting with the Incorporation of NN

The significance of PV power forecasting and the role of WD in this regard have been discussed but, the problem of error emergence persists with the stochastic techniques such as WD. The requirement of accurate forecasting emphasizes the inclusion of NN or ANN in the forecasting computations.

In [84], the authors reviewed the application of available ANN techniques on solar radiation prediction and identified the research gap. They also discussed the prediction accuracy of ANN models regarding dependency on different combinations of input parameters, training algorithms, and architectural configurations. Further research areas in ANN-based methodologies are also identified in the presented study.

In [85], the authors proposed a mechanism for PV-wind hybrid generation systems employed for the residential load. They also presented steps supportive in the enhancement of hybrid system penetration at the distribution voltage level. The equation used for the calculation of the output PV power is given in Equation (23).

$$P_{pv} = P_{n_{pv}} \times \left(\frac{G}{G_{STC}}\right) \times \left(1 + K(T_{cell} - T_{STC})\right) \tag{23}$$

where G and G_{STC} are the solar irradiance and the solar irradiance in standard testing conditions. The authors investigated different levels of penetrations in residential and commercial applications. Their analysis concluded that the proposed system improved the performance of residential distributed generation with a 1-min temporal resolution along-with the incorporation of active and reactive powers. The mechanism worked efficiently for low power applications, ignoring the requirement of power at large scale.

Reference [86] performed the prediction of global horizontal irradiation (GHI) for different locations in Zimbabwe. The proposed NN contained 10 neurons and a tensing transfer function for both input and output layers. The formulas for the calculation of inputs to the output neurons and final output are given in Equations (24) and (25), respectively.

$$B_j = \sum_{j=1}^{10} O_j W_{pj} \tag{24}$$

$$O_f = \frac{2}{1 + e^{-2B_j}} - 1 \tag{25}$$

The authors in [86] also considered statistical indices (RMSE, MPE, R^2, etc.) for achieving better accuracy of the forecast. The authors concluded that the pure linear transfer function emerged as the worst performer amongst the tested transfer functions, and the proposed model predicted the GHI for the specified period with relatively good accuracy.

In [87], the authors presented a new model formed by the integration of the advantages of non-linear artificial neural networks and the linear auto-regressive moving average (ARMA). The mathematical representation of the ARMA model used by the researchers is given in Equation (26).

$$Y_t = \sum_{I=1}^{p} \phi_i \, Y_{t-1} - \sum_{j=1}^{q} \theta_j e_{t-j} + e_t \tag{26}$$

where ϕ_i, θ_j and e_t are the auto-regressive parameter, average parameter, and white noise with variance, respectively. The authors also included the statistical indices for improving the accuracy of forecast by computing the percentage error in the calculation. The authors concluded that the proposed integrated model showed better results than the individual performances of ARMA and ANN, especially in terms of the statistical indices.

Reference [88] also presented a prediction model developed by the combination of Empirical Model Decomposition (EMD) and ANN for long-term prediction of the intensity of solar irradiance. The formula used to find the standard deviation (SD) is given by Equation (27).

$$SD = \frac{\sum_{k=1}^{T} \left(\left| h_{1(k-1)}(t) - h_{1k}(t) \right|^2 \right)}{h_{1(k-1)}^2(t)} \tag{27}$$

where T is the length of the sequence. The authors used daily historical data in the proposed system. They concluded that the predicted results showed the system to be more accurate with a simplified calculation model than the many available mechanisms.

Reference [89] proposed an ANN-based forecasting model for the PV generation system. The developed model was provided with available data for initialization, and this data was used in a vector having 146 network values. These values were labeled as training inputs, X, and network had just one single training target, T. The first input was taken as the season of the day is forecasted, the second input of the network was linked to the time of day, and the remaining 144 values were used to represent the solar irradiation values in an interval division of 10 min earlier 24 h. The single output of the presented ANN model was target "T' as the predicted irradiation value. The authors computed the generated power from the predicted solar irradiance values by Equation (28).

$$P_S = \eta SI(1 - 0.005(t_0 - 25)) \tag{28}$$

where P_S is generated electrical power, η is conversion efficiency co-efficient, S is the area of the module, I is the solar irradiance, and t_0 is the measured temperature. Through the results, the authors concluded that the developed model provided sufficiently high precision for the solar irradiation parameter when employed to micro-grids. They also remarked that the system would improve the instantaneous control of micro-grids, based on the fact that the uncertainty of solar irradiation was reduced and made the system more reliable.

The authors in [90] presented an algorithm based on deep neural networks, namely DeepEnergy to perform precise short-term load forecasting (STLF). The designed mechanism was initiated by flatting the pooling layer in 1D and build a structure with a fully connected output layer. The authors addressed the overfitting problem of NNs dropout technology was adopted in the fully connected layer. The authors evaluated the accuracy indexes by testing through Mean Average Percentage Error (MAPE) and Cumulative Variation of Root Mean Square Error (CV-RMSE). The experimental results were compared with five artificial intelligence algorithms commonly employed for forecasting, and the results were MAPE (9.77%) and CV-RMSE (11.66%), showing the system to be very accurate. The results concluded DeepEnergy to be a robust system with strong generalization ability. Table 5 given below summarizes the discussed schemes.

Table 5. Photovoltaic (PV) power forecasting with the incorporation of NN.

Model Used for Power Production or Resource/Power Forecasting	Prediction Error	Reference
$P_{pv} = P_{n_{pv}} \times \left(\frac{G}{G_{STC}}\right)$ $\times (1 + k(T_{cell} - T_{STC}))$	$\%VF_k = \sum_{i=1}^{n-1} \frac{\|V_{k,i+1} - V_{k,i}\|}{(n-1)} \times 100$ $\%VIF = \sqrt[2]{\frac{1 - \sqrt[2]{3-6\beta}}{1 + \sqrt[2]{3-6\beta}}} \times 100$	Here $P_{n_{pv}}$, G, G_{STC} and k are rated power of PVsystem, solar irradiance on PV surface, solar irradiance in standard test conditions, and efficiency temperature coefficient, respectively. Also, VF and *VIF* are voltage fluctuations and voltage imbalance factor, respectively [85].
$P = \gamma S\eta(1 - n\Delta t)$ $R = R_0\left(1 - 0.75n^{3.4}\right)$ $R_0 = 990\sin\phi - 30$ $\phi = \frac{\phi_{tp} + \phi_p}{2}$	$MAE = \frac{1}{n}\sum_{i=1}^{n}\left\|GHI_{(measured)} - GHI_{(predicted)}\right\|$ $MAPE = \left(\frac{1}{n}\sum_{i=1}^{n}\left\|GHI_{(measured)} - GHI_{(predicted)}\right\|\right)$ $RMSE = \left(\frac{1}{n}\sum_{i=1}^{n}\left\|GHI_{(measured)} - GHI_{(predicted)}\right\|^2\right)^{\frac{1}{2}}$ $R^2 = \left(1 - \left(\frac{\sum_{i=1}^{n}\left\|GHI_{(measured)} - GHI_{(predicted)}\right\|^2}{GHI_{(measured)}}\right)\right)$	Here R^2, n, R_0, ϕ_{tp} and ϕ_p are solar radiation, cloud cover, clear sky insolation, solar elevation angle and for previous and current hours, respectively. R^2 is the coefficient of determination. Also, $GHI_{(measured)}$ and $GHI_{(predicted)}$ are measured and predicted solar irradiations respectively [86].
$P = \gamma S\eta(1 - n\Delta t)$ $k_t = \frac{H}{H_0}$ $H_0 = I_{SC}E_0$ $\times(\sin\delta\sin\varphi + \cos\delta\cos\varphi\cos\omega)$	$RMSE = \sqrt{\sum_{i=1}^{N}\frac{(y_i - x_i)^2}{N}}$ $MBE = \sum_{i=1}^{N}\frac{(y_i - x_i)}{N}$ $MPE = \sum_{i=1}^{N}\left(\frac{(y_i - x_i)}{Nx_i}\right) \times 100$ $R^2 = \frac{\sum_{i=1}^{N}(y_i - x_i)^2}{\sum_{i=1}^{N}(y_i - \bar{y}_i)^2}$	Here k_t, H and H_0 are clearance index, global ground radiation, and extraterrestrial global radiation, respectively. Also, y_i and x_i, $\bar{y}$ and $\bar{x}$ are estimated, measured, and average estimated and measured values, respectively [87].
$P = \gamma S\eta(1 - n\Delta t)$	$RMSE = \sqrt{\frac{1}{n}\left(\sum_{i=1}^{n}\left(X_{hist,i} - X_{pred,i}\right)^2\right)}$ $MAPE = \left(\frac{1}{n}\sum_{i=1}^{n}\left\|\frac{X_{hist,i} - X_{pred,i}}{X_{hist,i}}\right\|\right) \times 100\%$ $R = \frac{\sum_{i=1}^{n}\left(X_{hist,i} - \bar{X}_{hist}\right)\left(X_{pred,i} - \bar{X}_{pred,i}\right)}{\sqrt{\sum_{i=1}^{n}\left(X_{hist,i} - \bar{X}_{hist}\right)^2\left(X_{hist,i} - \bar{X}_{hist,i}\right)^2}}$	*hist* and *pred* historical and predicted results [88]. P, γ, η, S, Δt and n stand for solar active power, amount of solar irradiance, efficiency, total area of PV modules, PV cell temperature's forecast error, and co-efficient of the temperature, respectively.

In addition to the above-reviewed techniques for solar irradiance forecasting, the authors in this paper also give consideration to machine learning for solar irradiance prediction and have presented a brief review of the research work based on machine learning-based wind forecasting.

Reference [91] introduces the mechanism for hourly forecasting of solar irradiance using machine learning algorithms. The developed prediction model was designed in two forms: the first step used the environmental parameters like temperature, pressure, wind speed, and relative humidity. This network was trained to find the new input values by using the data of the previous few months or years. This mechanism made the system capable of predicting the solar irradiance with relative accuracy by correcting the errors using ANNs; the second step used the time-series prediction of the solar irradiance. This system developed models to determine future values. This mechanism required a continuous database of solar irradiance to predict future values. The developed systems were assisted with multi-layer feed-forward neural networks (MLFFNN), radial basis function neural networks (RBFNN), support vector regression (SVR) and adaptive neuro-fuzzy inference system (ANFIS) for improving the precision of resources forecasting.

4. Optimal Economic Dispatching (OED) using PSO

Economic dispatch (ED) has been a significant concern for planners due to the rising fossil fuel prices and long-distance transmission from hydro-electric plants. With the growing penetration of RESs, the significance of ED gets re-emphasized because large scale power generation from RESs is possible only at distant locations where a number of wind turbines and/or PV panels can be installed, integrated, and utilized [92,93]. Considering the bulk amount of energy that can be extracted from large-scale RE farms, researchers and power system planners have presented many ED schemes to dispatch the generated power efficiency. PSO stands-out because of its fast convergence, simplified approach, and flexibility. Before presenting the review of the solution of the OED problem by PSO, a brief review of the PSO algorithm is presented here.

4.1. A Brief Review of PSO Algorithm

PSO is a stochastic algorithm used to obtain the most suitable solution for the optimization problems and was proposed by Kennedy in 1995. The algorithm of PSO is based on the simulation of behavior in which flock of birds flies together in a multi-dimensional search space, adjusting their movements and distances is to discover an optimum objective, subject to the constraints imposed [94]. PSO is described graphically in Figure 1.

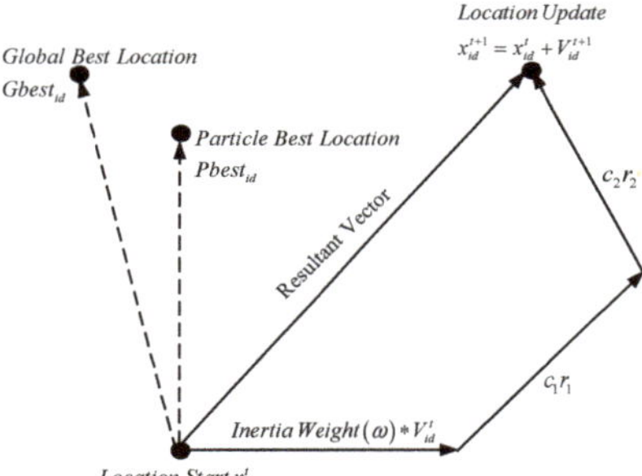

Figure 1. Particle swarm optimization (PSO) vector diagram.

Mathematically, PSO velocity and position formulas are expressed by Equations (29) and (30).

$$v_{id}^{t+1} = \overbrace{\omega v_{id}^{t}}^{Inertial\ Component} + \overbrace{c_1 r_1()(pbest_{id} - x_{id}^{t})}^{Cognitive\ Component} + \overbrace{c_2 r_2()(gbest_{gd} - x_{id}^{t})}^{Social\ Component}, \tag{29}$$

$$x_{id}^{t} = x_{id}^{t} + v_{id}^{t+1}; i = 1, 2, \ldots, n; d = 1, 2, \ldots, m \tag{30}$$

where i is particle's index, t discrete-time index, d dimension being considered, n number of particles in a group, m dimensions of a particle, ω inertia weight factor, and c_1, c_2 acceleration coefficient for the cognitive and social components, respectively [95–97].

4.2. Review of PSO Applied to OED Incorporating RESs

In [98], the authors proposed a linearized network model in the form of the DC power flow model while considering the thermal limits of transmission lines and real power constraints. The cost curves for generating units have been developed in the form of a piecewise linear model, and simulation was analyzed on HOMER software for intermittent RESs, specifically wind turbines. They concluded that the steady-state analysis of the power system with the inclusion of RESs is very important and results in significant cost saving.

Reference [99] presented a chaos PSO based algorithm for dispatch cost reduction of hybrid power system by treating wind and solar power generations as negative load. The proposed algorithm gives the best convergence competence and search performance in evaluation. They concluded that chaos PSO provides better results in convergence time and efficient mechanism for the use of RESs.

The incorporation of RESs in the power system reduces the emissions that pollute or environment. However, large-scale power generation using these resources is possible only at locations that are hundreds or thousands of kilometers away from the load centers. This emphasizes the use of a properly designed dispatch mechanism [100].

The authors in [101] designed the double-weighted PSO (DWPSO) to cater to the non-convexity in combined emission economic dispatch (CEED) when intermittent wind energy is used. Equation (31) shows the formula used for conversion double objective CEED problem into a single optimization problem.

$$\min\left\{C = \sum_{i=1}^{m} F_i(P_i) + h\sum_{i=1}^{m} E_i(P_i)\right\} \tag{31}$$

where h is the ratio between maximum fuel cost and maximum emission for each unit given in Equation (32).

$$h = \frac{F_i\left(P_i^{\mathrm{pmax}}\right)}{E_i\left(P_i^{\mathrm{pmax}}\right)} \tag{32}$$

The designed algorithm resulted in the successful reduction of the fuel cost by providing a solution to the non-convex wind penetration in the power system. The authors also concluded that the solution proposed also showed a decline in perilous emissions.

Reference [102] developed a mechanism of optimal operation for the distributed generation at the micro-grid (MG) level, in contrast to the algorithm developed for large-scale generation presented in [101]. The authors also performed the resource estimation for wind and calculated the power that could be extracted from wind and PV (without forecasting). They also focused on the formulations for the economic operation of small-scale energy zones. The designed cost function for generation and operation and maintenance are given in Equations (33) and (34), respectively.

$$Cost_{gen,si} = Cost_{gen,MT,si} + Cost_{gen,FC,si} + Cost_{gen,CHP,si} \tag{33}$$

$$Cost_{O\&M,s} = Cost_{O\&M,WT,s} + Cost_{O\&M,PV,s} + Cost_{O\&M,MT,s} + Cost_{O\&M,FC,s} + Cost_{O\&M,CHP,s} \tag{34}$$

where these functions are of generation cost, transaction powers, and O&M cost pollutant emission. The authors solved the dispatch problem in the imperialist competitive algorithm (ICA), and the results proved to be better than the Monte-Carlo simulation (MCS). In all the cases mentioned above, the requirement is to satisfy the load requirements as mentioned in Equation (35).

$$\begin{aligned} &\sum_{i}^{n} P_i = P_D + P_l \\ &P_{Gi}^{\mathrm{min}} \leq P_{Gi} \leq P_{Gi}^{Max}, i = 1, 2, \dots, N_G \end{aligned} \tag{35}$$

where P_t, P_D, and P_l are total power, power demand, and power losses respectively, and the second portion of Equation (35) presents the generation limits that must be met to keep the system balanced.

4.3. Constraints Handling by PSO

Despite many benefits, the application of PSO requires proper consideration of certain constraints such as frequency fluctuation, ramp-rate limits, compensation of load variation, and battery storage. Some prominent schemes to manage these constraints while performing OED of the generated energy are discussed in the following sub-sections.

4.3.1. Compensation for Load and Voltage Variation

The occurrence of load variation is a major limitation, as it disturbs the normal operation of the power system. The load variation is eminent, but the researchers and planners are interested in designing the mechanisms that have the capability of compensating the load variations. A model of Multi-Microgrid (MMG) for voltage regulation against load variations has already been discussed.

Reference [103] proposed an optimal generation rescheduling mechanism designed through dynamic PSO, for RESs to overcome load variations. Dynamic PSO worked by the variation in the acceleration co-efficient, and this resulted in fast convergence, improved global search capability, and achievement of the global optima at the end-stage. The calculation for the amount of solar and wind powers was performed using Equation (36).

$$P_S + P_W \leq \eta \times P_D^a \tag{36}$$

where P_S, P_W, and P_D^a are the solar power, wind power, and the actual load demand, respectively. Solar and wind powers generated have been taken as a negative load. The effectiveness of the proposed solar-wind system with the hybrid system emerged with a reduction in the fuel cost. The formula for calculation of the percentage reduction is given in Equation (37).

$$\Delta C = \left(1 - \frac{F_{fR}}{F_{fN}}\right) \times 100 \tag{37}$$

where F_{fR} and F_{fN} represent the fuel costs with and without RES. The authors concluded that the presented algorithm prevented premature convergence and also maintained the voltage by scheduling the generation as per load variation. But, the specific class of power generation value was ignored by the authors that is being discussed in the preceding discussions.

The authors in [104] presented Chaos PSO based optimization method for standalone MGs to achieve economic dispatch and voltage compensation. The proposed mechanism controlled not just the electricity but the cooling and heating systems as well. The power balance constraints are given in Equation (38).

$$\sum_{m=1}^{Q} P_m(t) = P_{E.L}(t) - P_{E.L,c}(t); P_H(t) = P_{H,L}(t) \& P_C(t) = P_{C.L}(t) \tag{38}$$

where $P_{E.L}(t)$, $P_{H,L}(t)$ and $P_{C.L}(t)$ are electricity, thermal and cooling load of the MG, respectively. The simulations performed by the authors proved the designed system to be capable of effectively solving the optimization problem for different situations while keeping the voltage variations in the limit and improving the overall economic and environmental efficiency.

4.3.2. Control of Frequency Fluctuations

The availability of RESs is never constant over a period of time. This uncertainty in the available supply demands proper consideration of the system frequency. This frequency control is very important in the sense that it enables the system to automatically and dynamically adjust generation to meet the load demand that improves load factor and consumption of energy. Frequency control mechanisms also increase the flexibility of the system components, which encourages the integration of a number of MGs to develop MMGs and supply good quality of energy to the load while keeping the frequency and other power system parameters in their specified limits [105]. Many researchers have investigated and presented different mechanisms to provide frequency control in RES based power system while having a proper plan for OED.

In [106], the authors presented a comprehensive review of the mitigating methods adopted to control the fluctuations of a power system containing PV cells. Load variations cause the system frequency to deviate, and the situation becomes even worse when the solar radiation causes the PV output to vary as well, as it was discussed in [107]. The authors also analyzed that the frequency fluctuations in PV based systems were less than the wind-based systems. However, this difference is of lesser significance as the modern grid system has interconnected hybrid solar and wind power systems. They also proposed that in order to compensate for the drooping frequency due to a rise in load, a battery storage system can be implemented. The authors presented simulation models to demonstrate the presented schemes for controlling power fluctuations especially variations in the frequency of PV sources.

Reference [108] presented a mechanism to control the frequency stability and ED of the power system networks. The authors presented asymptotic stability for integral frequency control to accommodate a decentralized power system. The authors then developed a distributed averaging-based integral (DAI) control that is designed to operate by sense and control system of local frequency. The formula for ED of the generated power is given in Equation (39).

$$\min_{\theta,u} \sum_{j \in N} \tfrac{1}{2} a_j u_j^2$$
$$\text{subject to}$$
$$p_j + u_j - \sum_{k \in N} B_{jk} \sin\left(\theta_j - \theta_k\right) = 0, \qquad j \in N$$
$$\left|\theta_j - \theta_k\right| \le \gamma_{jk} < \tfrac{\pi}{2}, \qquad j,k \in \varepsilon \tag{39}$$

The explicit synchronization frequency is given by Equation (40).

$$\omega^* = \frac{\sum_{j \in N} p_j + u_j}{\sum_{j \in N} D_j} \tag{40}$$

The results concluded that the proposed DAI control system provided the closed-loop stability to the system and achieved the desired ED. These conclusions were also validated by the simulations performed. However, the impact of back-up supply on frequency control was not considered.

The authors in [109] presented a droop control method for the frequency stability of RES based power systems. Unlike [108], the proposed system worked by controlling the charging power through the aggregated participation of the system frequency, which was instated as soon as the system frequency deviated from the rated frequency. The mathematical formula required to be satisfied for the power, while the system frequency remained regulated is given in Equation (41).

$$P_{agg,t,h} = P_{c,grant,t,h} + \Delta P; P_{agg,t,h} \ge P_{force,t,h} \tag{41}$$

The authors concluded through their results that the proposed system performed well in frequency regulation even when the penetration of RES increased. They also proved that the higher the RES penetration in the power system, the more significant the frequency regulation mechanism.

4.3.3. Regulation of Ramp-Rate Limits

The ramp rate (RR) is defined as the rate of change in power at the given time interval. If the change in power is positive, it is known as a ramp-up, and if the power change is negative, then it is known as a ramp-down event. Ramp rate accounts for the difference in power over the specified time interval [110]. The RR occurs due to the intermittent nature of RES as the wind speed or solar irradiation endures variation during different time intervals of a day. This variation in wind speed or solar radiation causes the power supply to change, thus impacting the overall performance of the power system as it may cause other imbalances to appear and disturb the normal operation of the power system. The wind energy is more prone to these RR limits (RRL) violation than the solar radiation, because of more variation of wind speed during different periods of the day. The overall impact of power generation variation should be considered by the planners to maintain the continuity and reliability of the supply. Many researchers have developed mechanisms to measure and sort-out the RRL violations.

Reference [111] developed an optimal operation mechanism for grid-connected hybrid RESs for residential applications. The designed system incorporated a variety of sources like PV, wind, fuel cell and solar thermal and supplied electricity and heating. The modified PSO performed the optimization problem. The ramp rate occurrence and the fuel cell (FC) assembly, start and stop cycles are given in Equation (42).

$$
\begin{aligned}
P_{\min} &\leq P_{FC,i} \leq P_{Max} \\
\left(T_{t-1}^{off} - MDT\right)&(U_i - U_{i-1}) \geq 0 \\
\left(T_{t-1}^{on} - MUT\right)&(U_{i-1} - U_i) \geq 0 \\
P_{FC,i-1} - P_{FC,i} &\leq \Delta P_d \\
P_{FC,i} - P_{FC,i-1} &\leq \Delta P_u
\end{aligned}
\tag{42}
$$

where T^{off} and T^{off} denote the FC off and on time respectively, and U denotes the on-off status of the FC following the binary system MUT and MDT represents the upper and lower ramp rate limits respectively. The simulation results obtained were after implementation of the proposed system for four different cases, and assessment of the results validated that the proposed hybrid energy system was more cost-effective and simpler than the stand-alone single-source systems, even when the system must supply energy to full load demand. A comparison between the optimization of modified PSO and genetic algorithm (GA) also concluded that modified PSO was more accurate with better convergence time.

In [112], designed a stochastic ED model for hybrid power systems with the wind, solar, and thermal power plants, to solve dynamic economic emission dispatch (DEED) problems while considering the environmental constraints. They used a weighted aggregation method for enabling PSO to solve multi-objective (MO) problems. The RR constraint incorporated in [112] is given in Equation (43).

$$
P_{ij} - P_{ij-1} \leq UR_i; P_{ij-1} - P_{ij} \leq DR_i
\tag{43}
$$

where UR_i and DR_i are ramp-up and ramp-down for the ith unit, respectively. The authors based on their simulation and results, concluded that the presented model provided an optimal solution to the DEED problem. The developed system tackled the uncertainty and system imbalance issues efficiently and operated the system securely and optimally.

The authors in [113] presented a model for combined economic emission dispatch (CEED) with a PV system integrated with serval thermal generating plants. The formulated problem was tackled through a decomposition framework that divided the problem into two sub-problems. PSO, Newton-Raphson method, and binary integer programming techniques were incorporated in the designed mechanism. The proposed solution for the optimization problem is given in Equation (44).

$$
\min_{P_i, Us_j, Ps_k} \sum_{i=1}^{n} (F_i(P_i) + E_i(P_i)) + \sum_{j=1}^{m} G_j - Us_j - \sum_{k}^{NB} Ps_k - \sum_{j=1}^{m} Us_j
\tag{44}
$$

The authors concluded that the presented mechanism could reduce overall fuel costs and had the potential to reduce emissions by enhancing the share of solar generation in the power system. The hybrid optimization scheme also provided an optimal solution for the ED of generation while satisfying RRL.

4.3.4. Storage Mechanism as a Solution

RESs based supply keeps on varying throughout the time period besides, their supply is also not available during a day as wind flow can vary, or a cloud cover may emerge at any time [114–116]. This intermittent nature of these sources also gives rise to certain issues such as frequency fluctuations and RRL that requires proper backup arrangement to keep up with the continuous load demand. The backup system can be of two categories:

(i)   Thermal Power Generation is dependable, but it presents major issues such as a rise in carbon emissions, an increase in fuel cost, and special consideration is required in system coordination. We cannot use thermal plants as backup generation only as they take a significant amount of time in their start-up, and fuel cost for spinning reserve contributes to disturbing the economic dispatch that is a major concern in current power system.

(ii)   Battery Storage compensation through batteries is a modern age replacement of backup thermal plants. It requires a properly designed storage system to provide adequate power to supply the load when power from RESs is lesser than the demand. The battery storage system also provides an additional benefit of peak shaving as RES starts supplying an excessive supply for storage [117,118]. The major concern is of designing a properly designed storage system to achieve the optimum cost-saving and stability of the power system.

Many researchers have presented mechanisms based on backup storage systems that maintain continuous supply according to the load demand while keeping the system constraints within their limits.

In [119], the authors discussed the economic allocation of energy storage systems for Wind Power Distribution. The authors argued that the improper size selection and wrong placement of the storage units cause voltage instability and an undesired increase in the cost of the power system. To solve this, issue the authors presented an algorithm based on the hybrid multi-objective PSO (HMOPSO) to improve the voltage profile and cost of the power system. The proposed algorithm was designed by the combination of MOPSO, non-dominant sorting GA, and probabilistic load flow techniques. The mathematical relation used for computation of the operational cost is given in Equation (45).

$$Cost_i = \sum_{j=1}^{NG} C(P_{G_i}) + C_w + C_s \tag{45}$$

where $C(P_{G_i})$, C_w and C_s are the fuel cost of the generator, cost of wind power generator, and cost of energy storage system, respectively. They concluded through the presented simulation results that the system provided proper placement and sizing of the Energy Storage System (ESS) as well as minimizing the total operational cost and improved the voltage profile. But, the power compensation extracted from batteries was slow.

The authors in [120] presented optimization of the battery ESS (BESS) through PSO implemented on stand-alone MGs. The proposed method was designed to install a battery of optimum size to compensate for the load demand when required and simultaneously control the frequency to avoid any instability occurrence in the power system. The proposed model was also compared on an economic basis with some of the available modern technologies. The authors also discussed some of the materials used in batteries to find out the optimum material for better battery storage. They proposed polysulfide-bromine based BESS to be cost-effective than the redox-based BESS for long-duration applications. The formulas used by the authors for BESS calculation are given in Equation (46).

$$E_{bac} = \frac{r_{bp}}{1 + ST_{bp}} I_{BESS}; \quad I_{BESS} = \frac{E_{bt} - E_{bac} - E_{b1}}{r_{bt} + r_{bs}}; \quad E_{b1} = \frac{r_{b1}}{1 + ST_{b1}} I_{BESS} \tag{46}$$

where $I_{BESS}, E_{b1}, E_{bt}, E_{bac}, r_{bp}$ and r_{b1} are current through the battery, battery resistance, V_{phase} of the battery side, $V_{open\text{-}circuit}$ of battery, self-discharge resistance, and over-voltage resistance, respectively.

Based on the presented results, the authors concluded that the presented system offered very fast compensation for active power that improved the dynamic stability of the power system. The results also validated the significance of the BESS based PSO mechanism for optimum sizing of the storage system.

Reference [121] emphasized the use of ESS to tackle the inherent uncertainty of the wind power system to make the power system more reliable. The authors discussed several technologies available for ESS of different stability purposes and concluded that the properly designed battery system is key to success. The formulas incorporated for ESS power output and remaining energy level (REL) is given in Equation (47).

$$P_{bess}^{ord} = \frac{-sT_f}{1 + sT_f} P_{wind}; REL = \frac{P_{bess}^{ord}}{s} = \frac{T_f}{1 + sT_f} P_{wind} \tag{47}$$

where T_f is time constant. It is also clear from Equations (46) and (47) that the larger the value of time-constant, higher will be smoothing effect and larger will be ESS power.

The authors concluded that recent models focused on the daily dispatch of the ESS operation and control to compensate for the power fluctuations [122]. The authors also discussed the technical constraints, some of which have already been discussed in this paper for the OED of RESs. The authors also emphasized the significance of time-constant in the smoothing effect along-with system power stability and capacity rating.

The authors in [123] designed a mechanism of optimal energy storage system (ESS) along-with a PV generation system. The mechanism presented by the authors was based on four steps, i.e., prediction of load and daily power generation, an optimization process for best battery power "P^b" and state of battery charge "SOC^b", power requests calculus, and E-broker auction algorithms. The value of P^b was computed using Equation (48).

$$P^b = 2 \cdot P_{cont} \cdot x_p \tag{48}$$

where x_p is the optimization variable. SOC^b was used to trigger the power requests from the system. The method avoided the scenario faced by researchers where they had to impose a division on the battery peak shaving and energy shifting as it optimized suitable variables and prevented the ESS from capacity wasting. The presented model also prevented power losses and unforeseen peaks occurring due to the supply or absorption of power by optimizing the boundaries of battery behavior. However, the system was not effective for active distribution systems (ADSs).

Reference [124] presented a fuzzy multi-objective bi-level problem for the planning of ESS in (ADS). The authors designed a model using PSO and differential evaluation (DE) for the solution to the mentioned problem. They considered two scenarios of ADSs like peak load shaving and failure status responsibility support system. The mechanism worked by dividing the yearly data into 365 intervals according to the load demand and power output by the REGs. The objective functions developed for peak load shaving, restraining voltage ESS reserve capability are given in Equations (49) and (50), respectively.

$$\min f_1 = \min \sum_{t=1}^{24} \left(P_{NL}(t) - P_{NL,average} \right)^2$$
$$\min f_2 = \min \sum_{t=2}^{24} \left(P_{NL}(t) - P_{NL}(t-1) \right)^2 \tag{49}$$
$$\min f_3 = \min \left(\frac{1}{\sum_{t=1}^{24} P_{ava}(t)} \right)$$

and

$$P_{ava}(t) = \min \left\{ P_{ESS}^R, \frac{\left(ESS(t-1) - E_{ESS}^{min} \right)\eta_D}{\Delta t} \right\} \tag{50}$$

where $P_{NL,average}$ is the average of the net load demand of the ADS, $ESS(t)$ is energy stored in battery bank at time t, and E_{ESS}^{min} is the minimum energy stored in the battery bank of the ESS.

Based on the simulation results, the authors concluded that the propped scheduling model contributes to obtaining a reasonable planning scheme by taking into consideration the ESS operation strategy. The system implemented in designing a proper ESS can help in tackling the time-varying nature of RESs and the load demand as well.

The authors in [125] presented a stochastic planning and scheduling model for ESSs to handle the congestion in the electric power systems consisting of RESs. The model provided a design mechanism for charging the dis-charging of ESSs to handle the intermittent nature of RESs. The output power of wind and solar using Gaussian probability density function (PDF) and Monte-Carlos simulation (MCS) along-with ESS to tackle unpredictability. The objective functions for congestion management and ESS cost minimization are mentioned in Equations (51), (52) and (53), respectively.

$$of_{cm} = C_n + IN_{ESS} \tag{51}$$

$$IN_{ESS} = (DV_E \times P_{ESS} \times IP_E) \times EAC \tag{52}$$

$$C_n = \left\{ \sum_{ll=1}^{mll} \left(\sum_{nl=1}^{NL} \left(S_{nl}^{ll} \times FTR_{nl}^{ll} \right) \right) \right\} \times T_{an} \tag{53}$$

where IN_{ESS} is annual installation cost of ESS, EAC is applied converter for life-cycle, C_n is power flow through all lines, and FTR is financial transmission right for daily congestion cost, respectively. The planning system requires three identical ESSs to manage the congestion and cost of the system. The network without ESS cannot meet the constraints, and simulation results concluded that ESS not only reduced the loss but also improved the voltage profile and stability margin. Table 6 summarizes the discussed solutions to the constraints along-with their presented models and objective functions.

Table 6. Constraints and their solutions for optimal economic (OED) of Renewable Energy Sources (RESs).

Constraints	Presented Model	Objective Function	Reference
Load and Voltage Variations	$P_D^a + P_L - \sum_{i=1}^{N_G} P_{Gi} = 0$ $P_D^a = P_D^t - (P_s + P_W)$	$\min F_f(P_{Gi})$ $= \sum_{i=1}^{N_G} \left(a_i + b_i P_{Gi} + c_i P_{Gi}^2 \right)$	Here P_D^a, P_L, P_s and P_W are the load demand, transmission losses, solar and wind powers, respectively [103,104].
Frequency Fluctuations	$P_{force,t,h} = \sum_{k=1}^{K_{t,h}} P_{rate,k}^{U_n} + \sum_{j=1}^{M_{t,h}} P_{rate,j}^{M}$ $P_{c,max,t,h} = \sum_{i=1}^{N_{t,h}} P_{rate,i}$	$\min \sum_{t=1}^{T} \left\{ \sum_{g=1}^{N_g} \left(a_g \cdot \left(P_{g,t}^{G,ref} \right)^2 + b_g \cdot P_{g,t}^{G,ref} + c_g \right) \right\}$	[105–109]
Ramp-Rate Limits	$F_i(P_i) = \alpha_i P_i^2 + \beta_i P_i + \gamma_i + \varepsilon_i \exp(\delta_i \times P_i)$	$\min_{P_i, Us_j, Ps_k} \sum_{i=1}^{n} (F_i(P_i) + E_i(P_i)) + \sum_{j=1}^{m} G_j - Us_j - \sum_{k}^{NB} Ps_k - \sum_{j=1}^{m} Us_j$	[110–113]
Storage Mechanism	$P_i - V_i \sum_{j=1}^{N} V_j \left(G_{ij} \cos \delta_{ij} + B_{ij} \sin \delta_{ij} \right) = 0$ $Q_i - V_i \sum_{j=1}^{N} V_j \left(G_{ij} \sin \delta_{ij} - B_{ij} \cos \delta_{ij} \right) = 0$	$\min f_1 = \sum_{i=1}^{5} Prob_i \cdot Cost_i$ $\min f_2 = \sum_{k=1}^{n} \left(\frac{V_k - V_k^{spec}}{\Delta V_k^{max}} \right)^2$	[114–125]

5. Conclusions

This paper reviews the constraints faced and their solutions for reliable, efficient, cost-effective, and sustainable RESs based power systems. The fluctuating RES input power categorized in time and space has different solutions. In the time domain, forecasting helps plan for the supply-demand mismatch. In the spatial domain, different regions across vast space have different generation capabilities and hence can be used to balance out load mismatch. The combination of time and space variations, both of which have different solutions integrated gives the best overall system solution. The intermittency of RESs poses a major hurdle in their large-scales implementations but, supply forecasting (i.e., wind speed and solar irradiance) helps power system planners to design and implement large RESs based on power generation farms. WD emerges as an accurate, reliable and fast technique for predicting resource availability during a specified time period. The forecasting mechanisms may also require the incorporation of NN for the correction of an error in the predicted value to make the system even more precise. The RESs dependent power systems also have to face some other constraints such as load and voltage variation, frequency fluctuations, ramp-rate limits, and energy storage mechanisms in OED. The proposed mechanisms and algorithms for the solution of these constraints have been discussed and summarized. The review showed that the consideration of these constraints improves the performance of the power system for optimal economic dispatching. This paper comprehensively provides a manuscript for investors and power system planners to be able to learn about constraints and their available solutions, and it can be beneficial for researchers by providing a broad source for their literature review.

Author Contributions: Conceptualization, M.E. and G.A.; data curation, A.R.; formal analysis, I.K., G.A. and P.M.K.; investigation, U.F.; methodology, M.E., G.A., A.R. and M.N.; resources, I.K., P.M.K. and M.N.; supervision, G.A.; validation, I.K. and P.M.K.; visualization, U.F.; writing—original draft, M.E. and G.A.; writing—review & editing, M.E., A.R., M.N., U.F. and G.A.

Funding: The open access publishing fees for this article have been covered by the Texas A&M University Open Access to Knowledge Fund (OAKFund), supported by the University Libraries and the Office of the Vice President for Research.

Conflicts of Interest: The authors declare no conflict of interest.

Abbreviations

ANN	Artificial Neural Networks
ANFIS	Artificial Neuro Fuzzy Inference System
CNN	Convolutional Neural Networks
CEED	Combined Emission Economic Dispatch
DED	Dynamic Economic Dispatch
DWPSO	Double Weighted Particle Swarm Optimization
DNI	Direct Normal Solar Irradiation
ED	Economic Dispatch
EDP	Economic Dispatch Problem
ERCOT	Electric Reliability Council of Texas
ESRMC	Energy and Spinning Reserve Market Cleaning
EMD	Empirical Mode Decomposition
GHI	Graphical Horizontal Solar Irradiation
GP	Genetic Programming
HANN	Hybrid Artificial Neural Network
ICA	Independent Component Analysis
LUBE	Lower-Upper Bound Estimation
MLFFNN	Multi-layer Feed-forward Neural Networks
MAPE	Model Predictive Control
MG	Micro-grid
MMG	Multi Micro-grid
MOM	Method of Moments
NN	Neural Networks
NWP	Numerical Weather Predictor
OED	Optimal Economic Dispatch
PSO	Particle Swarm Optimization
PV	Photovoltaics

PDEM	Part Density Energy Method
PI	Prediction Interval
PCA	Principal Component Analysis
RES	Renewable Energy Sources
RF	Reliability Factor
RBFNN	Radial Basis Function Neural Networks
RMSE	Root Mean Squared Error
SVM	Support Vector Machine
SD	Standard Deviation
SVR	Support Vector Regression
SSER	Small-Scale Energy Resource
WD	Weibull Distribution
WT	Wind Turbine
WTG	Wind Turbine Generator
WEC	World Energy Council

References

1. Hansen, J.; Kharecha, P.; Sato, M.; Masson-Delmotte, M.; Ackerman, F.; Beerling, J.D.; Hearty, P.J.; Hoegh-Guldberg, O.; Hsu, S.L.; Parmesan, C.; et al. Assessing dangerous climate change, required reduction of carbon emissions to protect young people, future generations and nature. *PLoS ONE* **2013**, *8*, e81648. [CrossRef] [PubMed]
2. Nuclear Power: Totally Unqualified to Combat Climate Change. Available online: https://worldbusiness.org/ (accessed on 21 December 2018).
3. Ela, E.; O'Malley, M. Studying the variability and uncertainty impacts of variable generation at multiple timescales. *IEEE Trans. Power Syst.* **2012**, *27*, 1324–1333. [CrossRef]
4. Widén, J.; Carpman, N.; Castellucci, V.; Lingfors, D.; Olauson, J.; Remouit, F.; Bergkvist, M.; Grabbe, M.; Waters, R. Variability assessment and forecasting of renewables: A review for solar, wind, wave and tidal resources. *Renew. Sustain. Energy Rev.* **2015**, *44*, 356–375. [CrossRef]
5. Milligan, M.; Kirby, B. *Calculating Wind Integration Costs: Separating Wind Energy Value from Integration Cost Impacts*; No. NREL/TP-550-46275; National Renewable Energy Lab.(NREL): Golden, CO, USA, 2009.
6. Rondina, J.M. Technology Alternative for Enabling Distributed Generation. *IEEE Lat. Am. Trans.* **2016**, *14*, 4089–4096. [CrossRef]
7. Distributed Generation. Environmental & Energy Study Institute. Available online: https://www.eesi.org/topics/distributed-generation/description (accessed on 27 December 2018).
8. Libra, M.; Beránek, V.; Sedláček, J.; Poulek, V.; Tyukhov, I.I. Roof photovoltaic power plant operation during the solar eclipse. *Sol. Energy* **2016**, *140*, 109–112. [CrossRef]
9. Distributed Generation. OFGEM. Available online: https://www.ofgem.gov.uk/electricity/distribution-networks/connections-and-competition/distributed-generation (accessed on 27 December 2018).
10. Microgrids. Grid Integration group, Energy Storage and Distributed Sources Division. Available online: https://building-microgrid.lbl.gov/microgrid-definitions (accessed on 29 December 2018).
11. Khan, I.; Zhang, Y.; Xue, H.; Nasir, M. A Distributed Coordination Framework for Smart Microgrids. In *Smart Microgrids*; Springer: Cham, Switzerland, 2019; pp. 119–136.
12. Wan, Q.; Zhang, W.; Xu, Y.; Khan, I. Distributed control for energy management in a microgrid. In Proceedings of the IEEE/PES Transmission and Distribution Conference and Exposition (T&D), Dallas, TX, USA, 3–5 May 2016; pp. 1–5.
13. Zhaoyun, Z.; Wenjun, Z.; Mei, Y.; Li, K.; Zhi, Z.; Yang, Z.; Guozhong, L.; Na, Y. Application of Micro-Grid Control System in Smart Park. *J. Eng.* **2019**, *2019*, 3116–3119. [CrossRef]
14. Weibull Analysis, An Overview of Basic Concepts. Available online: https://www.weibull.com/ (accessed on 22 December 2018).
15. More, A.; Deo, M.C. Forecasting wind with neural networks. *Mar. Struct.* **2003**, *16*, 35–49. [CrossRef]
16. Kirschen, D.S.; Ma, J.; Silva, V.; Belhomme, R. Optimizing the flexibility of a portfolio of generating plants to deal with wind generation. In Proceedings of the IEEE Power and Energy Society General Meeting, Detroit, MI, USA, 24–28 July 2011; pp. 1–7.

17. Gerven, M.V.; Bohte, S. Artificial neural networks as models of neural information processing. *Front. Comput. Neurosci.* **2018**, *11*, 114. [CrossRef]

18. Stergiou, C.; Siganos, D. Neural Networks. Available online: http://www.imperial.ac.uk/computing (accessed on 22 December 2018).

19. Nasle, A.; Nasle, A. Systems and Methods for Real-Time Forecasting and Predicting of Electrical Peaks and Managing the Energy, Health, Reliability, and Performance of Electrical Power Systems Based on an Artificial Adaptive Neural Network. U.S. Patent 9,846,839, 19 December 2017.

20. Jordehi, A.R. How to deal with uncertainties in electric power systems? A review. *Renew. Sustain. Energy Rev.* **2018**, *96*, 145–155. [CrossRef]

21. Banos, R.; Agugliarob, F.M.; Montoyab, F.G.; Gila, C.; Alcaydeb, A.; Gómezc, J. Optimization methods applied to renewable and sustainable energy: A review. *Renew. Sustain. Energy Rev.* **2011**, *15*, 1753–1766. [CrossRef]

22. Gamarra, C.; Guerrero, J.M. Computational optimization techniques applied to micro-grids planning: A review. *Renew. Sustain. Energy Rev.* **2015**, *48*, 413–424. [CrossRef]

23. Alizadeh, M.I.; Moghaddam, M.P.; Amjady, N.; Siano, P.; Eslami, M.K.S. Flexibility in future power systems with high renewable penetration: A review. *Renew. Sustain. Energy Rev.* **2016**, *57*, 1186–1193. [CrossRef]

24. García, F.; Bordons, C. Optimal economic dispatch for renewable energy microgrids with hybrid storage using Model Predictive Control. In Proceedings of the IECON 2013—39th Annual Conference of the IEEE Industrial Electronics Society, Vienna, Austria, 10–13 November 2013; pp. 7932–7937.

25. Bai, Q. Analysis of Particle Swarm Optimization Algorithm. *Comput. Inf. Sci.* **2010**, *3*, 180. [CrossRef]

26. Settles, M. *An introduction to particle swarm optimization*; University of Idaho: Moscow, Russia, 2005.

27. Rini, D.P.; Shamsuddin, S.M.; Yuhaniz, S.S. Particle swarm optimization: Technique, system and challenges. *Int. J. Comput. Appl.* **2011**, *14*, 19–26. [CrossRef]

28. Abbas, G.; Gu, J.; Farooq, U.; Asad, M.U.; El-Hawary, M. Solution of an Economic Dispatch Problem Through Particle Swarm Optimization: A Detailed Survey—Part I. *IEEE Access* **2017**, *5*, 15105–15141. [CrossRef]

29. Abbas, G.; Gu, J.; Farooq, U.; Raza, A.; Asad, M.U.; El-Hawary, M. Solution of an Economic Dispatch Problem Through Particle Swarm Optimization: A Detailed Survey—Part II. *IEEE Access* **2017**, *5*, 24426–24445. [CrossRef]

30. Jordehi, A.R. Particle swarm optimisation (PSO) for allocation of FACTS devices in electric transmission systems: A review. *Renew. Sustain. Energy Rev.* **2015**, *52*, 1260–1267. [CrossRef]

31. Sinha, S.; Chandel, S.S. Review of recent trends in optimization techniques for solar photovoltaic–wind based hybrid energy systems. *Renew. Sustain. Energy Rev.* **2015**, *50*, 755–769. [CrossRef]

32. Kirby, B.; Hirst, E. *Customer-Specific Metrics for the Regulation and Load-Following Ancillary Services*; ORNL/CON-474; Oak Ridge National Laboratory: Oak Ridge, TN, USA, 2000.

33. Kariyawasam, K.; Karunarathna, K.; Karunarathne, R.; Kularathne, M.; Hemapala, K. Design and Development of a Wind Turbine Simulator Using a Separately Excited DC Motor. *Smart Grid Renew. Energy* **2013**, *4*, 259–265. [CrossRef]

34. Genc, A.; Erisoglu, M.; Pekgor, A.; Oturanc, G.; Hepbasli, A.; Ulgen, K. Estimation of Wind Power Potential Using Weibull Distribution. *Energy Sources* **2005**, *27*, 809–822. [CrossRef]

35. Zhang, J.; Hodge, B.-M.; Miettinen, J.; Holttinen, H.; Lázaro, E.G.; Cutalulis, N.; L-Palima, M.; Sorensen, P.; Lovholm, A.L.; Berge, E.; et al. Analysis of Variability and Uncertainty in Wind Power Forecasting: An International Comparison. In Proceedings of the 12th International Workshop on Large-Scale Integration of Wind Power into Power Systems, London, UK, 22–24 October 2013.

36. Reza, S.E.; Zaman, P.; Ahammad, A.; Ifty, I.Z.; Nayan, M.F. A study on data accuracy by comparing between the Weibull and Rayleigh distribution function to forecast the wind energy potential for several locations of Bangladesh. In Proceedings of the 4th International Conference on the Development in the in Renewable Energy Technology (ICDRET), IEEE, Dhaka, Bangladesh, 7–9 January 2016; pp. 1–5.

37. Stoutenburg, E.D.; Jenkins, N.; Jacobson, M.Z. Variability and uncertainty of wind power in the California electric power system. *Wind Energy* **2014**, *17*, 1411–1424. [CrossRef]

38. Zhang, J.; Wang, J.; Wang, X. Review on probabilistic forecasting of wind power generation. *Renew. Sustain. Energy Rev.* **2014**, *32*, 255–270. [CrossRef]

39. Reliability Basics: Reliability Engineering Resources. The eMagzine for Reliability Professional. Issue 7. 2001. Available online: https://www.weibull.com/ (accessed on 22 December 2018).

40. Ela, E.; Kirby, B. *ERCOT Event on February 26, 2008, Lessons Learned*; National Renewable Energy Laboratory: Golden, CO, USA, 2008.
41. Papoulis, A.; Papoulis, S. *Probability, Random Variables and Stochastic Processes*, 4th ed.; McGraw-Hill: Boston, MA, USA, 2002; ISBN 0-07-366011-6.
42. Jiang, R.; Murthy, D.N.P. A study of Weibull shape parameter: Properties and significance. *Reliab. Eng. Syst. Saf.* **2011**, *96*, 1619–1626. [CrossRef]
43. Amari, S.; Nagaoka, H. *Methods of Information Geometry*; American Mathematical Society and Oxford University Press: Providence, RI, USA, 2000; p. 206. ISBN 0-8218-0531-2.
44. Galambos, J.; Simonelli, I. *Products of Random Variables: Applications to Problems of Physics and to Arithmetical Functions*; Marcel Dekker, Inc.: New York, NY, USA, 2004; Volume 1, pp. 139–140. ISBN 0-8247-5402-6.
45. Clarkson, E. *Comparison of Weibull and Normal Distributions*; NCAMP Documents/Publications, NCAMP (NASA), Technical Presentations; NASA: Washington, DC, USA, 2013.
46. Merovci, F.; Elbatal, I. Weibull Rayleigh Distribution: Theory and Applications. *Appl. Math. Inf. Sci.* **2015**, *9*, 2127–2137.
47. Pishgar-Komleh, S.H.; Keyhani, A.; Sefeedpari, P. Wind speed and power density analysis based on Weibull and Rayleigh distributions (A case study: Firouzkooh county of Iran). *Renew. Sustain. Energy Rev.* **2015**, *42*, 313–322. [CrossRef]
48. Ahmad, M.A.; Raqab, M.Z.; Kundu, D. Discriminating between the Generalized Rayleigh and Weibull Distributions: Some Comparative Studies. *Commun. Stat. Simul. Comput.* **2017**, *46*, 4880–4895. [CrossRef]
49. Fu, T.; Wang, C. A hybrid wind speed forecasting method and wind energy resource analysis based on a swarm intelligence optimization algorithm and an artificial intelligence model. *Sustainability* **2018**, *10*, 3913. [CrossRef]
50. Saxena, B.K.; Rao, K.V.S. Comparison of Weibull parameters computation methods and analytical estimation of wind turbine capacity factor using polynomial power curve model: Case study of a wind farm. *Renew. Wind Water Sol.* **2015**, *2*, 3. [CrossRef]
51. Carrillo, C.; Cidrás, J.; Díaz-Dorado, E.; Montaño, A.F.O. An Approach to Determine the Weibull Parameters for Wind Energy Analysis: The Case of Galicia (Spain). *Energies* **2014**, *7*, 2676–2700. [CrossRef]
52. Anuradha, M.; Keshavan, B.K.; Ramu, T.S.; Sankar, V. Probabilistic modeling and forecasting of wind power. *Int. J. Perform. Eng.* **2016**, *12*, 353–368.
53. Nikmehr, N.; Ravadanegh, S.N. Optimal Power Dispatch of Multi-Microgrids at Future Smart Distribution Grids. *IEEE Trans. Smart Grid* **2015**, *6*, 1648–1657. [CrossRef]
54. Reddy, S.S.; Bijwe, P.R.; Abhyankar, A.R. Joint Energy and Spinning Reserve Market Clearing Incorporating Wind Power and Load Forecast Uncertainties. *IEEE Syst. J.* **2015**, *9*, 152–164. [CrossRef]
55. Azada, A.K.; Rasul, M.G.; Alam, M.M.; Uddin, S.M.A.; Mondal, S.K. Analysis of wind energy conversion system using Weibull distribution. *Procedia Eng.* **2014**, *90*, 725–732. [CrossRef]
56. Kim, C.; Gui, Y.; Chung, C.C.; Kang, Y. Model predictive control in dynamic economic dispatch using Weibull distribution. In Proceedings of the IEEE Power & Energy Society General Meeting, Vancouver, BC, Canada, 21–25 July 2013; pp. 1–5.
57. Celik, A.K. A statistical analysis of wind power density based on the Weibull and Rayleigh models at the southern region of Turkey. *Renew. Energy* **2004**, *29*, 593–604. [CrossRef]
58. Sağbaş, A.; Karamanlıoğlu, T. The Application of Artificial Neural Networks in the Estimation of Wind Speed: A Case Study. In Proceedings of the 6th International Advanced Technologies Symposium (IATS'11), Elazığ, Turkey, 18 May 2011; 2011; pp. 16–18.
59. Ata, R. Artificial neural networks applications in wind energy systems: A review. *Renew. Sustain. Energy Rev.* **2015**, *49*, 534–562. [CrossRef]
60. Kadhem, A.A.; Wahab, N.I.A.; Aris, I.; Jasni, J.; Abdalla, A.N. Advanced Wind Speed Prediction Model Based on a Combination of Weibull Distribution and an Artificial Neural Network. *Energies* **2017**, *10*, 1744. [CrossRef]
61. Quan, H.; Srinivasan, D.; Khosravi, A. Short-Term Load and Wind Power Forecasting Using Neural Network-Based Prediction Intervals. *IEEE Trans. Neural Netw. Learn. Syst.* **2014**, *25*, 303–315. [CrossRef]
62. Quan, H.; Srinivasan, D.; Khosravi, A.; Nahavandi, S.; Creighton, D. Construction of Neural Network-Based Prediction Intervals for Short-Term Electrical Load Forecasting. In *2013 IEEE Computational Intelligence Applications in Smart Grid (CIASG)*; IEEE: Piscataway, NJ, USA, 2013; pp. 66–72.

63. Peng, H.; Liu, F.; Yang, X. A hybrid strategy of short term wind power prediction. *Renew. Energy* **2012**, *50*, 590–595. [CrossRef]

64. Mohammadi, K.; Shamshirband, S.; Yee, P.L.; Petkovi, D.; Zamani, M.; Chaudary, S. Predicting the wind power density based upon extreme learning machine. *Energy* **2015**, *86*, 232–239. [CrossRef]

65. Shamshirband, S.; Mohammadi, K.; Tong, C.W.; Petković, D.; Porcu, E.; Mostafaeipour, A.; Chaudary, S.; Sedaghat, A. Application of extreme learning machine for estimation of wind speed distribution. *Clim. Dyn.* **2016**, *46*, 1893–1907. [CrossRef]

66. Xu, Y.; Shi, L.; Ni, Y. Deep-learning-based scenario generation strategy considering correlation between multiple wind farms. *J. Eng.* **2017**, *13*, 2207–2210. [CrossRef]

67. Huang, C.J.; Kuo, P.H. A short-term wind speed forecasting model by using artificial neural networks with stochastic optimization for renewable energy systems. *Energies* **2018**, *11*, 2777. [CrossRef]

68. Zhang, Y.; Zhang, C.; Zhao, Y.; Gao, S. Wind speed prediction with RBF neural network based on PCA and ICA. *J. Electr. Eng.* **2018**, *69*, 148–155. [CrossRef]

69. Khan, G.M.; Ali, J.; Mahmud, S.A. Wind power forecasting—An application of machine learning in renewable energy. In Proceedings of the International Joint Conference on Neural Networks (IJCNN), Beijing, China, 6–11 July 2014; pp. 1130–1137.

70. World Energy Sources, Solar-2016. World Energy Council. Available online: https://www.worldenergy.org/wp-content/uploads/2017/03/WEResources_Solar_2016.pdf (accessed on 1 October 2019).

71. Shi, G. A Data Driven Approach to Solar Generation Forecasting. 2017. Thesis. ALL. 156. Available online: https://surface.syr.edu/thesis/156/ (accessed on 1 October 2019).

72. Bushong, S. *Advantages and Disadvantages of a Solar Tracker System*; Solar Power World: Cleveland, OH, USA, 2016.

73. Beránek, V.; Olšan, T.; Libra, M.; Poulek, V.; Sedláček, J.; Dang, M.Q.; Tyukhov, I. New monitoring system for photovoltaic power plants' management. *Energies* **2018**, *11*, 2495. [CrossRef]

74. Zhang, J.; Florita, A.; Hodge, B.; Lu, S.; Hamann, H.F.; Banunarayanan, V.; Brockway, A.M. A suite of metrics for assessing the performance of solar power forecasting. *Sol. Energy* **2015**, *111*, 157–175. [CrossRef]

75. Geoffrey, H. Reflections–The Economics of Renewable Energy in the United States. *Rev. Environ. Econ. Policy* **2009**, *4*, 139–154.

76. Kumara, S.; Sarkara, B. Design for reliability with weibull analysis for photovoltaic modules. *Int. J. Curr. Eng. Technol.* **2013**, *3*, 129–134.

77. Langella, R.; Proto, D.; Testa, A. Solar Radiation Forecasting, Accounting for Daily Variability. *Energies* **2016**, *9*, 200. [CrossRef]

78. Peruchena, C.M.F.; Ramı´rez, L.; Silva-Pe´rez, M.A.; Lara, V.; Bermejo, D.; Gasto´n, M.; Moreno-Tejera, S.; Pulgar, J.; Liria, J.; Macı´as, S.; et al. A statistical characterization of the long-term solar resource: Towards risk assessment for solar power projects. *Sol. Energy* **2016**, *123*, 29–39. [CrossRef]

79. Basri, M.J.M.; Abdullah, S.; Azrulhisham, E.A.; Harun, K. Higher order statistical moment application for solar PV potential analysis. In Proceedings of the AIP Conference Proceedings AIP Publishing, Penang, Malaysia, 10–12 April 2016; Volume 1782, p. 050004.

80. Munkhammar, J.; Widén, J.; Rydén, J. On a probability distribution model combining household power consumption, electric vehicle home-charging and photovoltaic power production. *Appl. Energy* **2015**, *142*, 135–143. [CrossRef]

81. Ayodele, T.R. Determination of Probability Distribution Function for Global Solar Radiation: Case Study of Ibadan, Nigeria. *Int. J. Appl. Sci. Eng.* **2015**, *13*, 233–245.

82. Maraj, A.; Londo, A.C.; Firat, C.; Karapici, R. Solar Radiation Models for the City of Tirana, Albania. *Int. J. Renew. Energy Res.* **2014**, *4*, 413–420.

83. Arias-Rosales, A.; Mejía-Gutiérrez, R. Optimization of V-Trough photovoltaic concentrators through genetic algorithms with heuristics based on Weibull distributions. *Appl. Energy* **2018**, *212*, 122–140. [CrossRef]

84. Yadav, A.K.; Chandel, S.S. Solar radiation prediction using Artificial Neural Network techniques: A review. *Renew. Sustain. Energy Rev.* **2014**, *33*, 772–781. [CrossRef]

85. Behravesh, V.; Keypour, R.; Foroud, A.A. Stochastic analysis of solar and wind hybrid rooftop generation systems and their impact on voltage behavior in low voltage distribution systems. *Sol. Energy* **2018**, *166*, 317–333. [CrossRef]

86. Chiteka, K.; Enweremadu, C.C. Prediction of global horizontal solar irradiance in Zimbabwe using artificial neural networks. *J. Clean. Prod.* **2016**, *135*, 701–711. [CrossRef]

87. Gairaa, K.; Khellaf, A.; Messlem, Y.; Chellali, F. Estimation of the daily global solar radiation based on Box Jenkins and ANN models: A combined approach. *Renew. Sustain. Energy Rev.* **2016**, *57*, 238–249. [CrossRef]

88. Li, F.F.; Wang, S.Y.; Wei, J.H. Long term rolling prediction model for solar radiation combining empirical mode decomposition (EMD) and artificial neural network (ANN) techniques. *J. Renew. Sustain. Energy* **2018**, *10*, 013704. [CrossRef]

89. Rodriguez, F.; Fleetwood, A.; Galarza, A.; Fontan, L. Predicting solar energy generation through artificial neural networks using weather forecasts for micro-grid control. *Renew. Energy* **2018**, *126*, 855–864. [CrossRef]

90. Kuo, P.H.; Huang, C.J. A High Precision Artificial Neural Networks Model for Short-Term Energy Load Forecasting. *Energies* **2018**, *11*, 213. [CrossRef]

91. Khosravi, A.; Koury, R.N.N.; Machado, L.; Pabon, J.J.G. Prediction of hourly solar radiation in Abu Musa Island using machine learning algorithms. *J. Clean. Prod.* **2018**, *176*, 63–75. [CrossRef]

92. Economic Dispatch and Technological Change. Report to Congress United States Department of Energy 2015. Washington, DC, USA. Available online: https://www.energy.gov/oe/articles/20142015-economic-dispatch-and-technological-change-report-congress-now-available (accessed on 26 December 2018).

93. Sigarchian, S.G.; Paleta, R.; Malmquist, A.; Pina, A. Feasibility study of using a biogas engine as backup in a decentralized hybrid (PV/wind/battery) power generation system–Case study Kenya. *Energy* **2015**, *90*, 1830–1841. [CrossRef]

94. Reddy, S.S.; Bijwe, P.R. Real time economic dispatch considering renewable energy resources. *Renew. Energy* **2015**, *83*, 1215–1226. [CrossRef]

95. Zhang, J.M.; Xie, L. Particle swarm optimization algorithm for constrained problems. *Asia-Pac. J. Chem. Eng.* **2009**, *4*, 437–442. [CrossRef]

96. Jeyakumar, D.N.; Jayabarathi, T.; Raghunathan, T. Particle swarm optimization for various types of economic dispatch problems. *Int. J. Electr. Power Energy Syst.* **2006**, *28*, 36–42. [CrossRef]

97. Gaing, Z.L. Particle swarm optimization to solving the economic dispatch considering the generator constraints. *IEEE Trans. Power Syst.* **2003**, *18*, 1187–1195. [CrossRef]

98. Elsaiah, S.; Benidris, M.; Mitra, J.; Cai, N. Optimal economic power dispatch in the presence of intermittent renewable energy sources. In Proceedings of the IEEE PES General Meeting|Conference & Exposition, National Harbor, MD, USA, 27–31 July 2014; pp. 1–5.

99. Huynh, D.C.; Nair, N. Chaos PSO algorithm based economic dispatch of hybrid power systems including solar and wind energy sources. In Proceedings of the IEEE Innovative Smart Grid Technologies—Asia (ISGT ASIA), Bangkok, Thailand, 3–6 November 2015; pp. 1–6.

100. Gholami, A.; Ansari, J.; Jamei, M.; Kazemi, A. Environmental/economic dispatch incorporating renewable energy sources and plug-in vehicles. *IET Gener. Transm. Distrib.* **2014**, *8*, 2183–2198. [CrossRef]

101. Kheshti, M.; Ding, L.; Ma, S.; Zhao, B. Double weighted particle swarm optimization to non-convex wind penetrated emission/economic dispatch and multiple fuel option systems. *Renew. Energy* **2018**, *125*, 1021–1037. [CrossRef]

102. Nikmehr, N.; Najafi-Ravadanegh, S. Probabilistic Optimal Power Dispatch in Multi-Microgrids Using Heuristic Algorithms. In *2014 Smart Grid Conference (SGC)*; IEEE: Piscataway, NJ, USA, 2014; pp. 1–6.

103. Huynh, D.C.; Ho, L.D. Optimal Generation Rescheduling of Power Systems with Renewable Energy Systems using a Dynamic PSO Algorithm. *Int. Adv. Res. J. Sci. Eng. Technol.* **2016**, *3*, 73–78. [CrossRef]

104. Wang, F.; Zhou, L.; Wang, B.; Wang, Z.; Shafie-khah, M.; Catalao, J.P. Modified Chaos Particle Swarm Optimization-Based Optimized Operation Model for Stand-Alone CCHP Micro-grid. *Appl. Sci.* **2017**, *7*, 754. [CrossRef]

105. Mudumbai, R.; Dasgupta, S.; Cho, B.B. Distributed Control for Optimal Economic Dispatch of a Network of Heterogeneous Power Generators. *IEEE Trans. Power Syst.* **2012**, *27*, 1750–1760. [CrossRef]

106. Shivashankar, S.; Mekhilef, S.; Mokhlis, H.; Karimi, M. Mitigating methods of power fluctuation of photovoltaic (PV) sources– A review. *Renew. Sustain. Energy Rev.* **2016**, *59*, 1170–1184. [CrossRef]

107. Senjyu, T.; Datta, M.; Yona, A.; Sekine, H.; Funabashi, T. A coordinated control method for leveling output power fluctuations of multiple PV systems. In Proceedings of the 7th International Conference on Power Electronics, Daegu, Korea, 22–26 October 2007; pp. 445–450.

108. Zhao, C.; Mallada, E.; Dorfler, F. Distributed frequency control for stability and economic dispatch in power networks. In Proceedings of the 2015 American Control Conference (ACC), Chicago, IL, USA, 1–3 July 2015; pp. 2359–2364.

109. Zhua, X.; Xiaa, M.; Chiang, H.-D. Coordinated sectional droop charging control for EV aggregator enhancing frequency stability of microgrid with high penetration of renewable energy sources. *Appl. Energy* **2018**, *210*, 936–943. [CrossRef]

110. Lee, D.; Kim, J.; Baldick, R. *Ramp Rates Control of Wind Power Output Using a Storage System and Gaussian Processes*; University of Texas at Austin, Electrical and Computer Engineering: Austin, TX, USA, September 2012.

111. Maleki, A.; Rosen, M.A.; Pourfayaz, F. Optimal operation of a grid-connected hybrid renewable energy system for residential applications. *Sustainability* **2017**, *9*, 1314. [CrossRef]

112. ElDesouky, A.A. Security and Stochastic Economic Dispatch of Power System Including Wind and Solar Resources with Environmental Consideration. *Int. J. Renew. Energy Res.* **2013**, *3*, 951–958.

113. Khan, N.A.; Sidhu, G.A.S.; Gao, F. Optimizing Combined Emission Economic Dispatch for Solar Integrated Power Systems. *IEEE Access* **2016**, *4*, 3340–3348. [CrossRef]

114. Solomon, A.A.; Child, M.; Caldera, U.; Breyer, C. ow much energy storage is needed to incorporate very large intermittent renewables? *Energy Procedia* **2017**, *135*, 283–293. [CrossRef]

115. Optimizing Battery Management in Renewable Energy Storage Systems. Pulse Power Electronics. Available online: https://www.power.pulseelectronics.com/optimizing-battery-management-in-renewable-energy-storage-systems (accessed on 31 December 2018).

116. Abdelrahman, A.; Lamont, L.; El-Chaar, L. *Energy Storage for Intermittent Renewable Energy Systems*; MDPI Sustainability Foundation: Basileia, Switzerland, 2012.

117. Pickard, W.F.; Abbott, D. Addressing the intermittency challenge: Massive energy storage in a sustainable future. *Proc. IEEE* **2012**, *100*, 317–321. [CrossRef]

118. Zsiborács, H.; Baranyai, N.H.; Vincze, A.; Zentkó, L.; Birkner, Z.; Máté, K.; Pintér, G. Intermittent Renewable Energy Sources: The Role of Energy Storage in the European Power System of 2040. *Electronics* **2019**, *8*, 729. [CrossRef]

119. Wen, S.; Lan, H.; Fu, Q.; Yu, D.C.; Zhang, L. Economic Allocation for Energy Storage System Considering Wind Power Distribution. *IEEE Trans. Power Syst.* **2015**, *30*, 644–652. [CrossRef]

120. Kerdphol, T.; Fuji, K.; Mitani, Y.; Watanabe, M.; Qudaih, Y. Optimization of a battery energy storage system using particle swarm optimization for stand-alone microgrids. *Int. J. Electr. Power Energy Syst.* **2016**, *81*, 32–39. [CrossRef]

121. Zhao, H.; Wua, Q.; Hu, S.; Xu, H.; Rasmussen, C.N. Review of energy storage system for wind power integration support. *Appl. Energy* **2015**, *137*, 545–553. [CrossRef]

122. IEC. Electrical energy storage white paper. Tech report. 2011.

123. Galván, L.; Navarro, J.M.; Galván, E.; Carrasco, J.M.; Alcántara, A. Optimal Scheduling of Energy Storage Using a New Priority-Based Smart Grid Control Method. *Energies* **2019**, *12*, 579. [CrossRef]

124. Rui, L.I.; Wei, W.; Zhe, C.; Xuezhi, W.U. Optimal planning of energy storage system in active distribution system based on fuzzy multi-objective bi-level optimization. *J. Mod. Power Syst. Clean Energy* **2018**, *6*, 342–355.

125. Hemmati, R.; Saboori, H.; Jirdehi, M.A. Stochastic planning and scheduling of energy storage systems for congestion management in electric power systems including renewable energy resources. *Energy* **2017**, *133*, 380–387. [CrossRef]

Article

A Doubly-Fed Induction Generator Adaptive Control Strategy and Coordination Technology Compatible with Feeder Automation

Peng Tian, Zetao Li and Zhenghang Hao *

College of Electrical Engineering, Guizhou University, Guiyang 550025, China; gs.ptian16@gzu.edu.cn (P.T.); gzulzt@163.com (Z.L.)
* Correspondence: haozhenghang@163.com

Received: 9 October 2019; Accepted: 21 November 2019; Published: 22 November 2019

Abstract: The extensive connection of distributed generation (DG) with the distribution network (DN) is one of the core features of a smart grid, but in case of a large number, it may result in problems concerning the DN-DG compatibility during fault isolation and service restoration, for which no efficient and economic solutions have been developed. This paper proposes a doubly-fed induction generator (DFIG) adaptive control strategy (ACS) and a coordination technology to be compatible with the typical feeder automation (FA) protection logics in the ring distribution system. First of all, an ACS simulating the inertia/damping characteristics and excitation principles of synchronous generators is developed to achieve seamless switching between DFIG grid-connection/island modes, and make distant synchronization possible. Next, a technology coordinating the DFIG islands controlled by ACS and the remote tie-switches based on local inspection of synchronization conditions for closing is developed to achieve the safety grid-connection of DFIG islands in the absence of DN-DG communication. At the last, a detailed simulation scenario with a ring DN accessed by five DFIGs is used to validate the effectiveness of ACS and coordination technology compatible with FA in various faults scenes.

Keywords: distribution network (DN); doubly-fed induction generator (DFIG); feeder automation (FA); compatibility; adaptive control strategy (ACS); coordination technology

---

1. Introduction

With more distributed generations (DGs) based on renewable energies integrating to the distribution network (DN), traditional DNs are evolving into active ones with more prominent DN-DG contradictions. Overall, the incompatibility and mutual repulsion between the DN and the DG have hindered the upgrading of active DN to smart DN [1–3]. Traditional DNs are challenged seriously by the compatibility between DG and DN in terms of topology structure, operating standards, control modes and protection configuration [1,4]. From the perspective of protection, traditional DN is generally a radiating structure from substations to users, and support fast fault isolation and service restoration based on equipment such as relays, circuit breakers, reclosing devices, sectionalizing switches and fuses [5–7]. Nevertheless, the application of DGs has changed the topology structure of traditional radical DN, which may lead to serious problems in the normal operation of protections to traditional DN, including false tripping of feeders, protection inaction, raised or reduced failure level, unintentional islanding, asynchronous closing and failed automatic reclosing [8,9]. Therefore, the reliability of traditional distribution systems are compromised by the installation of DGs as it results in failure of protection coordination. In the meantime, the safety of DGs is also challenged.

In recent years, researchers and institutions around the work have proposed various protection strategies and specific control technologies for active DNs, aiming to fast isolate the failures in the protection system at the grid side, and guarantee the safe operation of DGs at the DG side, while power authorities and international power organizations have formulated protection standards and specifications for DNs involving DGs to enhance the reliability of the DN [10,11]. Presently, the mainstream engineering DN-side protection is fast and non-selective connection of all DGs in case of any failure in the DN in any form, so that the existing protection will not be negatively affected [12]. Where an islanding status is detected by an existing DG, its internal control system will, in general cases, activate the anti-islanding protection mechanism to achieve disconnection and shutdown of DGs quickly [10,13], while DGs with fault rid-through capacity can be removed from the grid after a delay of a limited period in any failure [8]. For doubly-fed induction generator (DFIG), it will start crowbar protection immediately under fault scenario [7]. Obviously, the protection strategies at the DN and DG sides are comparatively conservative and weaken the advantage of DGs in improving power supply reliability. Furthermore, the disconnection and shutdown of DGs may affect stability of the grid, disadvantage service restoration, and result in additional maintenance costs [12,14].

Therefore, some new strategies are put forward successively and can be categorized generally into 2 types: ① DG improvements and reserved DG-side protection strategies; ② Improvements in DG-side protections and reserved DG strategies. In the first solution, DG output is limited and DG connecting position is optimized to avoid any impact on the DN protection [15,16]. Limiting DG output is easy to implement, but deviates from the initial intention of sufficient and flexible utilization of DG energies. Optimizing DG connecting position can reduce the power loss of the DN and improve voltage distribution. However, the best DG connecting position is subject to natural and geographical environments [17]. Therefore, optimizing the DG connecting position is not a preferred strategy in an established active DN. The second solution achieves adaptation to the large number of DGs by improving DG-side protection. It falls into six forms: voltage-based protection, improved current protection, differential protection, distance protection, adaptive protection and fault current compensation [2]. Those improved protection strategies have significantly improved the reliability of protective device whenever the DGs are in grid-connected or island modes. Essentially, they are developed against problems in the active DNs, such as fault identification, location and isolation, and fail to take into consideration the safe operation of DGs when the topology structure of the DN is changing dynamically. In particular, the specific strategies DG follow to withstanding the abnormal DN-side interruptions and service restoration in the continuous process from sudden tripping of circuit breakers to the closing of tie-switches are seldom referred. Reference [18] suggests closing the tie-switch 15 ms before circuit breakers trip to isolate the fault segments, to make sure DGs can maintain grid-connected power generation during and after faults, but, closing the tie-switch at the presence of faults may result in overvoltage or overcurrent, and the coordination between protection devices for actions in a ms places high requirements on the communication system and the circuit breakers. Reference [19] proposed a strategy of output matching of DGs through the local controllable load bank when they are disconnected, in order to maintain their status before the fault without immediate stopping, and realize fast reconnection by reclosing as the power is recovered. Instead of changing the structures and settings of existing DN-side protection system, this strategy further expands the application level of DGs, and directly reduces the costs of DG stopping and restart. However, it is a conservative disconnection strategy essentially that the DGs fail to continuously supply powers to local loads during a fault, and an additional controllable load bank is required.

It should be noted that in case of a fault, active DNs have achieved a lot in fault isolation, but the compatibility between DGs and DN-side protection devices is still a challenge in a true sense. The existing active DNs lack experience in technology docking and integration in terms of DN-DG coordination, and the strategies adopted are more conservative, while standards and operation guides concerning the deployments of DGs to the DN are based on the technical level at present, which will surely be removed through the technical route and utmost goals of smart DN to achieve the real integration of DGs and the DN.

Therefore, this paper gives a solution and the contributions of this paper which are emphasized in the following points:

① This paper proposes a doubly-fed induction generator (DFIG) adaptive control strategy (ACS) which possesses the capabilities of dual-mode operation, restraining sudden changes of rotor current and distant synchronization, making seamless switching between grid-connection and island can be achieved and making distant synchronization possible without switching control strategies.

② This paper proposes a coordination technology which combines the DFIG islanding controlled by ACS with remote tie-switches based on local inspection of synchronization conditions for closing to achieve the safety grid-connection of DFIG islands in absence of communication between DFIG and DN.

③ The ACS and coordination technology allows DFIG (not limited to a single wind turbine) continuously supply power to partial local loads during the dynamic process of feeder automation (FA) fault isolation and power recovery without shutdown and restart of DFIG, realizing the compatibility between DFIG and feeder automation (FA).

This paper consists of Section 2, which summarizes the active distribution system studied herein, and analyzes the contradictions between existing FA protection logics and DFIG, Section 3, which specifies the solution for DN-DFIG compatibility, Section 4, which describes and analyzes the proposed DFIG ACS and coordination technology in details, and Section 5, which exhibits the simulation results in various scenes and some discussions are given. Conclusions can be found in Section 6.

2. Contradictions in Active Distribution Network

2.1. System under Study

Modern DNs require flexible and reliable operations based on closed-loop design and open-loop operation [20]. Accordingly, an active DN as shown in Figure 1 is selected, in which, the primary system adopts a ring network structure with S1, S2, S3 and S4 at the outlet switches by both sides of the substation.

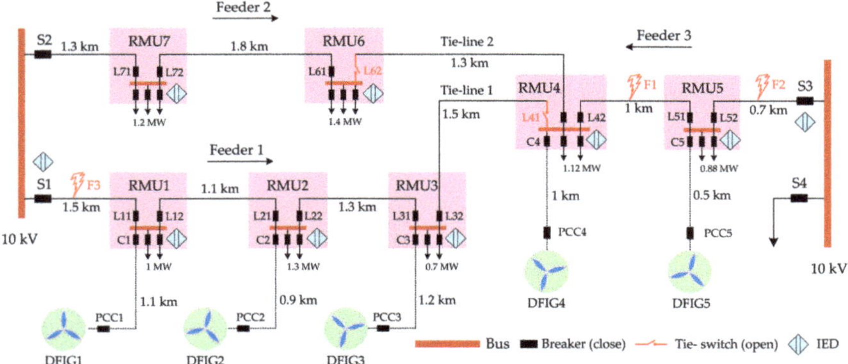

Figure 1. Active distribution network system under study.

Three feeders are led out via S1, S2 and S3, and interconnected through the ring main unit (RMU). All inlet and outlet switches of the RMU are circuit breakers, while the tie-switches L41 and L62 are normally opened. The tie-line 1 connects with feeders 1 and 3, and the tie-line 2 connects with the feeders 2 and 3. All feeders are underground cables. Five DFIGs connect to the DN through RMUs at different positions. The secondary system is configured as a typical intelligent distributed FA independent from the overall information from the master stations or slave stations, and exchanges fault information by peer-to-peer communication between intelligent electronic devices (IEDs), to realize fast isolation of feeder faults and service restoration [21–23]. The communication network supports optical fiber Ethernet structures according to IEC 61850, while the IEDs at the substations and the RMUs communicate with each other through the core switchers. The generic object oriented substation event (GOOSE), a fast message communication mechanism, is adopted for communications between IEDs [20].

2.2. FA Protection Logics

In Figure 1, permanent faults of cable feeders are designated as F1, F2 and F3. In case of a fault in the DN, FA controls the circuit breakers at both sides of the fault feeder through GOOSE message communication mechanism to isolate faults by tripping, after which, a GOOSE message of fault isolation successful (FIS) will be sent to the corresponding tie-switch for closing. For the relations between feeders and tie-switches, feeder 1 corresponds to the tie-switch L41, and feeder 3 to L62. Therefore, when a fault takes place on feeders 1 or 3, the tie-switches L41 or L63 is closed after fault isolation by FA accordingly.

2.3. Contradictions between FA and DFIG

Traditionally, DFIG is controlled by PQ (active and reactive power control) [24] for grid-connection and VF (constant-voltage constant-frequency control) for islanding [25]. In case of islanding a number of DFIGs, master-slave, peer-to-peer or hierarchical control is required [26]. Therefore, DFIG works on continuous grid-connection or pure islanding mode at present without switching between the two. There are few DFIGs capable of operating under the two modes, or supporting their seamless switching. As different modes demand different control strategies and operation rules, DFIG requires island detection technology to achieve switching between grid-connection and island modes. There are three island detection technologies: communication-based, active and passive [10], and the establishment of communication between DFIG and DN results in additional costs and higher complexity of DFIG penetration. Operated by different companies, the DFIG is not connected to the DN through secondary cables in real cases, while the active and passive island detection technologies are not reliable enough and a certain period of time elapses before an island is detected [10]. Furthermore, the DFIG is located comparatively remote from the DN (1-2 km in general cases), and the tie-switches are flexibly positioned, requiring a distant synchronous grid-connected process for DFIG islands in grid-connection through tie-switches, which makes this new mode essentially different from the traditional local synchronous grid-connection technology of DFIG. Again, the existing GOOSE fast communication mechanism, in essence, works by repeated transmission of state messages (e.g., circuit breaker tripping or closing) and fails to achieve the real-time transmission of remote sinusoidal quantity, while the SV message is not an economic and reliable solution for the DN.

For those reasons, when the DFIG is connected to the DN (Figure 1), 2 contradictions are expected between FA and DFIG to make sure the DFIG serves partial local loads without interruption, and safely close in coordination with the tie-switches. Contradiction 1: a series of switch state messages in the discretion of FA cannot be timely "notified" to DFIG, rendering the latter's failure to adapt to the topology changes of the grid; without supports to dual-mode operation, DFIG cannot adjust its control modes timely for the purpose of seamless switching between the grid-connection and islanding. Contradiction 2: the information channel and message specifications of the DN provide no supports to the distant synchronization function to transform the DFIG status from off-grid to grid-connection. The two contradictions account for the incompatibility and mutual repulsion between existing FA technologies and DFIGs, and hinder the upgrading from active DN to smart DN.

3. Solution for DN-DFIG Compatibility

The active distribution network system studied in this paper is 10 kv level, and DFIG is megawatt class (1.5 MW). It is assumed that FA can instantly isolate any fault. In addition, there is no any communication between DFIG and DN, and no support of energy storage widely used in microgrid [27,28]. With remarkable kinetic energy stored in its wind wheels and shafts, the megawatt DFIG can continuously supply power to partial local loads through advanced control strategies in case of an emergency. To sufficiently and flexibly leverage the potential of DFIG for enhanced resilience of the DN where no conditions for DN-DG communication are created, this paper proposes a DN-DG compatibility solution as shown in Figure 2, which contains two steps. Step 1: As the circuit breaker trips for fault isolation, the DFIG switches into the unintentional islanding mode seamlessly and operates stably. At the same time, DFIG islanding actively create distant synchronous conditions at remote tie-switch. Step 2: Remote tie-switch is permitted to close synchronously for the purpose of reconnection of DFIG island and recovering DFIG to the pre-fault status or to a new stable status.

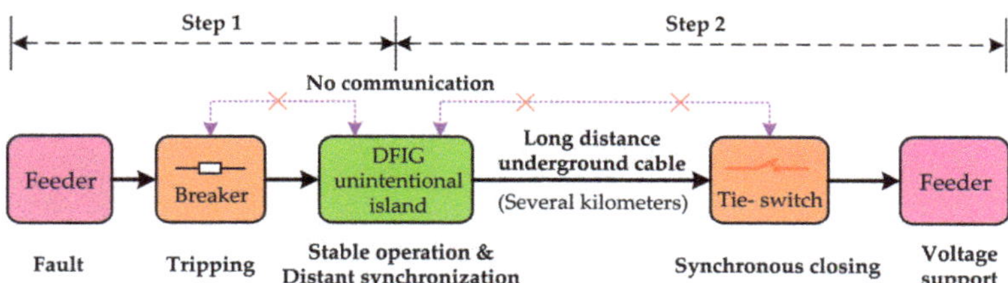

Figure 2. DN-DG compatibility solution.

For DFIG, the solution is to realize the continuous and seamless switching amongst unintentional stable operation of islands, distant island synchronization and island grid-connection. Such a process relies on advanced DFIG grid-connection/island dual-mode control strategy and coordination between DFIG and DG-side protection devices. Therefore, an adaptive control strategy and a coordination technology are proposed in the following Section.

4. The Proposed Adaptive Control Strategy and Coordination Technology

4.1. DFIG Adaptive Control Strategy

A typical DFIG system is illustrated in Figure 3. The wind wheels transform the wind energy captured into rotary mechanical energy which is delivered to DFIG via the gearbox shafts [29]. The stators of DFIG directly connect to the DN, while the rotators are through the back-to-back converter. The grid-side converter (GSC) is responsible for constant DC bus voltage, while the rotor-side converter (RSC) regulates the excitation voltage of rotors to realize DFIG grid-connection and power generation. This paper adopts GE 1.5 MW DFIG system [30] as an example, which contains mechanical and electrical controls. Mechanical control further divides into speed control, compensation control and torque control, and electric control consists of the GSC-side control and the RSC-side control. The nomenclatures of all variables used in this paper are shown in Table A1 (Appendix A). All quantities are in per unit except for the variables related to phase and special instructions in Section 4. Table A2 (Appendix A) and corresponding Figures shows the variable units in Section 5. The DFIG ACS proposed in this paper aims at RSC-side control and it has three capabilities, including dual-mode operation, restraining sudden changes of rotor current and distant synchronization. The basic control structure of the GE 1.5 MW DFIG [30] for the mechanical part is adopted, while the GSC-side control still follows the typical vector control strategy based on phase-locked loops (PLL) [31]. The maximum power point tracking (MPPT) of the wind turbine [30] is calculated as follows:

$$\omega_\mathrm{m}{}^* = -0.67P_e{}^2 + 1.42P_e + 0.51 \tag{1}$$

where $\omega_\mathrm{m}{}^*$ is the reference angular velocity of the wind turbine, and P_e is the actual output electromagnetic power. Therefore, in the MPPT, the speed of the wind turbine depends on the actual output electromagnetic power.

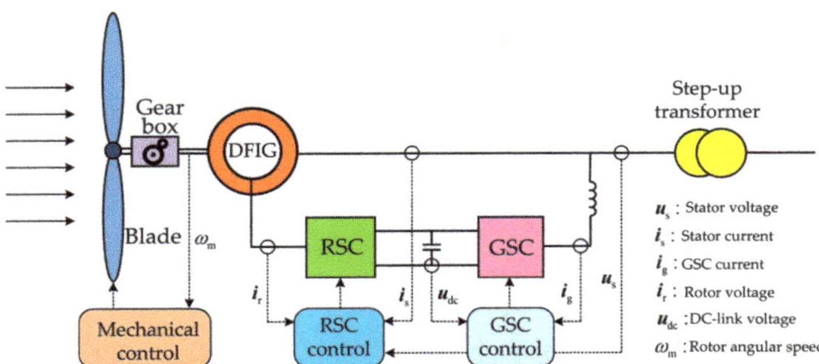

Figure 3. DFIG system.

When the rotor follows motor convention and the stator applies generator convention, the space vector model of the DFIG in the stator static reference frame can be expressed as:

$$\boldsymbol{u}_\mathrm{s} = -R_\mathrm{s}\boldsymbol{i}_\mathrm{s} + D(-L_\mathrm{s}\boldsymbol{i}_\mathrm{s} + L_\mathrm{m}\boldsymbol{i}_\mathrm{r}) \tag{2}$$

$$\boldsymbol{u}_\mathrm{r} = R_\mathrm{r}\boldsymbol{i}_\mathrm{r} - j\omega_\mathrm{m}(L_\mathrm{r}\boldsymbol{i}_\mathrm{r} - L_\mathrm{m}\boldsymbol{i}_\mathrm{s}) + D(L_\mathrm{r}\boldsymbol{i}_\mathrm{r} - L_\mathrm{m}\boldsymbol{i}_\mathrm{s}) \tag{3}$$

where all quantities are in per unit. D is the differential operator (d/dt); ω_m the electric angular frequency of rotor; R_s and R_r are resistances of stator and rotor, respectively; L_s, L_r and are stator self-inductance, rotor self-inductance and mutual inductance, respectively; $\boldsymbol{u}_\mathrm{s}$ and $\boldsymbol{u}_\mathrm{r}$, $\boldsymbol{i}_\mathrm{s}$ and $\boldsymbol{i}_\mathrm{r}$ are voltage and current space vectors of stator and rotor under stator static reference frame respectively.

The proposed DFIG ACS is shown in Figure 4 and consists of power control, voltage control and current inner loop, of which, the power control simulates the inertia / damping characteristics of the synchronous generator [32]. $1/(2H_\mathrm{A}s)$ is inertial link of power control; H_A inertia time constant of power control; K_A proportional coefficient of power control. The electromagnetic power reference P_e^* is produced by torque control in the mechanical control part. Where the actual power output losses its balance with the reference, ACS will regulate the inner potential angular frequency of the stator ω_si through the inertial link and the proportional link, based on which, the inner potential control phase of the stator θ_si is obtained through the integral. $\omega_\mathrm{b} = 50\mathrm{Hz}$ is the angular frequency of fundamental wave and the rotor's excitation current phase θ_Ir is obtained by θ_si minus the rotor phase $\theta_\mathrm{m} = \int \omega_\mathrm{m}dt$. In the meantime, damping negative feedback control is introduced to inhibit the frequency oscillations in the DFIG. The damping power P_D is determined by the difference between the inner potential angular frequency of the stator ω_si and the reference angular frequency of the grid $\omega_\mathrm{g}{}^*$ after the damping link. D_A is the power control damping coefficient.

The voltage control simulates the excitation principles of the synchronous generator. The reference of rotor's excitation current amplitude I_r^* is obtained through a typical PI controller based on the difference of U_s^*, the reference of stator voltage amplitude, and U_s, the actual stator voltage amplitude. $P_{Us,\mathrm{A}}$ and $I_{Us,\mathrm{A}}$ are respectively the voltage control proportion and integral coefficient.

A three-phase independent control mode is adopted for the current inner loop, in which, "r"(superscript) represents the value in the rotor reference frame. The space vector reference of the rotor excitation currents $\overrightarrow{\boldsymbol{i}}_r^{\,r*}$ is defined by the phase and amplitude obtained by power and voltage controls, and converted to *abc* coordinate to obtain the reference of three-phase rotor excitation currents

(i_{ra}^{r*}, i_{rb}^{r*} and i_{rc}^{r*}), which correspond to the actual excitation currents of rotor (i_{ra}^{r}, i_{rb}^{r} and i_{rc}^{r}), different between which is based on to obtain the three-phase excitation voltages of the rotor (u_{ra}^{r}, u_{rb}^{r} and u_{rc}^{r}) through the $G_{QPR}(s)$ control. $G_{QPR}(s)$, quasi-proportional resonant controller (QPR), better fits for precision control of sinusoidal signal as compared with traditional PI controllers, and the transfer function can be expressed as [33,34]:

$$G_{QPR}(s) = P_{QPR} + \frac{2K_{QPR}\omega_i s}{s^2 + 2\omega_i s + \omega_r^2} \tag{4}$$

where P_{QPR} is the proportional coefficient, K_{QPR} the resonance term gain, ω_i the resonance term bandwidth (actual value), and ω_r the slip angular frequency (actual value). The design of QPR parameters can be found in [33,34].

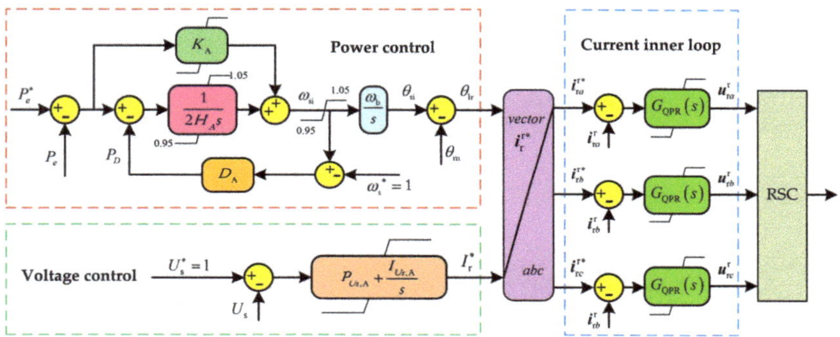

Figure 4. Structure diagram of proposed ACS.

4.1.1. Grid-connection/island Dual-mode Operation

According to Equation (2) and by neglecting the resistant of stator, the inner potential of stator e_s can be expressed as:

$$e_s = jX_m i_r = u_s + jX_s i_s, \tag{5}$$

where, $X_s = \omega_s L_s$, $X_m = \omega_s L_m$, $i_r = i_r^r e^{j\theta_m}$, and ω_s is the synchronous angular frequency. According to ACS control structure (Figure 4) and Equation (5), and by taking the current inner loop with rapid dynamic response as an ideal link, the inner potential of stator e_s can be further expressed as:

$$e_s = E_s e^{j\theta_{Es}} = jX_m i_r^{r*} e^{j\theta_m} = jX_m I_r^* e^{j\theta_{si}} \tag{6}$$

where E_s is the inner potential amplitude of the stator, and θ_{Es}, the inner potential phase of the stator. According to formula (6), the difference between θ_{Es} and θ_{si} is $\pi/2$, and the amplitude reference of rotor's excitation current I_r^* is proportional to the inner potential amplitude of the stator. Therefore, the inner potential of stator is subject to the direct control of ACS. When the DFIG is grid-connected, the stator voltage is clamped down by the grid voltage, and the power control will adjust the inner potential of stator based on the grid voltage, and transmit power to the grid. Thus power control plays a leading role, and the voltage control plays an auxiliary role. When the DFIG is suddenly islanded, and supply power to partial loads, the loss of supports from grid voltage will result in the voltage control at the dominating role and the power control automatically adjust the speed of the wind turbine and output energy to serve the island loads under the MPPT in formula (1). By damping negative feedback control and frequency amplitude limitation, the island frequency is controlled within a certain range. Instead of tracking the maximal wind energy, the MPPT under islanding mode guides the wind turbine to adjust its speed based on the loads. The particular case is that multiple DFIGs are suddenly islanded when the slow dynamic response of DFIG mechanical control will lead to the gradually balance between the mechanical energy absorbed by each wind turbine and the loads to jointly supply power for loads. Therefore, ACS can transmit the power captured to the grid during grid-connection mode, and supply power to loads by controlling the voltage of stators in the island mode. Furthermore, when the system switches between grid-connection/island modes, the ACS can directly produce the inner potential phase θ_{Es} of the stator by power control to avoid sudden change in the stator's voltage phase and facilitate the seamless switching between the two modes.

4.1.2. Equivalent Inertia/Damping Characteristics

The equivalent inertia/damping characteristics of DFIG take both mechanical control and electrical control into consideration. To facilitate analysis, it is assumed that the stator voltage, the wind speed v_w (actual value) and the pitch angle β (actual value) of blades remain constant when the mechanical energy capture by the wind wheel P_{wt} is out of balance with P_e, and the impact of speed ω_m on the wind turbine is neglected. By linear processing of the mechanical control of the wind turbine and the electrical control of the ACS, a small signal equivalent model as shown in Figure 5 is obtained. ω_{m0} is the stable speed of the wind turbine at a certain operation point, H_m the time constant of the wind turbine's inherent mechanical inertia, D_m the inherent mechanical damping coefficient of the wind turbine, P_v and I_v the proportional and integral coefficient of the torque control.

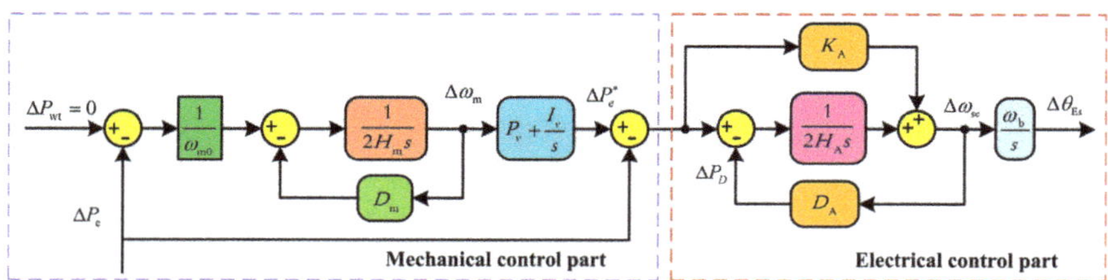

Figure 5. Small signal equivalent model of DFIG under ACS control.

The small signal equivalent model of DFIG in the Figure 5 is properly converted to obtain the equivalent inertia/damping model of DFIG as shown in Figure 6, where, T_{eq} and D_{eq} represent the equivalent inertia and equivalent damping of DFIG under ACS control. Formulas (7) and (8) are the expressions of T_{eq} and D_{eq} respectively.

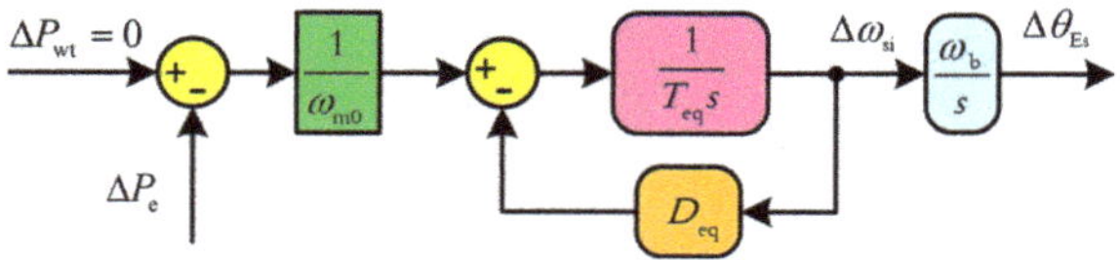

Figure 6. Equivalent inertia/damping model of DFIG under ACS control.

$$T_{eq}=\frac{4H_m H_A s^2+2D_m H_A s}{4H_m\omega_{m0}K_A H_A s^3+(2D_m\omega_{m0}K_A H_A+2P_v K_A H_A+2H_m\omega_{m0})s^2+(2I_v K_A H_A+D_m\omega_{m0}+P_v)s+I_v} \tag{7}$$

$$D_{eq}=\frac{4H_m D_A K_A H_A s^2+(2D_m D_A K_A H_A+2H_m D_A)s+D_m D_A}{4\omega_{m0}H_m K_A H_A s^2+(2D_m\omega_{m0}K_A H_A+2P_v K_A H_A)s+2I_v K_A H_A} \tag{8}$$

Formulas (7) and (8) reflect the equivalent inertia's association with the inertia time constant H_A and the proportional coefficient K_A, while the equivalent damping is related with the inertia time constant H_A, the proportional coefficient K_A and the damping coefficient D_A. It is assumed that the wind speed is 15 m/s, then the stable speed $\omega_{m0}=1.2$ pu, the inherent inertia/damping of the wind turbine are $H_\mathrm{m} = 4.96$ s, $D_\mathrm{m} = 1.5$, and the parameters of the torque control are $P_v = 3$ and $I_v = 0.6$ [32]. The maximal equivalent inertia of different inertia time constant H_A and proportional coefficient K_A are as shown in Figure 7a. When $K_\mathrm{A} \in (0, 0.5)$, the equivalent inertia significantly increases with H_A and when $K_\mathrm{A} \geq 0.5$, the increase slows down. Under the same operation conditions, the maximal equivalent damping of different inertia time constant H_A, damping coefficient D_A and proportional coefficient K_A are shown in Figure 7b. Overall, when the proportional coefficient satisfies the relation of $K_\mathrm{A} \in (0, 0.5)$, the equivalent damping significantly increases with the coefficient of damping D_A, and a smaller inertia time constant H_A corresponds to a larger equivalent damping. Obviously, ACS can have large equivalent inertia and damping by choosing appropriate parameters. In this paper, the parameters selected are $H_\mathrm{A} = 4.96$, $D_\mathrm{A} = 151$ and $K_\mathrm{A} = 0.11$.

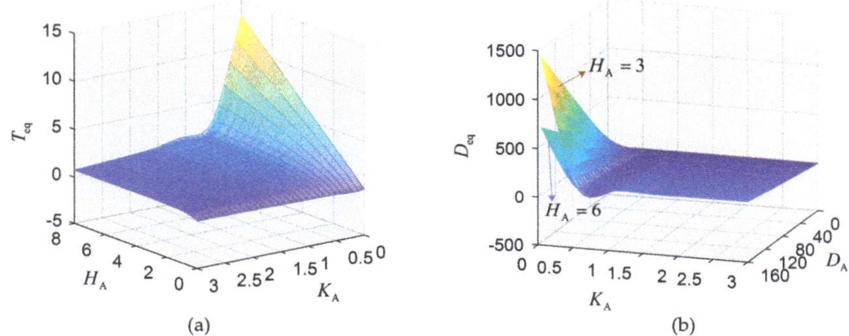

(a) (b)

Figure 7. ACS control parameter's impact on DFIG equivalent inertia/damping: (**a**) Maximal equivalent inertia with different inertia time constant and proportional coefficient; (**b**) Maximal equivalent damping with different inertia time constant, damping and proportional coefficient.

4.1.3. Features of Current Inner Loop Control

When the wind turbine accidentally enters into the island mode, the loads may be out of proportion to the outputs, resulting in fluctuation of stator's voltage amplitude and large induction voltage and current at the rotor side. Therefore, the switching between grid-connection/island modes requires proper control of the rotor current to avoid loss of control and damages to the rotor and the RSC. The current inner loop in ACS is designed as rapid dynamic response, and the voltage control is designed as slow dynamic response. In the transient of switching between grid-connection/island, it can be approximately taken as the reference of the rotor excitation current due to voltage control remains the same, and the current inner loop based on QPR can restrain the large current of the rotor to achieve stable and safe control.

4.1.4. Principles of Distant Synchronization

The absence of secondary cable connection between the DFIG and the tie-switch results in the fact that the grid-connection of DFIG islands via remote tie-switches cannot be solved by traditional DFIG local synchronous grid-connection technologies. Tie-switches cannot create conditions for synchronization actively. Instead, it is the DFIG responsible for this. The principles of distant synchronization of DFIG islands as proposed in this paper are shown in Figure 8.

The negative feedback control of damping in the ACS is equivalent to the proportional controller of the stator's voltage angular frequency ω_s. According to the principles of automatic control, proportional control is based on difference regulation, which will lead to inequality between the island's angular frequency $\omega_\mathrm{si} = \omega_\mathrm{s}$ and the grid's voltage angular frequency ω_g; furthermore, the given the slow dynamic response from the mechanical control of the DFIG, when the DFIG is accidentally islanded, its generation fails to reach a balance with the loads in a short period of time, leading to deviation in the island's voltage frequency. Henceforth, the difference regulation of ACS to the stator's voltage angular frequency and the power imbalance will result in relative movement between the DFIG's island voltage u_I and the grid voltage u_g that as time elapses, the synchronization conditions at any remote tie-switch can be satisfied automatically.

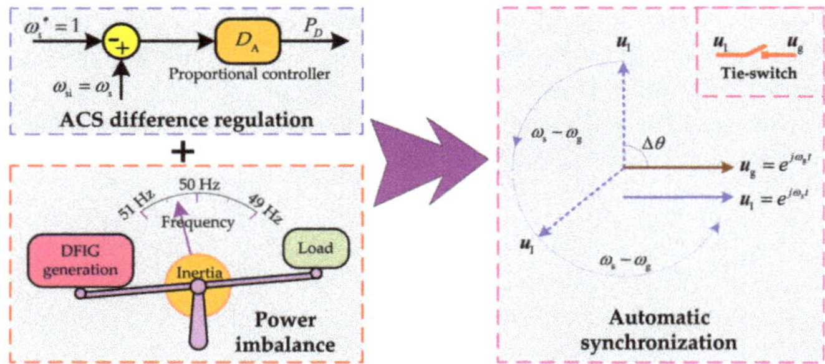

Figure 8. Principles of distant synchronization.

4.2. Coordination Technology

Traditional protection of passive DN focuses on the coordination of protection devices. In the active DN, if the tie-switch still follows the mechanism of closing upon receipt of the GOOSE message of fault isolation successful (FIS), the DIFG island is likely to be non-synchronized connection to grid, resulting in the failure of closing and even the destruction of DFIG and circuit breakers. Therefore, DN-DG coordination is also required in active DN. In this paper, a DN-DG coordination technology is proposed as shown in Figure 9. First is the synchronization check device at the tie-switch whose closing conditions are set as follows: in the process of automatically creating the distant synchronization conditions by the DFIG island under the control of ACS, the tie-switch closes immediately after the receipt of the FIS and detection of synchronization through the check device. As the DFIG controlled by ACS supports grid-connection/island dual-mode operation, DFIG will automatically adapt to the change from island to grid-connection after the tie-switch closes, and recover grid-connected operation. Therefore, in this paper, the technology of coordinating DFIG under the control of ACS with remote tie-switches for local check of synchronization conditions is proposed to realize the secure grid-connection of DFIG islands without an expensive and complex DN-DG communication system.

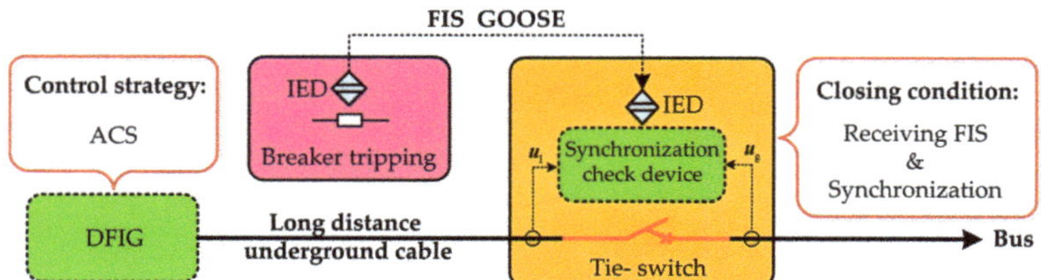

Figure 9. Coordination technology.

5. Case Analysis

To validate the effectiveness of the DFIG ACS and coordination technology compatible with FA, the simulation scenario as shown in Figure 1 and detailed models of DFIGs and converters [30,35] are established in MATLAB/Simulink. The selecting of five DFIGs in Figure 1 is to simulate the situation of multiple DGs connected to DN from different locations, the probabilities of multiple DGs unintentional island and multiple DGs reconnecting to grid. For parameters of DFIG, ACS and the DN, please refer to Table A2 (Appendix A). The first test target is the effectiveness of ACS in grid-connection mode, followed by the effectiveness of ACS and coordination technology after the tripping of circuit breakers under FA control in different DN fault scenes.

5.1. Grid-connection Test

The electromagnetic power waveforms of DFIG under traditional PQ control and ACS control at the same wind speed are shown in Figure 10, in which, the simulation time is 80 s, v_w is set as the typical wind speed, $P_{e,PQ}$ is the electromagnetic power waveform under traditional PQ control strategies, and $P_{e,ACS}$ is the electromagnetic power waveform under the ACS. It is obvious that when the wind speed varies sharply from 10 s to 50 s (7.9–15.45 m/s) in a jagged form, the waveform at $P_{e,PQ}$ can follow up the changes of the wind speed more quickly as compared with $P_{e,ACS}$ which changes more smoothly; when the wind speed changes slightly from 50 to 80 s (13 to 16.5 m/s), there is no huge difference between the waveforms of $P_{e,PQ}$ and $P_{e,ACS}$. The reason for this phenomenon is that the traditional PQ control does not have a large inertia characteristic, and its dynamic response to electromagnetic power is faster. However, the ACS control proposed in this paper has a large inertia, so its dynamic response to electromagnetic power is slower. In general, both waveforms have experienced the similar trends, and the DFIG under ACS control can track the power of wind turbine achieving effective operation in grid-connected mode.

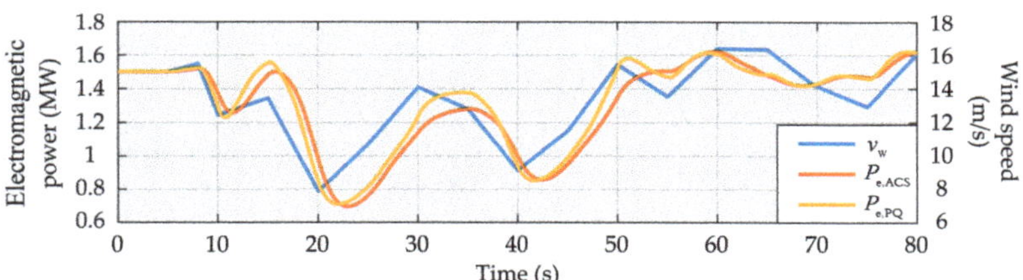

Figure 10. Electromagnetic power waveforms of DFIG under PQ control and proposed ACS control.

5.2. Tests under Different Fault Scenes

The DN in Figure 1 includes three permanent fault scenes of cable feeders, and has simulated the adaption and coordination process of different number of DFIGs with the DN at various wind speeds when they are islanded out of schedule. This paper ignores the effects of faults and assumes that the FA can isolate faults instantaneously. It is assumed that the maximal allowable transient frequency of the DN is 51 Hz, and the error between the voltage phases by both sides of the tie-switches is 180°.

Scene 1: Unintentional Island with One DFIG Reconnecting to Passive Feeder

Feeder 3 is designed with a F1 fault and the wind speed at DFIG4 is 15 m/s. FA controls the circuit breakers L42 and L51 tripping for fault isolation at 30 s, and the island load is 1.12 MW. The tie-switch L52 closes after detection of synchronization conditions. Figures 11 and 12 exhibit the changes in major electrical quantity at the grid side and the wind turbine side after a F1 fault.

①   Grid-connection/Island Switching: the power supply from feeder 3 is cut off at 30 s, DFIG4, RMU4, tie-line 2 and loads constitute single island. DFIG3 enters into the islanding mode under ACS control, while the voltage u_{T2} of the tie-line 2 jitters to 1.1 pu for within 1 cycle. No sudden change in phase is observed. However, mainly due to the sudden change of power (from 1.5 MW to 1.12 MW), the voltage frequency f_{T2} of tie-line 2 shows a peak of 50.6 Hz less than allowed (51 Hz); the rotor current amplitude of DFIG4, $I_{r\text{-}DFIG4}$ rapidly reduces from 0.8 pu to 0.65 pu without overcurrent; the electromagnetic power $P_{e\text{-}DFIG4}$ of DFIG4 also rapidly adapts to the island loads of 1.12 MW. It can be seen that the DFIG4 under ACS control can seamlessly enter into the single island mode.

② Distant Synchronization Process: when the DFIG4 enters into the islanding mode with imbalanced power, the amplitude of u_{T2} remains relatively stable while $I_{r\text{-DFIG4}}$ fluctuates at a frequency of 1.5 Hz with a decreasing tendency; $P_{e\text{-DFIG4}}$ matches with the island loads of 1.12 MW without significant fluctuation basically; f_{T2} fluctuates around 50.1 Hz to cause the error of voltage phase $\Delta\theta_{L62}$ between both sides of the tie-switch L62 reducing from 180° to 0°. It is obvious that the DFIG4 under ACS control can maintain the island stability and automatically create distant synchronization conditions.

③ Island Grid-connection Process: the tie-switch L62 receives the FIS GOOSE message through FA as the circuit breakers L42 and L51 trip. $\Delta\theta_{L62}$ reduces to 0° at 34.9 s, and L62 immediately closes to passive feeder 2 as the synchronization conditions are inspected locally through the synchronization check device, when the DFIG4 turns to the grid-connection status, and $I_{r\text{-DFIG4}}$ and $P_{e\text{-DFIG4}}$ rise gradually, during which, no vibrant fluctuation is experienced. The stable status before fault is recovered in about 10 s. In the meantime, u_{T2} and f_{T2} remain constant as they are clamped down by the voltage of feeder 2 after grid-connection, and $\Delta\theta_{L62}$ reserves 0°; Current amplitude I_{L62} of tie-switch L62 rises from 0pu after closing without dash current, and enters into a stable status about 10 s later. It is obvious that the DFIG4 island under ACS control can securely recover the grid-connection status through coordination with L62 for local synchronous closing.

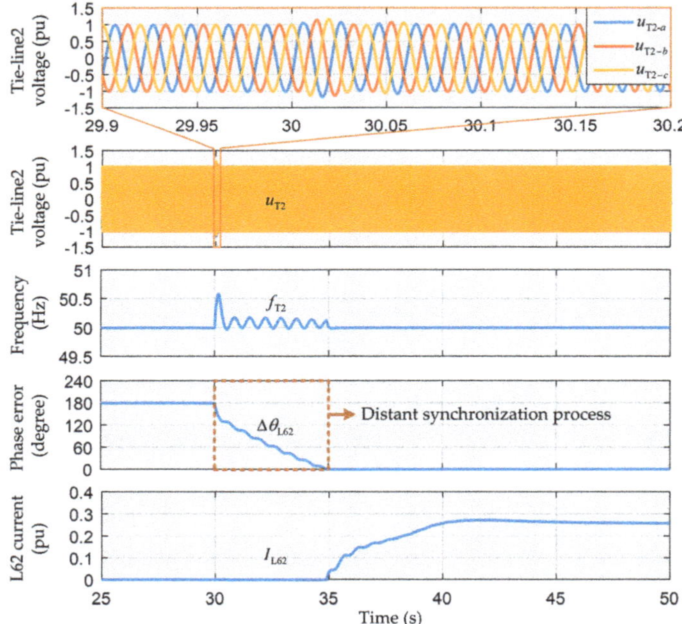

Figure 11. Changes in electrical quantity at the grid side after a F1 fault.

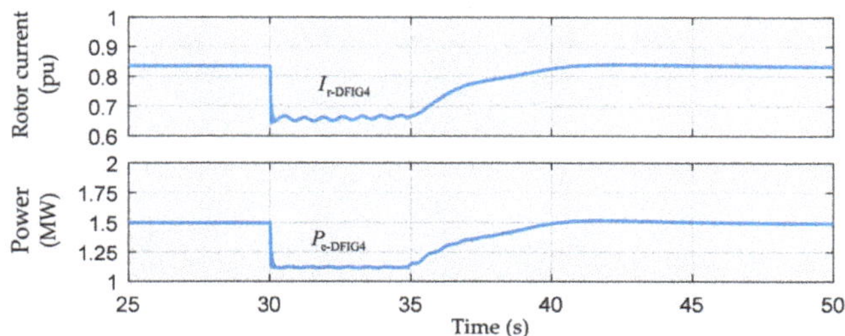

Figure 12. Changes in electrical quantity at the wind turbine side after a F1 fault.

Scene 2: Unintentional Island with Multiple DFIGs Reconnecting to Passive Feeder

Feeder 3 is designed with a F2 fault and the wind speed is 10m/s at DFIG4 and 15 m/s at DFIG5. FA controls the circuit breakers L52 and S3 tripping for fault isolation at 30 s, and the island load is 2 MW. The tie-switch L62 closes after detection of synchronization conditions. Figures 13 and 14 exhibit the changes in major electrical quantity at the grid side and the wind turbine side after a F2 fault.

① Grid-connection/Island Switching: as L52 and S3 trip, DFIG4, DFIG5, RMU4 and RMU5, tie-line 2 and loads constitute an island with two wind turbines. DFIG4 and DFIG5 enter into the islanding mode under ACS control, while the voltage u_{T2} of the tie-line 2 jitters slightly with a continuous waveform. Similarly, mainly due to the sudden change of power (from 2.3 MW to 2 MW), the voltage frequency f_{T2} of tie-line 2 shows a peak of 50.26 Hz below allowable; the rotor current amplitudes of DFIG4 and DFIG5 ($I_{r\text{-DFIG4}}$ and $I_{r\text{-DFIG5}}$) rapidly change without overcurrent; the total electromagnetic power of DFIG4 and DFIG5 ($P_C = P_{e\text{-DFIG4}} + P_{e\text{-DFIG5}}$) also rapidly adapts to the island loads of 2 MW. It can be seen that the DFIG4 and DFIG5 under ACS control can seamlessly enter into the multiple islands mode.

② Distant Synchronization Process: when the DFIG4 and DFIG5 enter into the islanding mode with imbalanced power, the amplitude of u_{T2} remains relatively stable while $I_{r\text{-DFIG4}}$ and $I_{r\text{-DFIG5}}$; DFIG4 and DFIG5 automatically allocate $P_{e\text{-DFIG4}}$ and $P_{e\text{-DFIG5}}$ to maintain P_C at 2MW constantly; f_{T2} fluctuates around 50.05 Hz and its undulatory property gradually weakens to cause the error of voltage phase $\Delta\theta_{L62}$ between both sides of the tie-switch L62 reducing to 0°. It is obvious that the island with DFIG4 and DFIG5 under ACS control can automatically create distant synchronization conditions.

③ Island Grid-connection Process: the tie-switch L62 receives the FIS GOOSE message through FA as the circuit breakers L52 and S3 trip. $\Delta\theta_{L62}$ reduces to 0° at 39.6 s, and L62 immediately closes to passive feeder 2 as DFIG4 and DFIG5 switch to grid-connection mode, $I_{r\text{-DFIG4}}$ and $I_{r\text{-DFIG5}}$ rise gradually. The pre-fault stable status is recovered in about 15 s for $P_{e\text{-DFIG4}}$ and $P_{e\text{-DFIG5}}$ (1.5 MW and 0.8 MW). In the meantime, u_{T2} and f_{T2} remain constant as they are clamped down by the voltage of feeder 2 after grid-connection, and $\Delta\theta_{L62}$ maintains at 0°; I_{L62} rises rapidly after closing without dash current, and enters into a stable status about 15 s later. It is obvious that the island with DFIG4 and DFIG5 under ACS control can securely recover the grid-connection status through coordination with L62 for local synchronous closing.

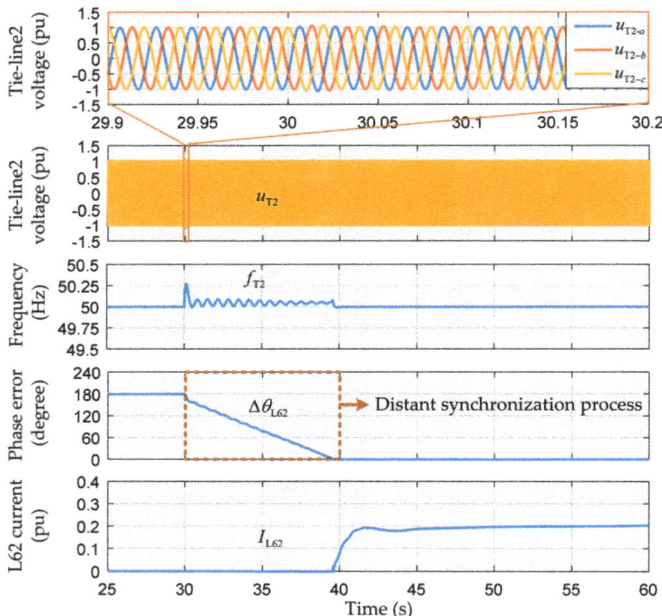

Figure 13. Changes in electrical quantity at the grid side after a F2 fault.

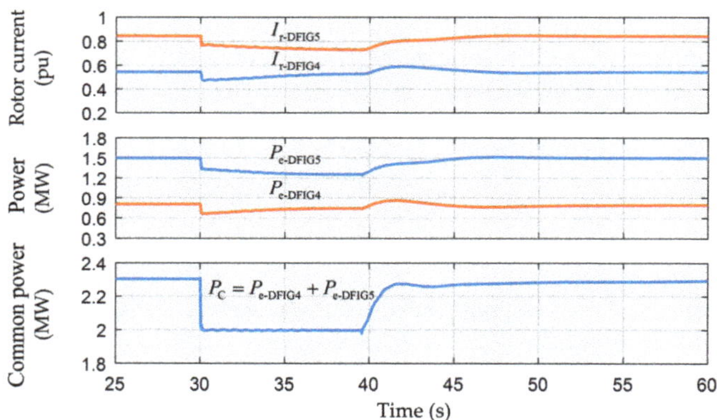

Figure 14. Changes in electrical quantity at the wind turbine side after a F2 fault.

Scene 3: Unintentional Island with Multiple DFIGs Reconnecting to Active Feeder

Feeder 3 is designed with a F3 fault and the wind speed is 15 m/s at DFIG1, 11.8 m/s at DFIG2 11.1 m/s at DFIG3, 12.1 m/s at DFIG4 and 11.4 m/s at DFIG5. FA controls the circuit breakers S1 and L11 tripping for fault isolation at 50 s, and the island load is 3 MW. The tie-switch L41 closes after detection of synchronization conditions with a F3 fault. Figures 15 and 16 exhibit the changes in major electrical quantity at the grid side and the wind turbine side after a F3 fault.

①   Grid-connection/Island Switching: the service is cut off at 50 s, DFIG1, DFIG2, DFIG3, RMU1, RMU2, RMU3, tie-line 1 and loads constitute an island with three wind turbines. DFIG1, DFIG2, and DFIG3 enter into the islanding mode under ACS control, while the voltage u_{T1} of the tie-line

1 jitters to 1.2 pu of the amplitude maximally for about 2.5 cycles, and the waveform transits flatly. Similarly, mainly due to the sudden change of power (from 3.9 MW to 3 MW), the voltage frequency f_{T1} of tie-line 1 shows a peak of 50.48 Hz, no more than the allowable value; the rotor current amplitudes of DFIG1, DFIG2, and DFIG3 ($I_{\text{r-DFIG1}}$, $I_{\text{r-DFIG2}}$ and $I_{\text{r-DFIG3}}$) suddenly change without overcurrent; the electromagnetic powers of DFIG1, DFIG2, and DFIG3 ($P_{\text{e-DFIG1}}$, $P_{\text{e-DFIG2}}$ and $P_{\text{e-DFIG3}}$) also rapidly change and their sum ($P_{\text{C}} = P_{\text{e-DFIG1}} + P_{\text{e-DFIG2}} + P_{\text{e-DFIG3}}$) matches with the island loads of 3 MW. It can be seen that the DFIG1, DFIG2, and DFIG3 under ACS control can seamlessly enter into the multiple wind turbine island mode.

②   Distant Synchronization Process: when the DFIG1, DFIG2, and DFIG3 enter into the islanding mode with imbalanced power, the amplitude of u_{T1} remains relatively stable while $I_{\text{r-DFIG1}}$, $I_{\text{r-DFIG2}}$ and $I_{\text{r-DFIG3}}$ close to each other; DFIG1, DFIG2, and DFIG3 automatically distribute $P_{\text{e-DFIG1}}$, $P_{\text{e-DFIG2}}$ and $P_{\text{e-DFIG3}}$ to maintain P_{C} at a relative stable value; f_{T1} fluctuates around 50.1 Hz to cause the error of voltage phase $\Delta\theta_{\text{L41}}$ between both sides of the tie-switch L41 reducing to 0°. It is obvious that the island with DFIG1, DFIG2, and DFIG3 under ACS control can automatically create distant synchronization conditions.

③   Island Grid-connection Process: the tie-switch L41 receives the FIS GOOSE message through FA as the circuit breakers L11 and S1 trip, and immediately closes to active feeder 3 as the synchronization conditions ($\Delta\theta_{\text{L41}} = 0°$) are detected locally through the synchronization check device at 54.7 s, when the DFIG1, DFIG2 and DFIG3 switch to the grid-connection status, and $I_{\text{r-DFIG1}}$, $I_{\text{r-DFIG2}}$ and $I_{\text{r-DFIG3}}$ recover in an undulatory manner, during which, $I_{\text{r-DFIG2}}$ and $I_{\text{r-DFIG3}}$ fluctuate significantly; $I_{\text{r-DFIG4}}$ and $I_{\text{r-DFIG5}}$ fluctuate transitorily (about 5 s) when L41 closes, and then stabilizes; $P_{\text{e-DFIG1}}$, $P_{\text{e-DFIG2}}$ and $P_{\text{e-DFIG3}}$ show the similar change tendency as the rotor currents, while $P_{\text{e-DFIG2}}$ and $P_{\text{e-DFIG1}}$ are subject to larger fluctuation as compared with the $P_{\text{e-DFIG1}}$ during grid-connection recovery, and $P_{\text{e-DFIG4}}$ and $P_{\text{e-DFIG5}}$ are shocked to a certain degree when L41 closes. In the meantime, u_{T1} and f_{T1} remain constant as they are clamped down by the voltage of feeder 3 after grid-connection, and $\Delta\theta_{\text{L41}}$ maintains at 0°; Current amplitude I_{L41} of tie-switch L41 rises rapidly without dash current, and enters into a stable status after slight fluctuation. It is obvious that the DFIG1, DFIG2 and DFIG3 islands under ACS control can securely recover the grid-connection status through coordination with L41 for local synchronous closing, while DFIG4 and DFIG5 stabilize rapidly after temporary and limited shocks.

It can be seen from three fault scenes: the magnitude and fluctuation of the island system frequency are closely related to the power variation rate. The greater the power changes, the bigger the frequency magnitude changes (including the overshoot), the larger the frequency fluctuates. However, the frequency fluctuation during island operation will gradually decrease over time (e.g., Figure 13). Moreover, the more the frequency of the island system changes, the faster the synchronization conditions are met, so the islanding time of DFIG is closely related to the power change rate too. In addition, the DFIG under ACS control is a temporary island during the coordination with FA, which, on the one hand, helps solve the problem of power quality arising from the long-term islanding of the DFIG, and on the other hand, upgrades the traditional one-way power supply recovery mechanism from feeders to loads to the two-way power supply recovery mechanism from DG and feeders to loads. Therefore, the solution proposed in this paper can make the DFIG truly non-off-grid and more economical than other existing solutions.

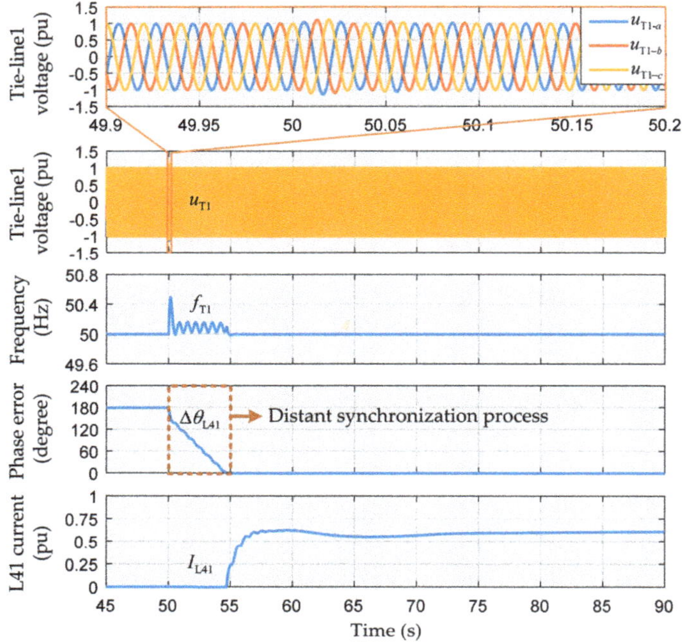

Figure 15. Changes in electrical quantity at the grid side after a F3 fault.

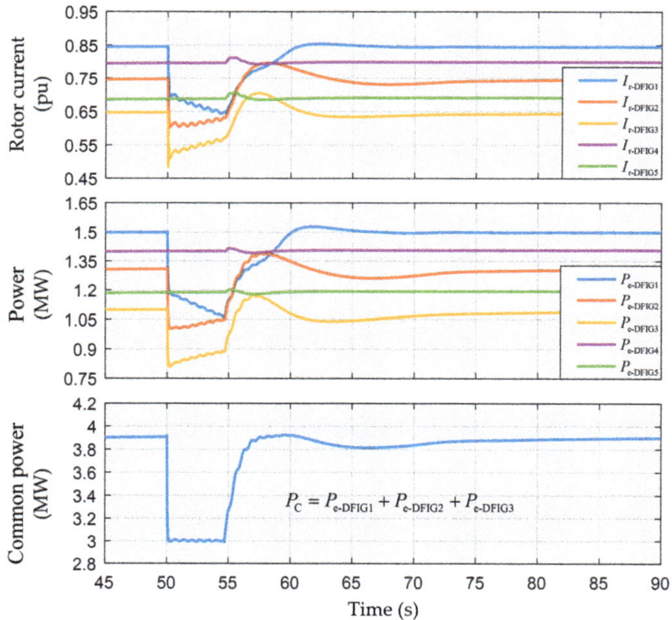

Figure 16. Changes in electrical quantity at the wind turbine side after a F6 fault.

It should be noted that, however, this paper mainly focuses on a situation that the total island loads are smaller than the total real-time capacity of all wind power generations in island, without consideration to other cases, including equivalent or larger relations. The underlying reasons shall be when the output of the wind turbine matches with the load, the wind turbine experiencing switching between grid-connection and island will not have obvious changes in status while the output of the wind turbine is smaller than the load, it has to coordinate with other power generation units (such as energy storage) to stabilize islands. Moreover, the compatibility between the FA technology and DFIG of the DN based on underground cables is another focus in this paper, while the FA technology for DN with overhead lines requires repeated reclosing, and different strategies to achieve compatibility with DFIG. In addition, this paper only studies DFIG, a type of DG. From the thought of DN-DG coordination, the solution proposed in this paper should have similar effect when applied to other types of DGs, such as photovoltaic, energy storage, permanent magnet synchronous generator, etc. However, if ACS proposed in this paper is to be applied to other DGs, appropriate modifications and adjustments may be required. Furthermore, for devices such as energy hub and energy router [36,37], which may be used in future smart grid, the idea in this paper should be helpful in solving the coordination problem between them and protection devices.

6. Conclusions

An adaptive control strategy of DFIG and coordination technology compatible with FA is proposed in this paper. The simulation is carried out based on MATLAB/Simulink and the results reveal that:

① The ACS proposed herein has similar power control characteristics as found in traditional PQ control that the DFIG can work effectively in grid-connection mode.

② As the FA isolates a feeder fault, one or multiple DFIGs under different wind speeds can achieve seamless switching between grid-connection/island modes under ACS control, and continuously supply power to partial local loads.

③ During DFIG islanding, ACS's differential regulation to frequency and power imbalance result in the automatic satisfaction of synchronization conditions at remote tie-switches.

④ After FA fault isolation, the tie-switches can close based on local detection of synchronization conditions and the DFIG under ACS control can be securely grid-connected without establishing an additional communication system with the DN.

In general, the ACS and coordination technology in this paper realize the compatibility between DFIG and FA without DFIG shutdown and restart.

Author Contributions: Z.H. contributed to the project idea, the results discussion and conclusions. P.T. contributed to specific strategy, theoretical analysis, simulation experiment design and data analysis. Z.L. reviewed the final manuscript and the results discussion.

Funding: This research was funded by the National Natural Science Foundation of China (51567005), in part by the Guizhou Province Science and Technology Platform and Talent Project of China ([2017]5788/LH Word [2017]7230), in part by the Guizhou Province Science and Technology Plan Project of China ([2018]5615), in part by the National Natural Science Foundation of China (61963009).

Conflicts of Interest: The authors declare that there is no conflict of interests regarding the publication of this paper.

Appendix A

Table A1. Nomenclature.

Variable	Description
P_e^*	Electromagnetic power reference
P_e	Actual output electromagnetic power
P_C	Total electromagnetic power of DFIG island
$P_{e,PQ}$	Electromagnetic power under traditional PQ control
$P_{e,ACS}$	Electromagnetic power under ACS control
P_D	Damping power
$P_{e\text{-DFIG}x}$	Electromagnetic power of DFIGx
ω_m	Electric angular frequency of rotor
ω_m^*	Reference angular velocity of wind turbine
ω_b	Angular frequency of fundamental wave
ω_g^*	Angular frequency reference of grid
ω_g	Grid voltage angular frequency
ω_r	Slip angular frequency
ω_{m0}	Stable speed of wind turbine at a certain operation point
ω_s	Synchronous angular frequency/ Stator voltage angular frequency
ω_i	Resonance term bandwidth of quasi-proportional resonant controller
ω_{si}	Island voltage angular frequency/ Inner potential angular frequency of stator
R_s & R_r	Resistances of stator and rotor
L_s & L_r	Self-inductance of stator and rotor
L_m	Mutual inductance
f_{T1}	Voltage frequency of tie-line 1
f_{T2}	Voltage frequency of tie-line 2
e_s	Inner potential of stator
E_s	Inner potential amplitude of stator
u_I	Space vectors of island voltage
u_g	Space vectors of grid voltage
u_{T1}	Voltage of the tie-line 1
u_{T2}	Voltage of the tie-line 2
u_s & u_r	Voltage space vectors of stator and rotor in stator static reference frame
u_{ra}^r, u_{rb}^r & u_{rc}^r	Three-phase excitation voltages of the rotor in rotor reference frame
U_s^*	Reference of stator voltage amplitude
U_s	Actual stator voltage amplitude
X_s	Synchronous reactance
X_m	Mutual reactance
i_s & i_r	Current space vectors of stator and rotor in stator static reference frame
i_{ra}^{r*}, i_{rb}^{r*} & i_{rc}^{r*}	Reference of three-phase rotor excitation currents in rotor reference frame
i_{ra}^r, i_{rb}^r & i_{rc}^r	Actual excitation currents of rotor in rotor reference frame
I_r^*	Reference of rotor excitation current amplitude
$I_{r\text{-DFIG}x}$	Rotor current amplitude of DFIGx
I_{L62}	Current amplitude of tie-switch L62
I_{L41}	Current amplitude of tie-switch L41
$\overrightarrow{i}_r^{r*}$	Space vector reference of rotor excitation currents in rotor reference frame
$\Delta\theta_{L62}$	Error of voltage phase between both sides of tie-switch L62
$\Delta\theta_{L41}$	Error of voltage phase between both sides of tie-switch L41
θ_{si}	Inner potential control phase of stator
θ_{Ir}	Excitation current phase of rotor
θ_{Es}	Inner potential phase of stator
θ_m	Rotor phase
P_{QPR}	Proportional coefficient of quasi-proportional resonant controller
K_{QPR}	Resonance term gain of quasi-proportional resonant controller
v_w	Wind speed
β	Pitch angle of blades

Table A1. *Cont.*

Variable	Description
H_A	Inertia time constant of power control
K_A	Proportional coefficient of power control
D_A	Damping coefficient of power control
P_{wt}	Mechanical energy capture by wind wheel
H_m	Inherent mechanical inertia time constant of wind turbine
D_m	Inherent mechanical damping coefficient of wind turbine
T_{eq} & D_{eq}	Equivalent inertia and damping of DFIG under ACS control
P_v & I_v	Proportional and integral coefficient of torque control
$P_{Us,A}$ & $I_{Us,A}$	Proportion and integral coefficient of voltage control

Table A2. Simulation system parameters.

DFIG Machine			
Parameter	**Value**	**Parameter**	**Value**
R_s	0.023 pu	U_{dc}	1150 V
R_r	0.016 pu	Rated power	1.5 MW
L_s	3.08 pu	Stator voltage	690 V
L_r	3.06 pu	Pole pairs	3
L_m	2.9 pu	Normal speed	1.2 pu
H_m	4.96	DC-link capacitor	10,000 uF
D_m	1.5	Rated frequency	50 Hz
ACS			
P_v	3	P_{QPR}	20
I_v	0.6	K_{QPR}	33
K_A	0.11	ω_i	π
H_A	4.96	$P_{Us,A}$	1
D_A	151	$I_{Us,A}$	40
DN			
Rated voltage	10 KV	Line inductance	0.9337×10^{-3} H/km
Rated frequency	50 Hz	Line capacitance	12.74×10^{-9} F/km
Line resistance	0.01273 Ω/km		

References

1. João, P.S.; Catalão, S.; Rahimi, E.; Javadi, M.S.; Nezhad, A.E.; Lotfi, M.; Shafie-khah, M.; Catalão, J.P.S. Impact of distributed generation on protection and voltage regulation of distribution systems: A review. *Renew. Sustain. Energy Rev.* **2019**, *105*, 157–167.
2. Blaabjerg, F.; Yang, Y.; Yang, D.; Wang, X. Distributed Power-Generation Systems and Protection. *Proc. IEEE* **2017**, *105*, 1311–1331. [CrossRef]
3. Roy, N.K.; Pota, H.R. Current Status and Issues of Concern for the Integration of Distributed Generation into Electricity Networks. *IEEE Syst. J.* **2015**, *9*, 933–944. [CrossRef]
4. Monadi, M.; Amin Zamani, M.; Ignacio Candela, J.; Luna, A.; Rodriguez, P. Protection of AC and DC distribution systems Embedding distributed energy resources: A comparative review and analysis. *Renew. Sustain. Energy Rev.* **2015**, *51*, 1578–1593. [CrossRef]
5. Kennedy, J.; Ciufo, P.; Agalgaonkar, A. A review of protection systems for distribution networks embedded with renewable generation. *Renew. Sustain. Energy Rev.* **2016**, *58*, 1308–1317. [CrossRef]
6. Nikolaidis, V.C.; Papanikolaou, E.; Safigianni, A.S. A Communication-Assisted Overcurrent Protection Scheme for Radial Distribution Systems with Distributed Generation. *IEEE Trans. Smart Grid.* **2016**, *7*, 114–123. [CrossRef]
7. Bansal, R. *Power System Protection in Smart Grid Environment*, 1st ed.; CRC Press: Boca Raton, FL, USA, 2019.

8. Norshahrani, M.; Mokhlis, H.; Abu Bakar, A.; Jamian, J.; Sukumar, S. Progress on Protection Strategies to Mitigate the Impact of Renewable Distributed Generation on Distribution Systems. *Energies* **2017**, *10*, 1864. [CrossRef]

9. Sajadi, A.; Strezoski, L.; Strezoski, V.; Prica, M.; Loparo, K.A. Integration of renewable energy systems and challenges for dynamics, control, and automation of electrical power systems. *Wires Energy Environ.* **2019**, *8*, e321. [CrossRef]

10. Vyas, S.; Kumar, R.; Kavasseri, R. Data analytics and computational methods for anti-islanding of renewable energy based Distributed Generators in power grids. *Renew. Sustain. Energy Rev.* **2017**, *69*, 493–502. [CrossRef]

11. Peixin, Y.; Peichao, Z. A Survey on Interconnection Protection of Distributed Resource. *Power Syst. Techn.* **2016**, *06*, 1888–1895.

12. Manditereza, P.T.; Bansal, R. Renewable distributed generation: The hidden challenges—A review from the protection perspective. *Renew. Sustain. Energy Rev.* **2016**, *58*, 1457–1465. [CrossRef]

13. Motter, D.; de Melo Vieira, J.C. The Setting Map Methodology for Adjusting the DG Anti-Islanding Protection Considering Multiple Events. *IEEE Trans. Power Deliv.* **2018**, *33*, 2755–2764. [CrossRef]

14. Zidan, A.; Khairalla, M.; Abdrabou, A.M.; Khalifa, T.; Shaban, K.; Abdrabou, A.; El Shatshat, R.; Gaouda, A.M. Fault Detection, Isolation, and Service Restoration in Distribution Systems: State-of-the-Art and Future Trends. *IEEE Trans. Smart Grid* **2017**, *8*, 2170–2185. [CrossRef]

15. Zhan, H.; Wang, C.; Wang, Y.; Yang, X.; Zhang, X.; Wu, C.; Chen, Y. Relay Protection Coordination Integrated Optimal Placement and Sizing of Distributed Generation Sources in Distribution Networks. *IEEE Trans. Smart Grid* **2016**, *7*, 55–65. [CrossRef]

16. Abdel-Ghany, H.A.; Azmy, A.M.; Elkalashy, N.I.; Rashad, E.M. Optimizing DG penetration in distribution networks concerning protection schemes and technical impact. *Electr. Power Syst. Res.* **2015**, *128*, 113–122. [CrossRef]

17. Badran, O.; Mekhilef, S.; Mokhlis, H.; Dahalan, W. Optimal reconfiguration of distribution system connected with distributed generations: A review of different methodologies. *Renew. Sustain. Energy Rev.* **2017**, *73*, 854–867. [CrossRef]

18. Xyngi, I.; Popov, M. An Intelligent Algorithm for the Protection of Smart Power Systems. *IEEE Trans. Smart Grid* **2013**, *4*, 1541–1548. [CrossRef]

19. Wheeler, K.A.; Elsamahy, M.; Faried, S.O. A Novel Reclosing Scheme for Mitigation of Distributed Generation Effects on Overcurrent Protection. *IEEE Trans. Power Deliv.* **2018**, *33*, 981–991. [CrossRef]

20. Zhu, Z.; Xu, B.; Brunner, C.; Yip, T.; Chen, Y. IEC 61850 Configuration Solution to Distributed Intelligence in Distribution Grid Automation. *Energies* **2017**, *10*, 528. [CrossRef]

21. Ling, W.; Liu, D.; Yang, D.; Sun, C. The situation and trends of feeder automation in China. *Renew. Sust. Energy Rev.* **2015**, *50*, 1138–1147. [CrossRef]

22. Chenghong, T.; Zhihong, Y.; Bin, S.; Yajun, Z. A Method of Intelligent Distributed Feeder Automation for Active Distribution Network. *Autom. Electr. Power Syst.* **2015**, *9*, 101–106.

23. Wanshui, L.; Dong, L.; Yiming, L.; Wenpeng, Y. Model of intelligent distributed feeder automation based on IEC 618510. *Autom. Electr. Power Syst.* **2012**, *36*, 90–95.

24. Tanvir, A.; Merabet, A.; Beguenane, R. Real-Time Control of Active and Reactive Power for Doubly Fed Induction Generator (DFIG)-Based Wind Energy Conversion System. *Energies* **2015**, *8*, 10389–10408. [CrossRef]

25. Shukla, R.D.; Tripathi, R.K. A novel voltage and frequency controller for standalone DFIG based Wind Energy Conversion System. *Renew. Sustain. Energy Rev.* **2014**, *37*, 69–89. [CrossRef]

26. Lin, L.; Xibin, S.; Zongxun, S. Operation mode analysis of micro-grid grid-connected and island based on DFIG control method. *Power Syst. Protect. Control* **2017**, *45*, 158–163.

27. Hosseini, S.M.; Carli, R.; Dotoli, M. Robust Day-Ahead Energy Scheduling of a Smart Residential User Under Uncertainty. In Proceedings of the 2019 18th European Control Conference, Naples, Italy, 25–28 June 2019; pp. 935–940.

28. Sperstad, I.; Korpås, M. Energy Storage Scheduling in Distribution Systems Considering Wind and Photovoltaic Generation Uncertainties. *Energies* **2019**, *12*, 1231. [CrossRef]

29. Sitharthan, R.; Karthikeyan, M.; Sundar, D.S.; Rajasekaran, S. Adaptive hybrid intelligent MPPT controller to approximate effectual wind speed and optimal rotor speed of variable speed wind turbine. *ISA. Trans.* **2019**, in press. [CrossRef]

30. Miller, N.W.; Sanchez-Gasca, J.J.; Price, W.W.; Delmerico, R.W. Dynamic modeling of GE 1.5 and 3.6 MW wind turbine-generators for stability simulations. In Proceedings of the IEEE Power Engineering Society General Meeting (IEEE Cat. No.03CH37491), Toronto, ON, Canada, 13–17 July 2003; pp. 1977–1983.

31. Pena, R.; Clare, J.C.; Asher, G.M. Doubly fed induction generator uising back-to-back PWM converters and its application to variable- speed wind-energy generation. *IEE Proc. Electr. Power Appl.* **1996**, *3*, 231–241. [CrossRef]

32. Wang, S.; Hu, J.; Yuan, X. Virtual Synchronous Control for Grid-Connected DFIG-Based Wind Turbines. *IEEE J. Emerg. Sel. Top. Power Electron.* **2015**, *3*, 932–944. [CrossRef]

33. Ye, T.; Dai, N.; Lam, C.; Wong, M.; Guerrero, J.M. Analysis, Design, and Implementation of a Quasi-Proportional-Resonant Controller for a Multifunctional Capacitive-Coupling Grid-Connected Inverter. *IEEE Trans. Ind. Appl.* **2016**, *52*, 4269–4280. [CrossRef]

34. Holmes, D.G.; Lipo, T.A.; McGrath, B.P.; Kong, W.Y. Optimized Design of Stationary Frame Three Phase AC Current Regulators. *IEEE Trans. Power Electr.* **2009**, *24*, 2417–2426. [CrossRef]

35. Clark, K.; Miller, N.W.; Sanchez-Gasca, J.J. *Modeling of GE Wind Turbine-Generators for Grid Studies*; General Electric International, Inc.: Schenectady, NY, USA, 2009.

36. Carli, R.; Dotoli, M. Decentralized control for residential energy management of a smart users microgrid with renewable energy exchange. *IEEE/CAA J. Autom. Sin.* **2019**, *6*, 641–656. [CrossRef]

37. Geidl, M.; Koeppel, G.; Favre-Perrod, P.; Klockl, B.; Andersson, G.; Frohlich, K. Energy hubs for the future. *IEEE Power Energy Mag.* **2007**, *5*, 24–30. [CrossRef]

Article

Numerical Investigation on the Influence of Mechanical Draft Wet-Cooling Towers on the Cooling Performance of Air-Cooled Condenser with Complex Building Environment

Jun Fan [2,†], Haotian Dong [1,†], Xiangyang Xu [3], De Teng [3], Bo Yan [3] and Yuanbin Zhao [1,*]

[1] School of Energy and Power Engineering, Shandong University, Jinan 250061, China; donghaotian2019@sdu.edu.cn

[2] School of Water Conservancy and Civil Engineering, Shandong Agricultural University, Taian 271000, China; fanjun2019@sdau.edu.cn

[3] Energy Engineering Excellence (ENEXIO) Energy Technology (Beijing) Co., Ltd, Beijing 100600, China; xuxiangyang2019@enexio.com (X.X.); Tengde2019@enexio.com (D.T.); Yanbo2019@enexio.com (B.Y.)

* Correspondence: zhyb@sdu.edu.cn; Tel.: +86-531-8801-5631

† These authors contributed equally to this work and should be considered co-first authors.

Received: 17 September 2019; Accepted: 5 November 2019; Published: 29 November 2019

Abstract: In air-cooled power units, an air-cooled condenser (ACC) is usually accompanied by mechanical draft wet-cooling towers (MCTs) so as to meet the severe cooling requirements of air-cooling auxiliary apparatuses, such as water ring vacuum pumps. When running, both the ACC and MCTs affected each other through their aerodynamic fields. To make the effect of MCTs on the cooling performance of the ACC more prominent, a three-dimensional (3D) numerical model was established for one 2 × 660 MW air-cooling power plant, with full consideration the ACC, MCTs and adjacent main workshops, which was validated by design data and published test results. By numerical simulation, we obtained the effect of hot air recirculation (HAR) on the cooling performance of the ACC under different working conditions and the effect of MCTs on the cooling performance of the ACC. The results showed that as the ambient wind speed increases, the hot recirculation rate (HRR) of the ACC increased and changed significantly with the change of wind directions. An increase in ambient temperature can cause a significant rise in back pressure of the ACC. The exhaust of the MCTs partially entered the ACC under the influence of ambient wind, and the HRR in the affected cooling units was higher than that of the nearby unaffected cooling units. When the MCTs were turned off, the overall HRR of the ACC decreased. The presence of MCTs had a local influence on the cooling performance of only two cooling units, and then slightly impacted the overall cooling performance of the ACC, which provides a good insight into the arrangement optimization of the ACC and the MCTs.

Keywords: air-cooled condenser; mechanical draft wet-cooling towers; hot recirculation rate; complex building environment; numerical simulation

---

1. Introduction

The air-cooled condenser (ACC) uses air as a cooling medium to exchange heat and has obvious water conservation benefits; therefore, in arid regions, the ACC has been widely used [1,2]. Apart from thermal power plants, the ACC is also diffusely used in the cold ends of the refrigeration and air conditioning industries [3,4]. Inevitably, ambient wind and surrounding structures can have unpredictable effects on the cooling performance of the ACC. Efficient and economical operation of the entire energy network is inseparable from the ACC's adequate and predictable cooling performance [5].

In air-cooled power units, the exhausted steam is cooled in the ACC, but the auxiliary critical apparatus, such as the water ring vacuum pump, oil cooler, etc., should be cooled by mechanical draft wet-cooling towers (MCTs) due to their severe cooling requirements, so MCTs are usually built nearby ACCs.

Hot air recirculation (HAR) is an unavoidable phenomenon during the operation of ACCs. Ambient wind and adjacent structures play a leading role in the ACC performance reduction. Gu et al. [6] experimentally studied the influence of wind directions, wind speed and ACC platform height on the HAR of an air-cooling system. Liu [7] and Wang [8] simulated the phenomenon of hot air reflow and proposed valuable suggestions to reduce these adverse effects. Meyer and Kroger [9,10] studied the influence of the characteristics of fan and heat exchangers and the geometry of the static pressure chamber on the flow loss of ACCs by experimental measurements.

The cooling performance of ACCs is affected by many factors, and many measures and methods have been developed to enhance the cooling performance of ACCs. Owen et al. [11] numerically investigated the influencing factors of the performance of ACCs under windy conditions and compared the results of the numerical simulation with experimental data. It was found that they have good correlation. He also pointed out that the performance degradation of the fan due to the distortion of the airflow at the upstream fan inlet is the main reason for the decline in the ACC's performance, and the accumulation of hot plume also led to a reduction in the ACC's performance. Alan O'Donovan et al. [12] experimentally studied that the increase in fan speed causes the temperature and pressure of the ACC to decrease, and vice-versa. Wang et al. [8] proposed setting side panels below or above the platform to prevent plume recirculation because of the flow separation at the inlet of the fan. Zhang et al. [13] proposed to set up a windproof mesh outside the steel supporting structure and under the ACC platform in order to enhance the cooling performance of the ACC. Meyer [14] recommended setting an aisle outside the edge of the platform of the fan and removing the inlet portion of the peripheral fan to restore the ACC's cooling performance under ambient wind. Gao et al. [15] recommended installing deflecting plates below the platform of the ACC to reduce the influence of strong wind on the ACC by introducing sufficient cold air into the windward area of the fan to achieve a uniform air mass flow rate, thereby improving ACCs cooling performance. Huang et al. [16] optimized the baffle geometry under the platform of the fan to make the ACC have better heat flow performance in a windy situation. Jin et al. [17] proposed to replace the traditional arrangement with a square array of ACC structures to improve the thermal performance of ACCs in a windy situation. Yang et al. [18] studied three different types of windproof wall and took into account the effects of the steam engine room, boiler room and chimney on the ACC. The results showed that increasing the width of the inner or outer channels can enhance the performance of the ACC.

MCTs are diffusely used for deep cooling of industrial water [19–21]. MCTs' performance is constrained by many factors, but at the same time, the plume generated by the MCTs will have an adverse effect on the environment [22]. Many studies [23–26] have made detailed analyses of the environmental impact of plumes discharged from MCTs. Ambient air enters the MCTs through the air inlet. After a series of heat exchange with the high-temperature steam, it is discharged into the environment by the fan at the air outlet. The temperature of this part of the exhausted air is often higher than the ambient air, and the exhausted air always entrains high-temperature steam. Under the influence of ambient wind, this part of the exhausted air will pass through the air inlet of the ACC, causing the air temperature sucked by the ACC to increase. Similarly, the air discharged from the ACC will also influence the performance of the MCTs. In power plants, MCTs and ACCs are coexisting, and there is bound to be an interaction between them. Therefore, it is very necessary to research the influence of MCTs on the cooling performance of ACCs.

The aforementioned research all showed that wind speeds, wind directions and even the installation of various equipment around ACCs had some impacts on the cooling performance of the ACC. In a power plant, MCTs were usually built nearby the ACC, in this process, the MCTs were bound to have some impacts on the cooling performance of the ACC. However, the investigation on the effect of MCTs on the performance of the ACC with a complex building environment is less. Taking a representative 2 × 600 MW power plant as an example, and taking into account factors such as boiler rooms, steam engine rooms and chimneys, the specific effects of MCTs on the cooling performance of the ACC were mainly studied, which fills in the blanks of this aspect. The final results showed that the presence of MCTs had a local influence on the cooling performance of only two cooling units, and then slightly impacted the overall cooling performance of the ACC, which provides a good insight into the arrangement optimization of ACCs and MCTs. Meanwhile, it is beneficial to the optimal design and operation of the ACC of the power plant.

2. Numerical Simulation

2.1. Physical Model

Figure 1 shows a typical 2 × 660 MW air-cooled power plant, which is composed of a chimney, boiler room, turbine room, support structure, an ACC and MCTs. The size of this ACC system is 100 m at width and is about 184 m at length. The structure consists of 124 (16 × 8) fans with a diameter of 9.6 m. The MCTs of the auxiliary cooling water system of the project are a square, open-type exhaust, counter flow combined-type and have a water distribution system with tubular pressure distribution. There are three MCTs with the water-cooling area of a single cooling tower of 196 m^2 and the height of 14 m arranged side by side. The cooling water volume is 3000 m^3/h, and the fan diameter is 8.53 m with the fan shaft power of 132 KW. Generally speaking, these MCTs are conventional tower-types, which have typical representative meanings. This paper focuses on the analysis of mutual influences. The tower-type will be discussed in subsequent articles. The main dimensions of the studied MCTs with S-wave fill are shown in Table 1.

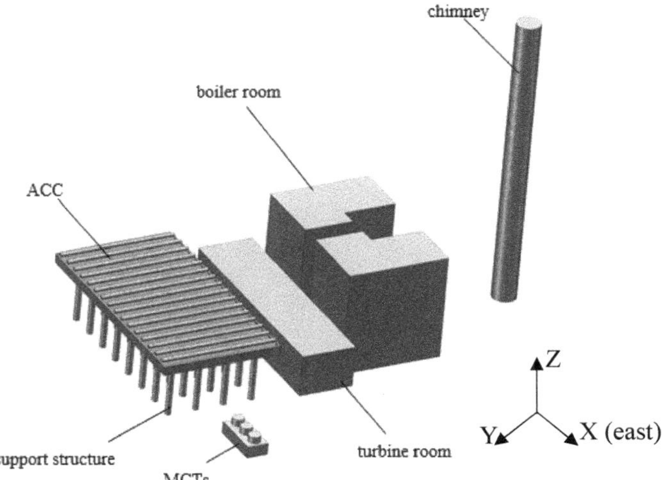

Figure 1. Physical model.

Table 1. The main dimensions of the studied wet cooling towers (mechanical draft wet-cooling towers (MCTs)).

Items	Value	Unit
Water drenching area	196	m^2
Cooling fill height	1.5	m
Tower height	14	m
Tower length and width	14	m
Air inlet height	2.7	m
Fan diameter	8.53	m
Fill bottom height	2.7	m
Tower throat inner diameter	8.60	m

Based on the information provided, the main affected buildings were identified (the buildings and terrain were all taken into consideration). In order to determine the calculation area range, in the case of maximum wind speed and maximum temperature rise disturbance, when the surrounding environment space was large enough, a mathematical model with an enlarged area diameter of 2600 m and the height of 2000 m was used to predict the range of the wind field to reach a stable area. According to the design drawing, the mathematical model of the plant was generated for the main buildings of the power plant. Except for the ACC and the MCTs, other buildings in the model area were treated as obstacles in the flow, which only affected the flow, regardless of whether it was hot or endothermic. The ACC was divided into heat transfer units corresponding to the wind turbine, considering the structure of the ACC and the complexity of heat exchange, the model was simplified appropriately. The details of the air-cooled units were neglected. It was proposed to adopt the radiator model in Computational Fluid Dynamics (CFD) software under the premise of keeping the macroscopic resistance and heat transfer characteristics of the ACC unchanged.

2.2. Numerical Model

Under the given environmental meteorological conditions and operating conditions, the ACC and the MCTs can be considered to be in stable operation; therefore, the Reynolds-averaged Navier–Stokes conservation equations are adopted to describe the flow of air in the ACC and its building environment. The Reynolds stress average term is turbulently closed by adopting the RNG k-ε turbulence model. The first-order upwind schemes are used for convection terms in discrete momentum, energy and turbulence equations, and the diffusion term adopts the central difference format. The ACC radiator is simulated by adopting the RADIATOR model in FLUENT software in which the thermal characteristics and resistance characteristics of the radiator are strictly on the basis of the data provided by the manufacturer. The flow field outside the ACC and the MCTs is considered to be incompressible, and the momentum equation and the continuous equation in the basic governing equation are solved by the SIMPLE algorithm for pressure-velocity coupling.

Grid modeling is performed strictly according to the actual model to ensure that the calculated flow field reflects the actual flow field. Important buildings in the factory area, such as boiler rooms, turbine rooms, chimneys, and surrounding mountains, have been reflected in the model. The partitioning of the solid mesh adopted the partitioned meshing technique, and different regions are divided by different mesh types and sizes. A structured grid with a small size was used near the finned tube heat sink, the grid near the air-cooled platform area also selected a smaller spacing, and the other areas used a larger spacing. There are three sets of grids of the numerical for grid independence testing. The resulting grid number is 3,379,659 for simulation modeling of the entire project.

In this study, a lumped radiator model was used to simplify the ribbed tube bundles of the ACC to a non-thickness surface. The radiator simulated in FLUENT is supposed to be infinitely thin and the pressure drop across it is also supposed to be in proportion to the dynamic head of the fluid, and has an empirical formula of the loss factor provided. That is, the pressure drop and the normal velocity component across the radiator can be written as:

$$\Delta P = K_L \frac{1}{2} \rho v^2 \tag{1}$$

where, ρ and v represent the air density and the magnitude of the local fluid velocity perpendicular to the radiator, respectively. K_L is the flow loss coefficient, which can be expressed as a polynomial function, a piecewise linear function or a piecewise polynomial function [27].

Polynomial functions are expressed as:

$$K_L = \sum_{n=1}^{3} r_n v^{n-1} \tag{2}$$

where, r_n are the polynomial coefficients with $r_1 = 70.665$, $r_2 = -15.86$ and $r_3 = 2$, in the light of the calculation results of the pressure drop across finned tube bundles.

The heat flux from the radiator to the surrounding fluid can be written as:

$$q = h(T_{HX} - T_{exit}) \tag{3}$$

where, q, T_{HX} and T_{exit} denote the heat flux, the radiator temperature and the temperature of the outlet fluid, respectively. The convection heat transfer coefficient h is a piecewise polynomial function, which are expressed as:

$$h = \sum_{n=0}^{N} h_n v^n; 0 \leq N \leq 7 \tag{4}$$

where, h_n and v represent the polynomial coefficient and the magnitude of the local fluid velocity perpendicular to the radiator, respectively. The heat transfer and resistance characteristics of the radiator in this study are provided by the manufacturer.

According to the size parameters of the radiator structure provided by the manufacturer, the length of the single tube bundle is 10 m, and the angle between the bundles is 60°, the width is 2.22 m, the thickness is 0.214 m, and the finned ratio is 122.69, as shown in Figure 2 below.

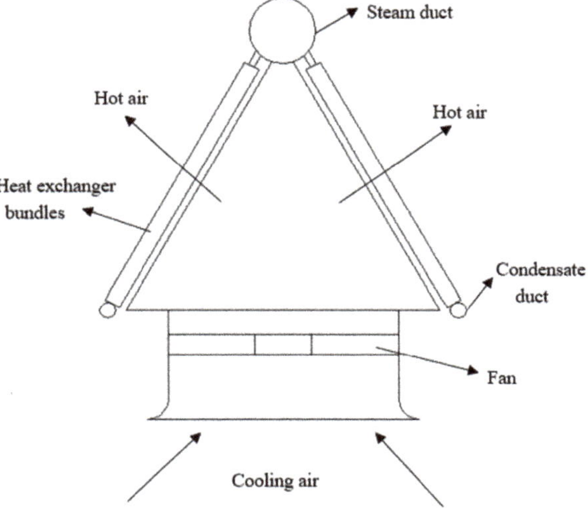

Figure 2. The structure of the air-cooled condenser (ACC).

2.3. Boundary Conditions

2.3.1. Fan Boundary Conditions

Injecting the surrounding cooling air into the heat exchanger is an important function of the axial fan, so this study chose the fan model in the FLUENT software because it can achieve this purpose by increasing the fluid pressure flowing through the fan. The FAN model is a lumped model that can be adopted to predict the influence of a fan with known deterministic properties on the flow field in a large basin. The fan characteristic curve expresses the relationship between the pressure-flow of the fan at a given known rotational speed. Taking the FLUENT software as an example, in the FAN model of FLUENT, simplifying the fan into an infinitely thin layer causes the discontiguous pressure to rise to overcome the flow resistance, and the fan's discontiguous pressure rise is designated as a function of fan speed. The axial flow fan is simplified to a pressure jump surface, and the pressure rise ΔP is calculated as a polynomial form of the axial velocity v. The relationship between them can be a piecewise polynomial function and is expressed as [27]:

$$\Delta P = \sum_{n=1}^{6} f_n v^{n-1} \tag{5}$$

where, ΔP and v represent the pressure jump and the size of the local fluid velocity perpendicular to the fan, respectively. Both the MCTs and the ACC contain fans, and the coefficients are different. f_{n1} are the pressure-jump polynomial coefficients of MCTs, with $f_1 = 459.2698, f_2 = -28.93151, f_3 = 5.4779, f_4 = -0.89391, f_5 = 0.04158$ and $f_6 = -0.000750621$. f_{n2} are the pressure-jump polynomial coefficients of the ACC, with $f_1 = 34.07611, f_2 = 178.7873, f_3 = -61.21099, f_4 = 9.15276, f_5 = -0.66468$ and $f_6 = 0.01831$, which are obtained by fitting the change of the static pressure with the volume flow rate.

2.3.2. Inlet Boundary Conditions

The boundaries of the air-inlet surfaces are set as the velocity inlet and the calculation should use the atmospheric boundary function, which is expressed as:

$$\frac{U_i}{U_\infty} = \left(\frac{Z_i}{Z_\infty} \right)^\alpha \tag{6}$$

where, Z_∞ is the speed at which the airflow reaches a uniform, U_∞ and U_i represent the average speed at the Z_∞ and Z_i, respectively, α is the wall roughness coefficient, the greater the roughness, the larger the value of α.

2.3.3. Outlet Boundary Conditions

The boundary of the outlet is set to a pressure outlet, and the static pressure at the flow outlet boundary is given according to the relationship between the barometric pressure and the elevation of the ocean level, as is the inlet pressure.

$$P = 760 e^{-\frac{a}{7926}} \tag{7}$$

where, a is the altitude and the unit is m, P is atmospheric pressure and the unit is mmHg.

2.3.4. Other Boundary Conditions

Using K-ε to describe the turbulence model and ambient air turbulence is very weak, with a recommended turbulence intensity of 10% and a viscosity ratio of 5 [28]. Surfaces such as ground and buildings are set as wall boundary conditions, regardless of their heat dissipation or heat absorption, only considering the impact on the flow. In this paper, it is assumed that air is an ideal gas, and its density is also calculated as an ideal gas. All parameters such as specific heat, heat transfer coefficient, etc., are constant. Figure 3 shows the computation domain and its air flow boundaries for the ACC and MCTs.

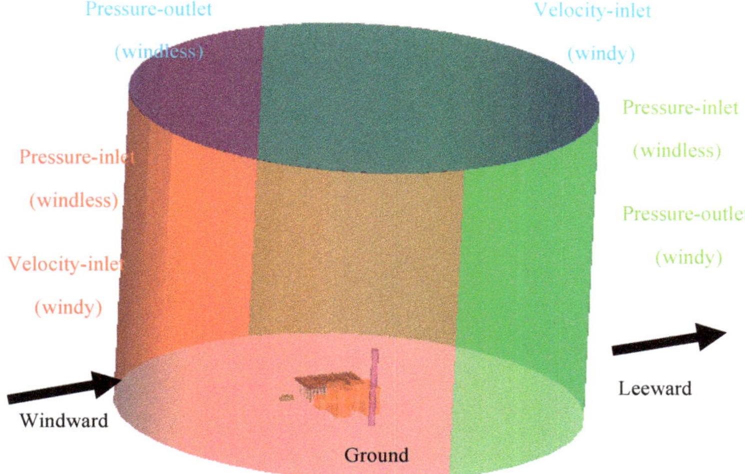

Figure 3. Computation domain and its air flow boundaries for the ACC and MCTs.

3. Results and Discussion

3.1. Effect of Hot Air Recirculation (HAR) on Heat Transfer Performance of the Air-Cooled Condenser (ACC)

HAR means that the hot air discharged from the ACC is returned to the air-cooling fan suction port, which causes the air temperature of the inlet of the ACC to rise. The hot recirculation rate (HRR) represents the strength of the HAR, the greater the HRR, the stronger the HAR. On account of the complexity of the direct ACC itself and the overall arrangement of its surrounding buildings, the flow field embracing them is very complicated; therefore, it is very necessary to study the wind or confirm the influence law of different ambient winds on the HRR. We selected the ambient temperatures of 34 °C and 38 °C, different incident wind directions of west-north-west (WNW), south-south-east (SSE) and wind speeds of 5 m/s and 7 m/s to complete the numerical simulation of the impact of ambient wind on HRR.

For quantitative analysis of thermal reflow, the HRR is defined as follows:

$$R = \frac{T_i - T_a}{T_o - T_a} \times 100\% \tag{8}$$

where, T_i is the average temperature at the inlet of the fan and T_a and T_o represent the ambient temperature and the average temperature at the outlet of the ACC, respectively.

For the ACC, the wall heat conduction thermal resistance and the steam condensation heat resistance are neglected. It is considered that the exhaust steam temperature is equal to the tube bundle average wall temperature, that is, the temperature of the exhaust steam and the pressure of the exhaust steam are calculated according to the tube bundle average wall temperature.

$$T_s = \frac{q}{h} + \frac{T_{a1} + T_{a2}}{2} \tag{9}$$

where, T_s is the exhaust steam temperature and q and h represent the heat flux per square meter and the heat transfer coefficient, respectively. T_{a1} and T_{a2} represent the average temperature of the inlet air and outlet air of the ACC, respectively.

The heat transfer capacity of a single unit under a design condition is about 820.86 MW, and the calculated heat flow per square meter is 47,400.47 W/m^2. T_{a1}, T_{a2} and h were obtained by three-dimensional (3D) model calculation. The HRR and the back pressure of the ACC can represent the level of the cooling performance of the ACC, a higher HRR and back pressure will have a lower cooling performance of the ACC. Combining different wind speeds, ambient temperature and wind directions, a total of eight working conditions were obtained. The calculation and analysis of these working conditions were carried out, and the corresponding back pressure and HRR were obtained to further study the effect of different environmental conditions on the heat transfer performance of the ACC. The calculated data is shown in Table 2 below.

Table 2. Combined working condition calculation data table.

Design Work Conditions	1	2	3	4	5	6	7	8
Ambient temperature (°C)	34	34	34	34	38	38	38	38
Wind direction	SSE	SSE	WNW	WNW	SSE	SSE	WNW	WNW
Wind speed (m/s)	5	7	5	7	5	7	5	7
Fan inlet average temperature (°C)	34.107	35.983	35.093	35.642	38.733	39.821	39.510	40.241
Exhaust steam temperature (°C)	63.629	65.809	63.538	64.925	67.413	69.397	68.442	69.855
Back pressure (kPa)	23.55	25.96	23.45	24.96	27.87	30.40	29.16	31.01
Hot recirculation rate (HRR) (%)	0.375	7.248	4.079	5.856	2.777	6.614	5.529	7.862

It can be seen from Table 2 that the change of wind direction affects the average temperature at the fan inlet and the exhaust temperature at the fan outlet. The wind direction of WNW increases the average temperature at the fan inlet. From Figure 4, we can see that under the effect of HAR, the back pressure of the unit was higher than 24 kPa under the ambient temperature of 38 °C. Under the conditions of the ambient temperature of 34 °C, wind speed of 7 m/s and wind direction of SSE and WNW, the back pressure of the unit was also higher than 24 kPa, which exceeded the safety back pressure (24 kPa) at the steam exhaust port of the turbine. At the same ambient temperature and the same wind direction, HRR and back pressure increased as the wind speed increased from 5 m/s to 7 m/s. Similarly, an increase in ambient temperature can cause a significant increase in back pressure. Among them, the increase of ambient temperature led to the performance degradation of the ACC, and the increase of wind speed also caused the same situation to occur, but the influence was less than the increase of ambient temperature and greater than the change of wind direction.

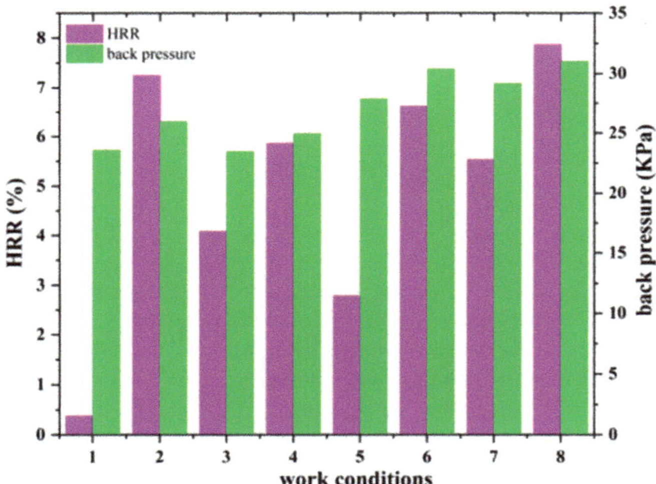

Figure 4. Hot recirculation rate (HRR) and back pressure under various working conditions.

3.2. Influence of Summer Extreme Conditions on Heat Transfer Performance of the Direct ACC

As the ambient temperature rose, especially in the extreme high-temperature conditions in summer, the heat transfer performance of the direct ACC would be correspondingly reduced, resulting in an increase of turbine back pressure. The output of the units was not guaranteed, and even affected the safe operation of the units. Therefore, the heat transfer performance of the direct ACC under high temperature and unfavorable wind direction in summer was studied to understand the law of the influence of extreme meteorological conditions on the ACC's cooling performance. Summer extreme conditions and unfavorable wind direction and wind speed were corresponding to the ambient temperature of 38 °C, the summer-dominant wind direction of SSE and the wind speed of 7 m/s. For the ACC, under the extreme summer conditions, the average temperature at the inlet of the fan was 39.82 °C, and the cooling triangle outlet average temperature was 65.534 °C. According to the formula, the back pressure of the ACC can be calculated to be 30.40 kPa, and the HRR of the ACC was 6.614%.

The ACC has a total of 128 cooling units with 16 columns and 8 rows. We represented X as a column and Y as a row, for example, $X8Y2$ represents the heat exchange unit in the eighth row and the second column, and so on. Figure 5 shows the transverse section temperature distribution of the ACC and it can be seen that the temperature gradient decreased from the windward side to the leeward side. The temperature on the windward side is significantly higher than the temperature on the leeward side. We can use Equation (8) to calculate the HAR of the ACC; therefore, the HAR in the upstream was significantly higher than the downstream of the wind field, so the heat transfer efficiency of the ACC heat exchange units on the windward side was reduced. Figure 6 showed the ACC and MCTs transverse section velocity distribution. Because of the influence of HAR, the heat exchange units of the ACC on the windward side upstream of the wind field produced significant eddy currents. It can be seen in Figure 6 that there was no obvious eddy current around the wind field of the MCTs, so there was no mutation in the velocity at the outlet of the MCTs. Therefore, it can be concluded that the MCTs were not obviously affected by the ambient wind. Besides, it can be indicated from Figure 6 that the influence of ambient wind on the ACC was significantly greater than that of MCTs. Figure 7 shows the cooling triangle temperature distribution of the ACC by comparing the $X15Y4$ and $X16Y4$ cooling units. Due to the influence of eddy currents, the volume flow of the units decreased from 327.622 m³/s to 242.931 m³/s and the fan power dropped from 76.95 KW to 49.12 KW. Therefore, under summer extreme conditions, due to the effect of HAR, hot air intrusion occurred at the edge of the ACC, resulting in the reduced cooling performance of the ACC.

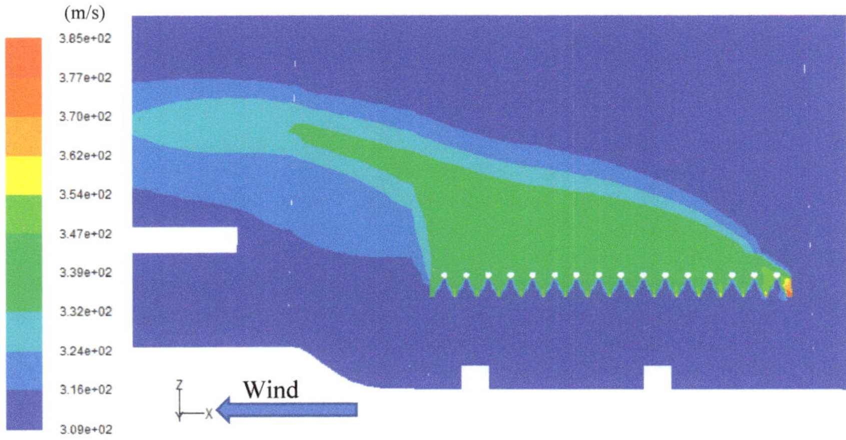

Figure 5. Transverse section temperature distribution of the ACC.

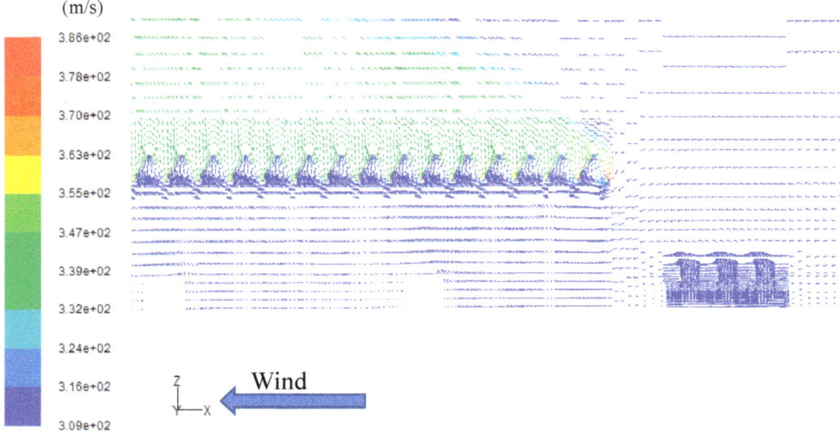

Figure 6. Transverse section velocity distribution of the ACC and MCTs.

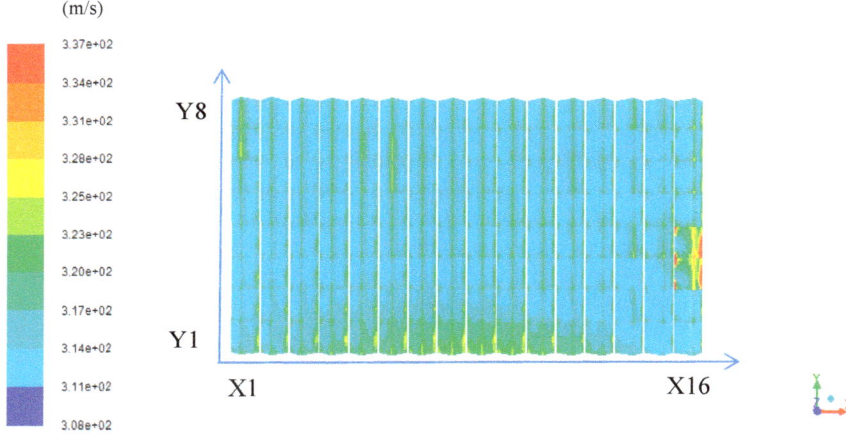

Figure 7. Cooling triangle temperature distribution of the ACC.

3.3. Effect of MCTs on the Cooling Performance of the ACC

In power plants, MCTs and the ACC are coexisting, and there is bound to be an interaction between them. The exhaust of the MCTs will partially enter the ACC under the effect of the ambient wind, which will affect the cooling performance of the ACC. Under different environmental conditions, MCTs will have different effects on the cooling performance of the ACC. Ambient air enters the MCTs through the air inlet. After a series of heat exchange with the high-temperature steam, it is discharged into the environment by the fan at the air outlet. The temperature of this part of the exhausted air is often higher than the ambient air, and the exhausted air always entrains high-temperature steam. The increase in temperature causes the air density to become smaller, so the discharged air will rise. Under the influence of ambient wind, this part of the exhausted air will pass through the air inlet of the ACC, causing the temperature of the air sucked by the ACC to increase. According to the heat exchange formula, the heat exchange temperature difference is reduced, and the ACC needs to do more work to complete the same heat transfer quantity, which greatly affects the heat transfer performance of the ACC, especially the entrained water mist, which may have serious consequences for the ACC.

Similarly, the air discharged from the ACC will also affect the heat transfer performance of the MCTs. Therefore, when MCTs and the ACC work together in the same power plant, it is very indispensable to study the interaction between them. According to the numerical simulation, the effect of the MCTs on the cooling performance of the ACC and the influence of the presence or absence of the MCTs on the cooling performance of the ACC were studied, which provides a feasible scheme for the research on the ACC. In order to better analyze the effect of MCTs on the cooling performance of the ACC, we chose to be in summer extreme conditions (ambient temperature of 38 °C, wind speed of 7 m/s, wind direction of SSE), which can weaken other influencing factors, focusing on the impact of MCTs on the cooling performance of the ACC.

Two typical cooling units, $X10Y7$ and $X11Y7$, were selected to investigate the area where the ACC was affected by the MCTs. In the area where the ACC was not affected by the MCTs, we selected two typical cooling units, $X7Y7$ and $X8Y7$, for analysis. Figure 8a,b shows the velocity and temperature distribution of the influence of MCTs on the ACC, respectively. We can see that under the condition that the ambient temperature of 38 °C, the wind speed of 7 m/s and the wind direction of SSE, the exhaust of the MCTs partially entered the ACC under the effect of the ambient wind. In the affected cooling unit $X10Y7$, the HRR was calculated to be 0.593%, and in the cooling unit $X11Y7$, the HRR was 0.583%. In the nearby unaffected cooling unit $X7Y7$, the HRR was 0.036%, and the cooling unit $X8Y7$ had a HRR of 0.280%. Focusing on Figure 9, the HRR of the cooling units affected by MCTs increased, which meant that MCTs had a great effect on the cooling performance of the ACC. However, in the cooling units not affected by MCTs, the increase of HRR was not large, which indicated that the presence of MCTs had a local influence on the cooling performance of only two cooling units, and then slightly impacted the overall cooling performance of the ACC, which provides a good insight into the arrangement optimization of the ACC and MCTs.

In the above discussion, we separately discussed the effects of MCTs on the affected and non-affected areas of the ACC. For a better study, the influence of the overall cooling performance of the ACC, under the same environmental conditions, the HRR of the ACC both in the absence and presence of MCTs were calculated separately. Through such a comparative study, theoretical support was provided for the study of the effect of the MCTs on the cooling performance of the ACC.

Figure 10a,b shows the cooling triangle temperature distribution of the ACC in the presence and absence of MCTs respectively, under the condition that the ambient temperature is 38 °C, the wind speed is 7 m/s and the wind direction is SSE, the exhaust of the MCTs partially entered the ACC under the effect of ambient wind. The temperature at the outlet of the ACC was 65.534 °C, the back pressure of the ACC was 30.396 kPa and the HRR was 6.614%. If the MCTs were turned off at this time, the temperature at the outlet of the ACC would change to 65.488 °C, the back pressure would be 30.335 kPa and the HRR would drop to 6.477%. In order to analyze the influence of the presence and absence of MCTs on the heat transfer performance of the ACC, we obtained the total average heat transfer coefficient and the average heat transfer coefficient of each column in the presence and absence of MCTs. From Table 3, we can see the comparison of the heat transfer performance of the ACC in the absence and presence of MCTs. It was very clear to see that in the absence of MCTs, the total average heat transfer coefficient of the ACC and most of the average heat transfer coefficient per column are slightly higher, while the back pressure and the HRR of the ACC were lower than in the presence of MCTs; however, the magnitude of the reduction was relatively small, so the presence of MCTs had a minor impact on the overall cooling performance of the ACC. The main reason was that after the ambient air entered the MCTs, the two phases of gas and water were directly contacted, which was mainly based on latent heat transfer, and the dry bulb temperature was not greatly increased, while the ACC was dominated by sensible heat transfer, and the oncoming flow air dry bulb temperature played a leading role in its thermal performance. Secondly, the cooling air volume (1089.55 kg/s) of the MCTs was significantly smaller than that (59,017.36 kg/s) of the ACC; therefore, the presence of MCTs had a relatively small effect on the overall thermal characteristics of the ACC.

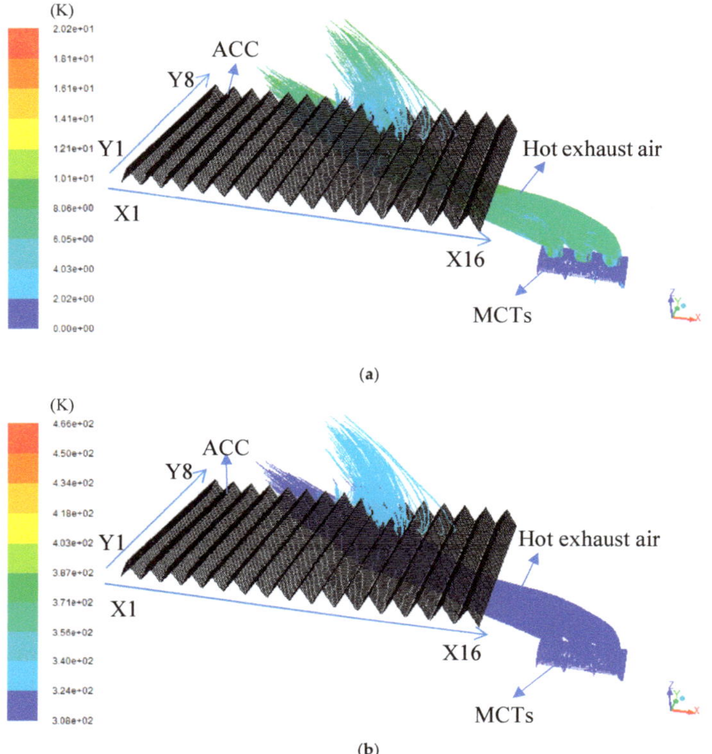

Figure 8. The influence of MCTs on the ACC. (**a**) Velocity distribution of the influence of MCTs on the ACC and (**b**) Temperature distribution of the influence of MCTs on the ACC.

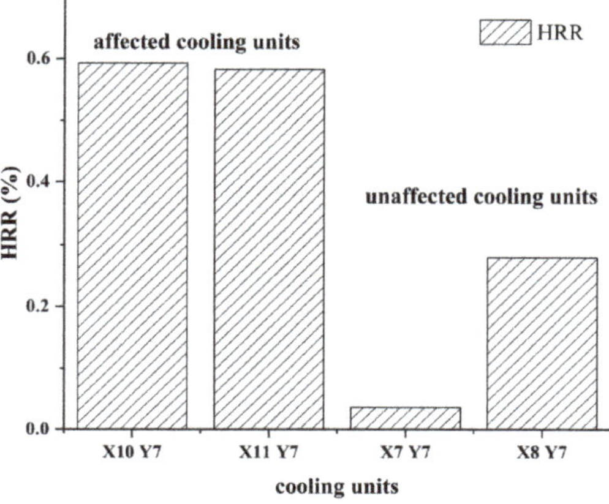

Figure 9. Relationship between the hot recirculation rate (HRR) and cooling units.

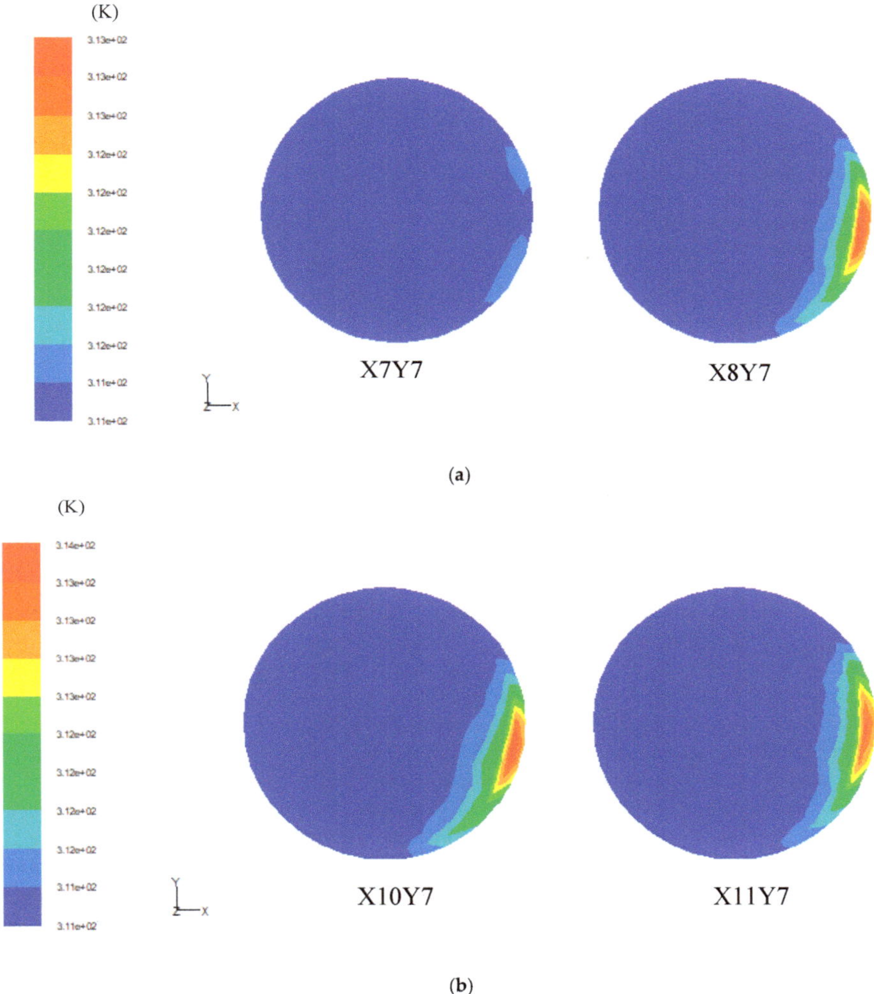

Figure 10. Fan temperature profile of the ACC in the presence and absence of MCTs. (**a**) Fan temperature profile of the ACC in the absence of MCTs. (**b**) Fan temperature profile of the ACC in the presence of MCTs.

Table 3. Comparison of heat transfer performance of the ACC in the absence and presence of MCTs.

Parameters	MCTs Effect	No MCTs Effect
HRR (%)	6.614	6.477
Back pressure (kPa)	30.396	30.335
h Total average heat transfer coefficient (W·m^{-2}·K^{-1})	3551.545	3552.097
h_1 Average heat transfer coefficient of X1 (W·m^{-2}·K^{-1})	3570.499	3570.685
h_2 Average heat transfer coefficient of X2 (W·m^{-2}·K^{-1})	3567.984	3567.887
h_3 Average heat transfer coefficient of X3 (W·m^{-2}·K^{-1})	3567.040	3567.048
h_4 Average heat transfer coefficient of X4 (W·m^{-2}·K^{-1})	3567.790	3567.998
h_5 Average heat transfer coefficient of X5 (W·m^{-2}·K^{-1})	3566.222	3566.620
h_6 Average heat transfer coefficient of X6 (W·m^{-2}·K^{-1})	3565.509	3565.957
h_7 Average heat transfer coefficient of X7 (W·m^{-2}·K^{-1})	3562.848	3563.341
h_8 Average heat transfer coefficient of X8 (W·m^{-2}·K^{-1})	3562.626	3563.008
h_9 Average heat transfer coefficient of X9 (W·m^{-2}·K^{-1})	3558.128	3558.397
h_{10} Average heat transfer coefficient of X10 (W·m^{-2}·K^{-1})	3558.433	3558.926
h_{11} Average heat transfer coefficient of X11 (W·m^{-2}·K^{-1})	3554.801	3555.224
h_{12} Average heat transfer coefficient of X12 (W·m^{-2}·K^{-1})	3554.407	3554.876
h_{13} Average heat transfer coefficient of X13 (W·m^{-2}·K^{-1})	3546.935	3547.624
h_{14} Average heat transfer coefficient of X14 (W·m^{-2}·K^{-1})	3542.062	3542.211
h_{15} Average heat transfer coefficient of X15 (W·m^{-2}·K^{-1})	3529.595	3529.525
h_{16} Average heat transfer coefficient of X16 (W·m^{-2}·K^{-1})	3345.026	3344.195

4. Conclusions

Using the mature and reliable CFD software FLUENT, the air-cooling system and surrounding buildings are accurately modeled, which can accurately calculate the air flow field around the ACC and MCTs, and the effect of the MCTs on the cooling performance of the ACC. Based on the above research, the conclusions are summarized as follows:

(1) By calculating the HRR and back pressure of the ACC under eight working conditions, the influence of HAR, ambient temperature, wind direction and wind speed on the cooling performance of the ACC was obtained. An increase of ambient temperature led to the performance degradation of the ACC, and the increase of wind speed also caused the same situation to occur, but the influence was less than the increase of ambient temperature and greater than the change of wind direction.

(2) Under the summer extreme conditions, the HRR of the ACC was 6.614%, and the back pressure of the ACC was 30.40 kPa. The influence of ambient wind on the ACC was significantly greater than that of MCTs.

(3) The influence of MCTs on the cooling performance of the ACC was analyzed emphatically. According to the results, the exhaust of MCTs partially entered the ACC under the effect of ambient wind, and the HRR of the affected cooling units was higher than that of the nearby unaffected cooling units. When the MCTs were turned off, the overall HRR of the ACC decreased. The presence of MCTs had a local influence on the cooling performance of only two cooling units, and then slightly impacted the overall cooling performance of the ACC, which provides a good insight into the arrangement optimization of the ACC and MCTs.

Author Contributions: Conceptualization, J.F.; Formal analysis, J.F.; Investigation, H.D.; Project administration, Y.Z.; Resources, X.X., D.T. and B.Y.

Funding: The financial supports for this research, from the National Natural Science Foundation of China (Grant No. 51606112) and the Key Technology Research and Development Program of Shandong Province (Grant No. 2017GGX40122), are both gratefully acknowledged.

Acknowledgments: I wish to thank Yajin Liu for advice on my English writing.

Conflicts of Interest: In my opinion, there is no conflict of interest in the submission of this manuscript, and all authors agree to publish the manuscript.

Nomenclature

a	Altitude (m)
f_n	Pressure-jump polynomial coefficients
h	Heat transfer coefficient (W·m^{-2}·K^{-1})
k	Turbulent kinetic energy per unit mass (m^2·s^{-2})
K_L	Flow loss coefficient
ΔP	Pressure jump (pa)
P	Atmospheric pressure (mmHg)
q	Heat flux (W/m^2)
r_n	Polynomial coefficients
T_i	Average temperature at the inlet of the fan (K)
T_a	Ambient temperature (K)
T_{a1}	Average temperature of the inlet air of the ACC (K)
T_{a2}	Average temperature of the outlet air of the ACC (K)
T_o	Outlet temperature of the ACC (K)
T_s	Exhaust steam temperature (K)
T_{HX}	Radiator temperature (K)
T_{exit}	Temperature of the outlet fluid (K)
v	Velocity (m/s)
X	Column
Y	Row

Greek Letters

ρ	Air density (kg/m^3)
ε	Turbulence dissipation rate (m^2·s^{-3})
α	Wall roughness coefficient

Subscripts

i	Inlet
a	Ambient
o	Outlet
N	Number
s	Steam
L	Loss

References

1. Duvenhage, K.; Kröger, D.G. The influence of wind on the performance of forced draught air-cooled heat exchangers. *J. Wind Eng. Ind. Aerod.* **1996**, *62*, 259–277. [CrossRef]
2. Wilber, K.R.; Zammit, K. Development of Procurement Guidelines for Air-Cooled Condensers. In Proceedings of the Advanced Cooling Strategies/Technologies Conference, Sacramento, CA, USA, 1–2 June 2005; pp. 1–24.
3. Ge, Y.; Cropper, R. Air-cooled condensers in retail systems using R22 and R404A refrigerants. *Appl. Energy* **2004**, *78*, 95–110. [CrossRef]
4. Yu, F.W.; Chan, K.T. Modeling of a condenser-fan control for an air-cooled centrifugal chiller. *Appl. Energy* **2007**, *84*, 1117–1135. [CrossRef]
5. Maulbetsch, J.S.; DiFilippo, M.N. *Effects of Wind on Air-cooled Condenser Performance*; Paper No. TP07-04; Cooling Technology Institute: Houston, TX, USA, 2007.
6. Gu, Z.; Chen, X.; Lubitz, W.; Li, Y.; Luo, W. Wind tunnel simulation of exhaust recirculation in an air-cooling system at a large power plant. *Int. J. Therm. Sci.* **2007**, *46*, 308–317. [CrossRef]
7. Liu, P.; Duan, H.; Zhao, W. Numerical investigation of hot air recirculation of air-cooled condensers at a large power plant. *Appl. Therm. Eng.* **2009**, *29*, 1927–1934. [CrossRef]
8. Wang, Q.W.; Zhang, D.J.; Zeng, M.; Lin, M.; Tang, L.H. CFD simulation on a thermal power plant with air-cooled heat exchanger system in north China. *Eng. Comput.* **2008**, *25*, 342–365. [CrossRef]

9. Meyer, C.; Kröger, D. Plenum chamber flow losses in forced draught air-cooled heat exchangers. *Appl. Therm. Eng.* **1998**, *18*, 875–893. [CrossRef]

10. Meyer, C.J.; Kroger, D.G. Numerical investigation of the effect of fan performance on forced draught air-cooled heat exchanger plenum chamber aerodynamic behavior. *Appl. Therm. Eng.* **2004**, *24*, 359–371. [CrossRef]

11. Owen, M.T.F.; Kröger, D.G. An Investigation of Air-Cooled Steam Condenser Performance Under Windy Conditions Using Computational Fluid Dynamics. *J. Eng. Gas Turbines Power* **2011**, *133*, 604502. [CrossRef]

12. O'Donovan, A.; Grimes, R. A theoretical and experimental investigation into the thermodynamic performance of a 50 MW power plant with a novel modular air-cooled condenser. *Appl. Therm. Eng.* **2014**, *71*, 119–129. [CrossRef]

13. Zhang, X.; Chen, H. Effects of windbreak mesh on thermo-flow characteristics of air-cooled steam condenser under windy conditions. *Appl. Therm. Eng.* **2015**, *85*, 21–32. [CrossRef]

14. Meyer, C. Numerical investigation of the effect of inlet flow distortions on forced draught air-cooled heat exchanger performance. *Appl. Therm. Eng.* **2005**, *25*, 1634–1649. [CrossRef]

15. Gao, X.; Zhang, C.; Wei, J.; Yu, B. Performance prediction of an improved air-cooled steam condenser with deflector under strong wind. *Appl. Therm. Eng.* **2010**, *30*, 2663–2669. [CrossRef]

16. Huang, X.; Chen, L.; Kong, Y.; Yang, L.; Du, X. Effects of geometric structures of air deflectors on thermo-flow performances of air-cooled condenser. *Int. J. Heat Mass Transf.* **2018**, *118*, 1022–1039. [CrossRef]

17. Jin, R.; Yang, X.; Yang, L.; Du, X.; Yang, Y. Square array of air-cooled condensers to improve thermo-flow performances under windy conditions. *Int. J. Heat Mass Transf.* **2018**, *127*, 717–729. [CrossRef]

18. Yang, L.J.; Du, X.Z.; Yang, Y.P. Influences of wind-break wall configurations upon flow and heat transfer characteristics of air-cooled condensers in a power plant. *Int. J. Therm. Sci.* **2011**, *50*, 2050–2061. [CrossRef]

19. Berman, L.D. *Evaporative Cooling of Circulating Water*; Pergamon press: Oxford, UK, 1961; p. 710.

20. Kroger, D.G. *Air-Cooled Heat Exchangers and Cooling € Towers*; Begell House: New York, NY, USA, 1998.

21. Thulukkanam, K. *Heat Exchanger Design Handbook*; Hemisphere Publishing Corporation: New York, NY, USA, 1983.

22. Davis, L.R. *Fundamentals of Environmental Discharge Modeling*; CRC Press: Boca Raton, FL, USA, 1998.

23. Buss, J.R. *Control of Fog from Cooling Towers*; Cooling Tower Institute Annual Meeting: New York, NY, USA, 1967.

24. Campbell, J.C. *The Prevention of Fog from Cooling Towers*; Annual Meeting of the Cooling Tower Institute: Houston, TX, USA, 1976.

25. Michioka, T.; Sato, A.; Kanzaki, T.; Sada, K. Wind tunnel experiment for predicting a visible plume region from a wet cooling tower. *J. Wind. Eng. Ind. Aerodyn.* **2007**, *95*, 741–754. [CrossRef]

26. Xu, X.; Wang, S.; Ma, Z. Evaluation of plume potential and plume abatement of evaporative cooling towers in a subtropical region. *Appl. Therm. Eng.* **2008**, *28*, 1471–1484. [CrossRef]

27. ANSYS Inc. *Fluent User's Guide*; ANSYS Inc.: Canonsburg, PA, USA, 2015.

28. Architectural Institute of Japan (AIJ). *AIJ Recommendations for Loads on Buildings*; Architectural Institute of Japan: Tokyo, Japan, 1996. (In Japanese)

Article

An Impedance Network-Based Three Level Quasi Neutral Point Clamped Inverter with High Voltage Gain

Muhammad Aqeel Anwar [1], Ghulam Abbas [1,*], Irfan Khan [2,*], Ahmed Bilal Awan [3], Umar Farooq [4] and Saad Saleem Khan [5]

[1]   Department of Electrical Engineering, The University of Lahore, Lahore 54000, Pakistan;
      aqeel.anwar@tech.uol.edu.pk
[2]   Marine Engineering Technology Department in a Joint Appointment with the Electrical and Computer
      Engineering Department, Texas A&M University, Galveston, TX 77554, USA
[3]   Department of Electrical Engineering, College of Engineering, Majmaah University,
      Al-Majmaah 11952, Saudi Arabia; a.awan@mu.edu.sa
[4]   Department of Electrical Engineering, University of the Punjab, Lahore 54590, Pakistan;
      engr.umarfarooq@yahoo.com
[5]   Department of Electrical Engineering, UAE University, Al-Ain 64141, UAE; saad.khan@uaeu.ac.ae
*   Correspondence: ghulam.abbas@ee.uol.edu.pk (G.A.); irfankhan@tamu.edu (I.K.);
    Tel.: +1-409-740-4549 (I.K.)

Received: 27 December 2019; Accepted: 28 February 2020; Published: 9 March 2020

Abstract: Due to the impediments of voltage source inverter and current source inverter, Z-Source Inverter (ZSI) has become notorious for better power quality in low and medium power applications. Several modifications are proposed for impedance source in the form of Quasi Z-Source Inverter (QZSI) and Neutral Point Clamped Z-Source Inverter (NPCZSI). However, due to the discontinuity of the source current, NPCZSI is not suitable for some applications, i.e., fuel cell, UPS, and hybrid electric vehicles. Although in later advancements, source current becomes continuous in multilevel QZSI, low voltage gain, higher shoot-through duty ratio, lesser availability of modulation index, and higher voltage stress across switches are still an obstacle in NPCZSI. In this research work, a three-level high voltage gain Neutral Point Clamped Inverter (NPCI) that gives three-level AC output in a single stage, is proposed to boost up the DC voltage at the desired level. At the same time, it detains all the merits of previous topologies of three-level NPCZSI/QZSI. Simulations have been done in the MATLAB/Simulink environment to show the effectiveness of the proposed inverter topology.

Keywords: Neutral Point Clamped Z-Source Inverter (NPCZSI); shoot-through duty ratio; modulation index; voltage gain; power quality

---

1. Introduction

Z-source inverter (ZSI) has the capability for buck/boost operation; this unique feature is not available in a traditional Current Source Inverter (CSI) and Voltage Source Inverter (VSI). Regardless of this unique feature, Z-source inverter has some curtailments such as lower voltage gain, discontinuous current, and high voltage stresses across the switches [1–3]. In the previous literature, several topologies in the form of Quasi Z-Source Inverter (QZSI), Neutral Point Clamped Z-Source Inverter (NPCZSI), and Neutral Point Clamped Quasi Z-Source (NPCQZSI) Inverter were presented to overcome these limitations. For instance, a voltage doubler, in conjunction with an isolation transformer, was utilized in [4] for quasi ZSI for distributed generation applications. Although the topology ensures the continuity in the input current, it offers a limited boosting ability. Theoretical results for four topologies of QZSI inverter having less element count and simplified control techniques were presented in [5,6].

The topologies ensure a continuous current without including a major advancement in boosting ability and voltage stresses across switches. A traditional ZSI integrated with a bridge rectifier was proposed in [7] for adjustable speed drive applications. A QZSI was proposed in [8] that included two inductors, two capacitors, and one diode in the QZS network. This topology offers continuity in the input current, however, with the same boosting ability as that of the traditional ZSI. To increase the modulation index range with the continuity in current, a switched inductor QZSI was proposed in [9]. No drastic increase in the boost factor of the inverter was observed. By utilizing a predictive control method, a grid-connected closed-loop QZSI topology was introduced [10] that claimed an improvement in the inverter performance with the same boosting ability as that of traditional ZSI and high voltage stresses across switches. Similarly, different ZS networks integrated with VSI and CSI were investigated in [11]. These are indeed different topologies with the same boost factor as that of traditional ZSI. To make a smoother DC input current, a family of embedded ZSI was suggested with the traditional boosting ability and high stresses across switches [12]. In all the aforementioned references, improvement in boosting ability is still a challenge.

In [13,14], a three-level NPCZSI was proposed that deployed the two conventional impedance sources that included four capacitors and four inductors with two separate dc sources. This proposed topology of the inverter provides an additional state (0 V) in the output voltage to yield the staircase waveform and reduces the harmonic distortion in output voltage. Due to the deployment of two separate impedance sources, the design of this topology becomes very bulky; additionally, the voltage gain of the inverter does not increase too much. This limitation was addressed in [15], where a single impedance source was deputed with the NPCI to reduce the cost and volume of the inverter while retaining all the advantages of the previous topology.

To achieve the optimal results from NPCZSI in the form of enhanced output voltage, waveform quality, and boosting ability, a detailed comparison between two modulation techniques, Phase Disposition Sinusoidal Pulse Width Modulation (PDSPWM) and Phase Opposite Disposition Sinusoidal Pulse Width Modulation (PODSPWM) were analyzed in [16]. A continuous model was suggested in [17] to balance the voltage between two capacitors in NPCI. Instead of standard inductors, the use of coupled transformers was introduced in [18] to enhance the voltage gain ability of NPCZSI by adjusting the turns ratio of transformers. It also overcomes the dependence on modulation index, and hence, makes the stresses lower across switches. However, with this topology, a colossal rise in voltage gain is restricted due to the shoot-through duty ratio and turns ratio of coupled transformers affecting each other. Said constraint is addressed in [19] by modifying the impedance network with the deployment of coupled inductors. A Coupled Quasi Z-Source Inverter (CQZSI) is suggested in [20] to diminish the input current ripples. A single-phase five-level hybrid NPCI was proposed in [21], which integrated the NPCI with an H-bridge inverter. The proposed inverter contends the accomplishment to lower the total harmonic distortion (THD) through the Selective Harmonic Elimination (SHE) technique.

An NPCI, with a Z-source, that was composed of two diodes, two inductors, two switches, and two capacitors was suggested in [22] that acquired all the benefits of QZSI and provided the three-level output voltages, but due to its lower voltage gain ability, it was not an appropriate choice where a higher boost was required. In [23], two topologies named Diode Assisted QZSI and Capacitor Assisted QZSI were proposed, which were extendable, but the repetition of a greater number of units increased the size of the inverter and made it bulky. From the invention of ZSI [24], a lot of attempts have been made to ameliorate this topology to accomplish the optimal benefits from it, but a small voltage gain, lower value of modulation index, and higher voltage stress are still present in previous topologies suggested in the literature.

This research work introduces a new topology for ZSI that overcomes the drawbacks such as lower boosting ability, utilization of higher shoot-through duty ratio, lesser availability of modulation index, and higher voltage stresses across switches of the previous topologies. This topology offers a remarkable boosting ability and provides a high voltage gain by utilizing a lower shoot-through ratio to make available a higher modulation index and keeps the lower stress across switches.

The paper is organized in the following way. The proposed inverter topology is presented in Section 2, with its complete working modes of operation. A detailed mathematical analysis is performed in Section 3. Section 4 covers the PWM technique, and the Boost Control Method applied to the proposed topology. Simulations and results are provided in Section 5, whereas Section 6 presents a detailed comparison with the previous topologies. Section 7 includes the conclusion.

2. The Proposed Inverter Topology

The schematic diagram of the proposed inverter topology, which can enhance the applied DC voltage to the desired level and transform it into a three-phase three-level output, is illustrated in Figure 1. The applied DC voltage can be procured from a DC battery or some other DC source such as a fuel cell or PV applications.

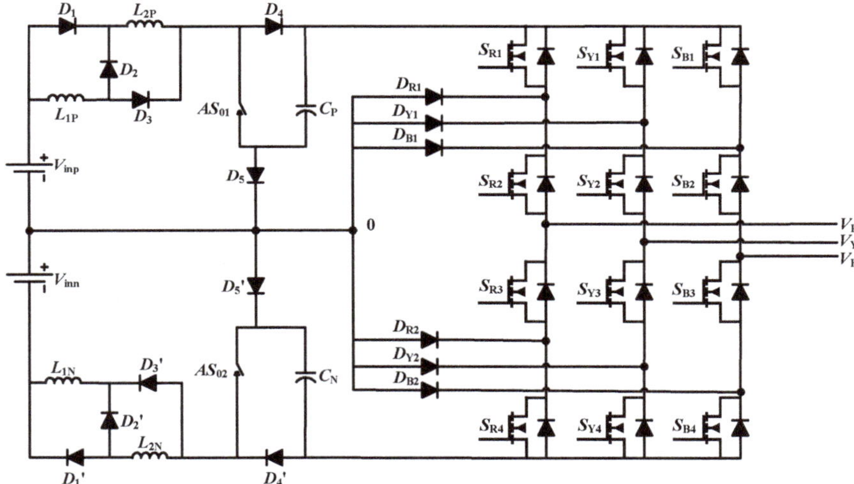

Figure 1. The proposed topology of the inverter.

The inverter is configured with two symmetrical impedance networks, where each network is comprised of two inductors, two capacitors, four diodes, and one active switch to provide the numerous advantages over the conventional inverters. Two alike dc sources are utilized to energize these networks. The conventional NPCI [25] can operate only in two types of states, active-states, and zero-states. However, the proposed inverter includes one more state of operation that is a Shoot Through (ST) state (that allows all switches to turn on at a time).

2.1. Active State

Here, the DC voltage applied to the inverter is exposed to the three-phase AC load after conversion to a three-phase three-level AC, at the desired level. In this mode of operation, $+V_{dc}$ or $-V_{dc}$ voltage appear at the poles of the inverter.

To achieve $+V_{dc}$, diodes D_2 and D_4 operate in the conduction mode and D_1 and D_3 remain in the nonconduction mode; L_{1P} and L_{2P} come in series. DC source V_{inp} and both inductors L_{1P} and L_{2P} of the upper impedance network energize the capacitor C_P. The direction of the current is shown in the equivalent circuit of the active state in Figure 2. In the active state, the current adopts the two paths, one from voltage source V_{inp} to inductor L_{1P}, diode D_2, inductor L_{2P}, diode D_4, and capacitor C_P, and completes its path through D_5. In this path, capacitor C_P is charged by voltage source V_{inp} and by inductors L_{1P} and L_{2P}, while in the second path, the current approaches to ac load after passing through L_{1P}, D_2, L_{2P}, S_{K1}, S_{K2} (where $K = R, Y, B$) and completes its path through the load to neutral point N and back to V_{inp}.

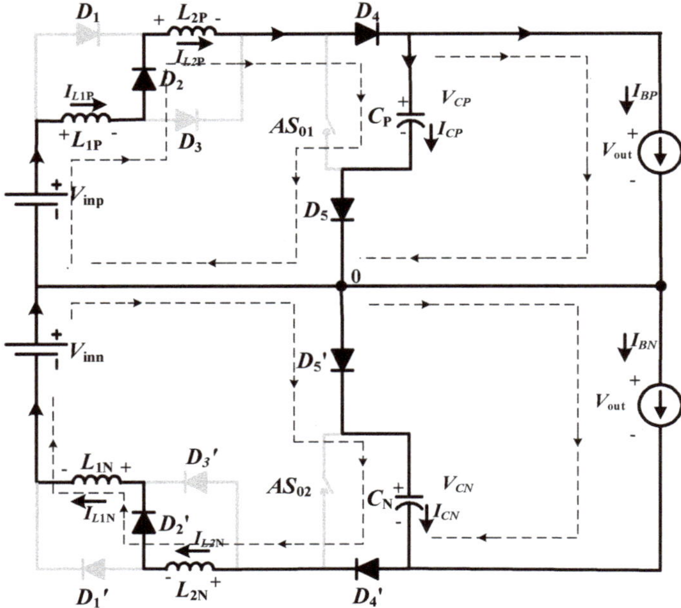

Figure 2. Equivalent circuit for the active state.

To achieve $-V_{\text{dc}}$, diodes $D_2{}'$ and $D_4{}'$ are in conduction mode, while $D_1{}'$ and $D_3{}'$ remain in nonconduction mode. DC source voltage V_{inn} and both inductors L_{1N} and L_{2N} of lower impedance network energize the capacitor C_N. The direction of the flow of current is shown in the equivalent circuit of the active state in Figure 2. In the active state, the current follows the two paths, one from voltage source V_{inn} to capacitor C_N through $D_5{}'$ and completes its path after passing through inductor L_{2N}, diode $D_2{}'$, and inductor L_{1N}. In this path, the capacitor is charged by voltage source V_{inn} and by inductors L_{1N} and L_{2N}. In the second path, the current approaches to AC load after passing through S_{K3} and S_{K4} (where $K = R, Y, B$), and completes its path through the inductor L_{2N}, diode $D_2{}'$, inductor L_{1N}, and back to V_{inn}. In the active state, both active switches (AS_{01} and AS_{02}) remain in off state and play no role.

2.2. Zero-State

During the zero-state of operation, no voltage appears across the load terminals. In the zero-state, two intermediate switches of each leg of the inverter are in on state, whereas the topmost and lowermost switch of each leg of the inverter remains in the off state. During this mode of operation, the diodes D_2 and D_4 are in the conduction mode, while D_1 and D_3 remain in the nonconduction mode. DC source V_{inp} and both inductors L_{1P} and L_{2P} of the upper impedance network energize the capacitor C_P. Similarly, for the lower network, the diodes $D_2{}'$ and $D_4{}'$ are in conduction mode, while $D_1{}'$ and $D_3{}'$ remain in nonconduction mode and DC source V_{inn} and both inductors L_{1N} and L_{2N} of lower impedance network energize the capacitor C_N.

The direction of the flow of current in the equivalent circuit of zero-state is shown in Figure 3, where it traverses voltage source V_{inp}, inductor L_{1P}, diode D_2, inductor L_{2P}, and capacitor C_P and reaches back to V_{inp} through diode D_4. In this path, capacitor C_P is charged by V_{inp} and by both inductors L_{1N} and L_{2N}. Similarly, for the lower network capacitor, C_N is charged by V_{inn} and by the combination of L_{1N} and L_{2N}, no power is transferred to load. Like in the active state, both AS_{01} and AS_{02} remain in off state.

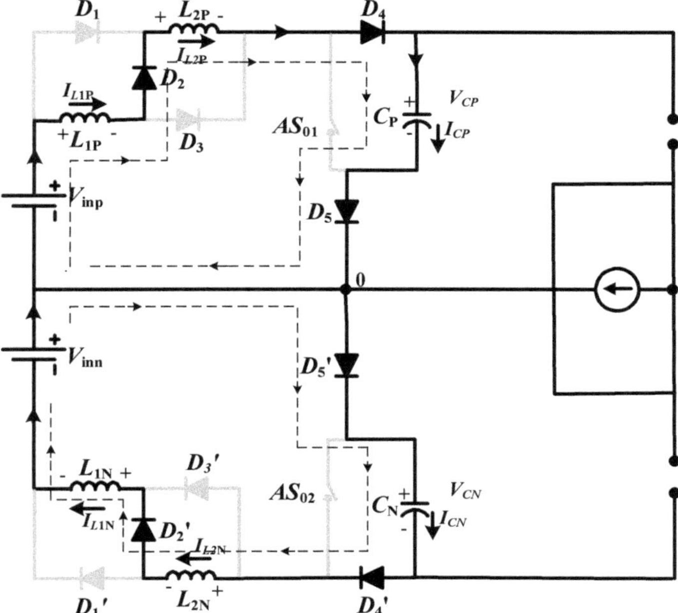

Figure 3. Equivalent circuit for the zero-state.

2.3. Shoot-Through (ST) State

In ST state, active switches, along with all switches of one or more legs of an inverter, go to on state simultaneously, which elicits 0 V across the load (see Figure 4). Seven different approaches that are summarized in Table 1 can be adopted to attain this state.

On closing the active switches, inductors L_{1P} and L_{2P} come in parallel in the upper impedance network and turns diodes D_2, D_4, and D_5 in reverse bias mode. Similarly, L_{1N} and L_{2N} come in parallel in the lower impedance network and configure diodes D_2', D_4', and D_5' in reverse bias mode. In ST state, V_{inp} and capacitor C_P charge inductors L_{1P} and L_{2P}; similarly, in the lower network, V_{inn} and capacitor C_N charge inductors L_{1N} and L_{2N}.

The span of the ST state is limited up to the premises of zero-state and is maximum when it occupies the time duration of the zero-state. This state enables the inverter to perform the buck/boost operation. Thus, by choosing its appropriate value, desired results can be achieved.

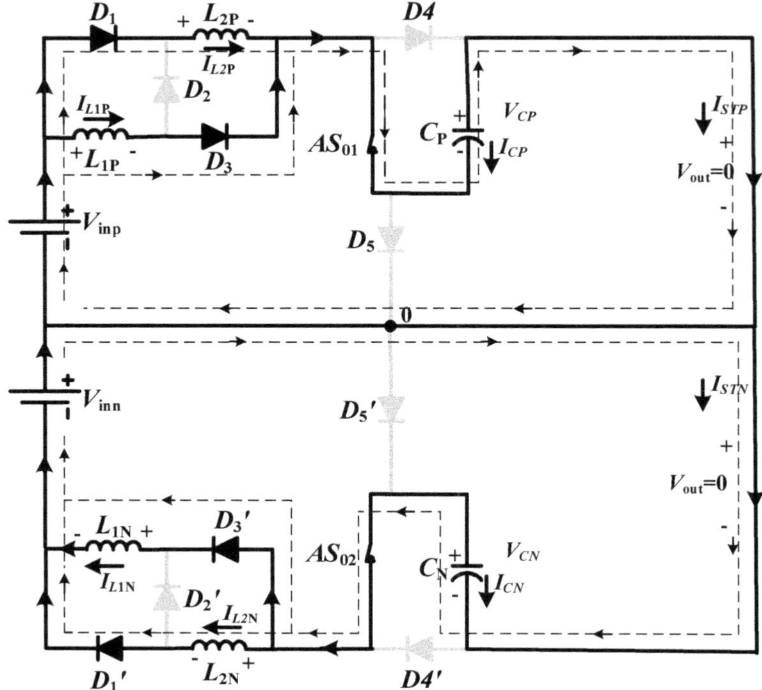

Figure 4. Equivalent circuit for the ST state.

Table 1. Different approaches for ST state.

Sr. No.	ON Switches	OFF Switches
1	$S_{R1}, S_{R2}, S_{R3}, S_{R4}, AS_{01}, AS_{02}$	$S_{Y1}, S_{Y2}, S_{Y3}, S_{Y4}, S_{B1}, S_{B2}, S_{B3}, S_{B4}$
2	$S_{Y1}, S_{Y2}, S_{Y3}, S_{Y4}, AS_{01}, AS_{02}$	$S_{R1}, S_{R2}, S_{R3}, S_{R4}, S_{B1}, S_{B2}, S_{B3}, S_{B4}$
3	$S_{B1}, S_{B2}, S_{B3}, S_{B4}, AS_{01}, AS_{02}$	$S_{R1}, S_{R2}, S_{R3}, S_{R4}, S_{Y1}, S_{Y2}, S_{Y3}, S_{Y4}$
4	$S_{R1}, S_{R2}, S_{R3}, S_{R4}, S_{Y1}, S_{Y2}, S_{Y3}, S_{Y4}, AS_{01}, AS_{02}$	$S_{B1}, S_{B2}, S_{B3}, S_{B4}$
5	$S_{R1}, S_{R2}, S_{R3}, S_{R4}, S_{B1}, S_{B2}, S_{B3}, S_{B4}, AS_{01}, AS_{02}$	$S_{Y1}, S_{Y2}, S_{Y3}, S_{Y4}$
6	$S_{Y1}, S_{Y2}, S_{Y3}, S_{Y4}, S_{B1}, S_{B2}, S_{B3}, S_{B4}, AS_{01}, AS_{02}$	$S_{R1}, S_{R2}, S_{R3}, S_{R4}$
7	$S_{R1}, S_{R2}, S_{R3}, S_{R4}, S_{Y1}, S_{Y2}, S_{Y3}, S_{Y4}, S_{B1}, S_{B2}, S_{B3}, S_{B4}, AS_{01}, AS_{02}$	Nil

3. Mathematical Analysis of the Proposed Inverter Topology

In this section, we perform the necessary mathematical calculations for the proposed inverter topology.

3.1. Non-ST-State

Applying the Kirchhoff's Voltage Law (KVL) to Figure 2, the voltage across inductors L_{1P} and L_{2P} are:

$$\begin{cases} V_{L1P} = V_{inp} - V_{CP} - V_{L2P} \\ V_{L2P} = V_{inp} - V_{CP} - V_{L1P} \end{cases} \tag{1}$$

where:

$$V_{inp} = V_{L1P} + V_{L2P} + V_{CP} \tag{2}$$

$$V_{CP} = V_{out}$$

To find the inductor and capacitor currents during the non-ST state, apply the Kirchhoff's Current Law (KCL) on upper impedance network:

$$I_{L1P} = I_{L2P} = I_{CP} + I_{BP} \tag{3}$$

$$I_{CP} = I_{L1P} - I_{BP} \tag{4}$$

or:

$$I_{CP} = I_{L2P} - I_{BP} \tag{5}$$

Similarly, for the lower network during the non-ST state:

$$\begin{cases} V_{L1N} = V_{inn} - V_{CN} - V_{L2N} \\ V_{L2N} = V_{inn} - V_{CN} - V_{L1N} \end{cases} \tag{6}$$

where:

$$V_{inn} = V_{L1N} + V_{L2N} + V_{CN} \tag{7}$$

$$V_{out} = V_{CN}$$

Furthermore, the inductor and capacitor currents are:

$$I_{L1N} = I_{L2N} = I_{CN} + I_{BN} \tag{8}$$

$$I_{CN} = I_{L1N} - I_{BN} \tag{9}$$

or:

$$I_{CN} = I_{L2N} - I_{BN} \tag{10}$$

3.2. ST-State

Apply KVL to Figure 4, the voltage across the inductors L_{1P} and L_{2P} are given as:

$$V_{L1P} = V_{L2P} = V_{inp} + V_{CP} \tag{11}$$

$$V_{out} = 0$$

For inductor and capacitor currents during the ST state, apply the KCL on upper impedance network:

$$I_{STP} = I_{L1P} + I_{L2P} = -I_{CP} \tag{12}$$

or:

$$I_{CP} = -(I_{L1P} + I_{L2P}) = -I_{STP} \tag{13}$$

$$V_{L1N} = V_{L2N} = V_{inn} + V_{CN}, \; V_{out} = 0 \tag{14}$$

$$I_{STN} = I_{L1N} + I_{L2N} = -I_{CN} \tag{15}$$

or:

$$I_{CN} = -(I_{L1N} + I_{L2N}) = -I_{STN} \tag{16}$$

3.3. Calculations of Current, Voltage, Boost Factor and Gain Factor

As per the Volt-Second Balance Principle (VSBP), the net voltage across the inductor remains zero during a period of one switching cycle. Apply the VSBP at upper impedance network across both inductors L_{1P} and L_{2P} during a complete switching time period T_{osc}:

$$(V_{inp} - V_{CP} - V_{L2P})(1 - D)T_{osc} + (V_{inp} + V_{CP})DT_{osc} = 0 \tag{17}$$

$$(V_{\text{inp}} - V_{\text{CP}} - V_{\text{L1P}})(1 - D)T_{\text{osc}} + (V_{\text{inp}} + V_{\text{CP}})DT_{\text{osc}} = 0 \tag{18}$$

Solve (17) to find out the voltage across inductor L_{2P} during the non-shoot-through state as:

$$\begin{aligned} V_{\text{L2P}} &= \frac{V_{\text{inp}} + V_{\text{CP}}(2D-1)}{(1-D)} \\ &= \frac{V_{\text{inp}}}{(1-D)} + \frac{V_{\text{CP}}(2D-1)}{(1-D)} \end{aligned} \tag{19}$$

Put the above value V_{L2P} into (18):

$$\left(\frac{V_{\text{inp}}}{(1-D)} + \frac{V_{\text{CP}}(2D-1)}{(1-D)} \right)(1-D)T_{\text{osc}} + (V_{\text{inp}} + V_{\text{CP}})DT_{\text{osc}} = 0 \tag{20}$$

By solving (20), the voltage across capacitor C_P is given as:

$$V_{\text{CP}} = \frac{V_{\text{inp}}(D+1)}{1-3D} \tag{21}$$

Similarly, apply the VSBP in lower impedance network across inductors L_{1N} and L_{2N} during a complete switching time period T_{osc}:

$$(V_{\text{inn}} - V_{\text{CN}} - V_{\text{L2N}})(1 - D)T_{\text{osc}} + (V_{\text{inn}} + V_{\text{CN}})DT_{\text{osc}} = 0 \tag{22}$$

$$(V_{\text{inn}} - V_{\text{CN}} - V_{\text{L1N}})(1 - D)T_{\text{osc}} + (V_{\text{inn}} + V_{\text{CN}})DT_{\text{osc}} = 0 \tag{23}$$

After solving (22) and (23), the voltage across capacitor C_N is given as:

$$V_{\text{CN}} = \frac{V_{\text{inn}}(D+1)}{1-3D} \tag{24}$$

As both the dc input sources are identical $V_{\text{inp}} = V_{inn} = V_{\text{IN}}$, (21) and (24) can be re-written in a general form:

$$V_{\text{CP}} = V_{\text{CN}} = \frac{V_{\text{IN}}(D+1)}{1-3D} \tag{25}$$

As per the Ampere Second Balance Principle (ASBP), the net current through the capacitor remains zero during a period of one switching cycle. Apply the ASBP to capacitor C_P in the upper network to find out the current through both inductors L_{1P} and L_{2P}:

$$(I_{\text{L1P}} - I_{\text{IBP}})(1 - D)T_{\text{osc}} - (I_{\text{L1P}} + I_{\text{L2P}})DT_{\text{osc}} = 0 \tag{26}$$

$$(I_{\text{L2P}} - I_{\text{IBP}})(1 - D)T_{\text{osc}} - (I_{\text{L1P}} + I_{\text{L2P}})DT_{\text{osc}} = 0 \tag{27}$$

Similarly, for the lower network, apply the ASBP to capacitor C_N to find out the current through both inductors L_{1N} and L_{2N}:

$$(I_{\text{L1N}} - I_{\text{IBN}})(1 - D)T_{\text{osc}} - (I_{\text{L1N}} + I_{\text{L2N}})DT_{\text{osc}} = 0 \tag{28}$$

$$(I_{\text{L2N}} - I_{\text{IBN}})(1 - D)T_{\text{osc}} - (I_{\text{L1N}} + I_{\text{L2N}})DT_{\text{osc}} = 0 \tag{29}$$

Solve (26) and (27), the average current through L_{1P} and L_{2P} is:

$$I_{\text{L1P}} = I_{\text{L2P}} = \frac{(1-D)I_{\text{IBP}}}{(1-3D)} \tag{30}$$

Similarly, the current through inductors L_{1N} and L_{2N} is found as:

$$I_{L1N} = I_{L2N} = \frac{(1-D)I_{IBN}}{(1-3D)} \tag{31}$$

The boost factor B is given as:

$$B = \frac{V_{out}}{V_{inp}} = \frac{V_{CP}}{V_{inp}} = \frac{V_{out}}{V_{inn}} = \frac{V_{CN}}{V_{inn}} \tag{32}$$

or:

$$B = \frac{(D+1)}{1-3D} \tag{33}$$

The overall gain factor G for the proposed inverter topology is given as:

$$G = BM = \frac{(D+1)M}{1-3D} \tag{34}$$

This completes the mathematical calculations for the proposed inverter topology.

4. PWM and Boost Control Techniques

4.1. PWM Signals

The proposed topology is designed for the three-phase three-level inverter; thus, three sine waves with a phase difference of 120 degrees are utilized in the PODSPWM technique [26–30], to generate the Gating Signals (GS) for 12 switches used in the main inverter circuit. The simulation model of the proposed inverter topology is carried out with a 50 Hz frequency for each sinusoidal signal, and the frequency of each triangular signal is kept at 5 kHz. GS for switches in the main inverter circuit is shown in Figure 5.

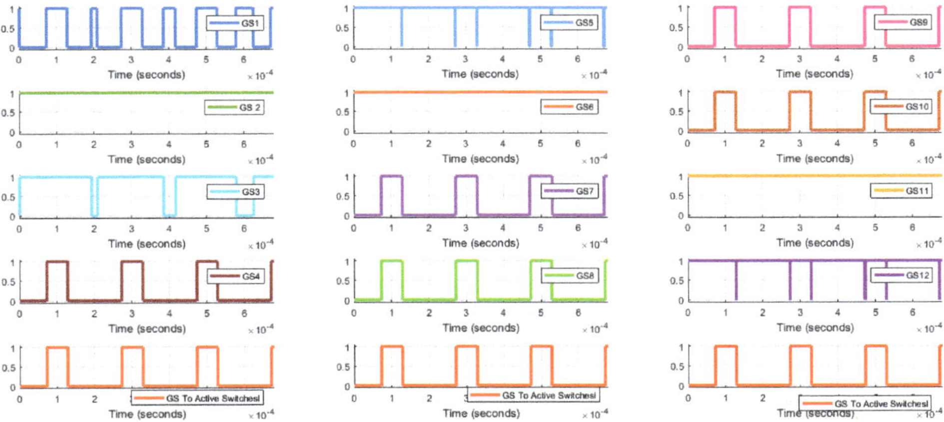

Figure 5. Gating signals for the main inverter circuit.

The GS applied to active switches in the impedance network are illustrated in Figure 6.

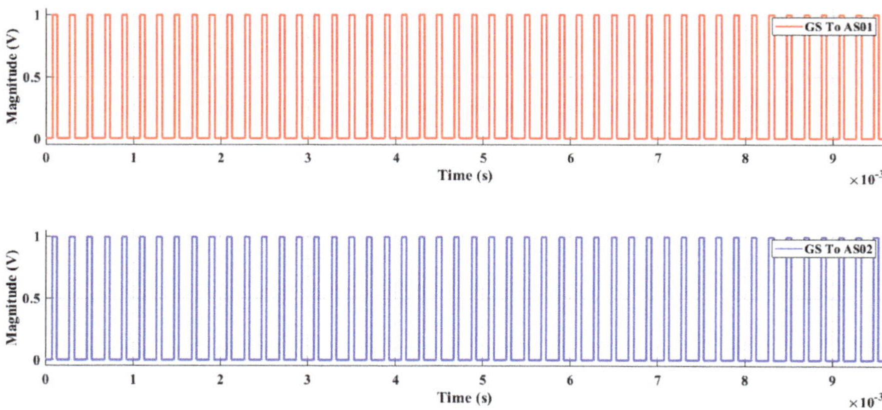

Figure 6. GS to AS_{01} and AS_{02}.

4.2. Maximum Constant Boost Control Method (MCBCM)

Although the Maximum Boost Control Method [31] provides a higher value of G with a smaller value of stress across switches V_s; however, due to diversity in the value of D for each switching cycle, it generates the low-frequency ripples in inductor current, which is not desirable [32]. To overcome this situation, and to achieve a higher value of G at lower values of V_s, MCBCM was introduced to provide a constant value of D by using two envelope signals V_P and V_N. Here, the third harmonic component with a magnitude of 1/6 of the fundamental component is dumped with the sine waves to enhance the range of M. The value of D is given as [32]:

$$D = \left(\frac{2 - \sqrt{3}M}{2}\right) = 1 - \frac{\sqrt{3}M}{2} \tag{35}$$

The value of M can be increased up to $2/\sqrt{3}$. An increase in the range of M causes a reduction in V_S [32]. Due to the numerous advantages of MCBCM over the other boost control methods, MCBCM with PODSPWM is utilized in the Simulink model. The situation is depicted in Figure 7.

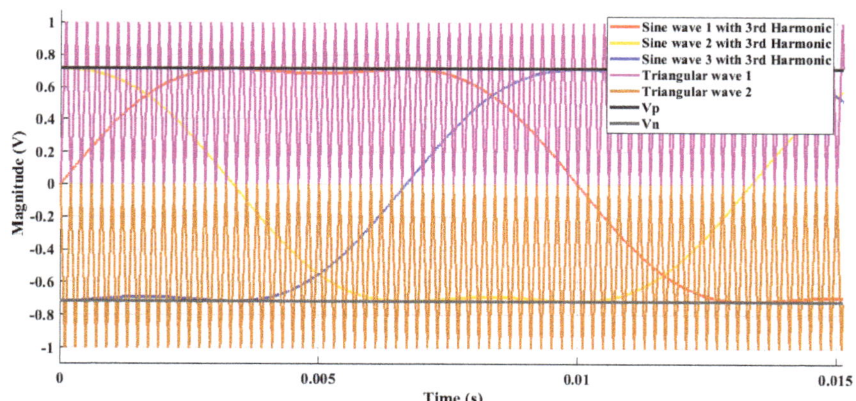

Figure 7. MCBCM with PODSPWM.

The boost factor, overall voltage gain, and stress across the switches for the proposed inverter topology are given as:

$$B = \frac{4 - \sqrt{3}M}{3\sqrt{3}M - 4} \tag{36}$$

$$G = \frac{M(4 - \sqrt{3}M)}{3\sqrt{3}M - 4} \tag{37}$$

$$V_S = BV_{IN} = \frac{(4 - \sqrt{3}M)V_{IN}}{3\sqrt{3}M - 4} \tag{38}$$

5. Simulations Results and Discussion

For solid validation of proposed topology, the inverter is simulated with a detailed switching model in discrete time simulations by using the SimPowerSystems toolbox in MATLAB/Simulink, where the conducting and switching losses are considered for the components used in the impedance network and the main inverter circuit. All the simulation results have a strong agreement with the theoretical results. The details of all components and parameters used in the simulation model for the proposed inverter topology are provided in Tables 2 and 3.

Table 2. Parameters specifications of the proposed inverter topology.

Parameters/Component	Value
Applied DC Voltage	40 V
Capacitor	1000 µF
Inductor	2 mH
Load	250 Ω
Frequency of Reference Signal(s)	50 Hz
Frequency of Carrier Signal(s)	5000 Hz
Modulation Index, M	0.825
Shoot Through Duty Ratio, D	0.2855291
Boost Factor	8.96
Overall Voltage Gain, G	7.392

Table 3. Device parameters in the proposed inverter topology.

Device	Parameter	Value
Diode	Internal Resistance	0.001 Ω
	Forward Voltage Drop	0.7 V
	Snubber Resistance	500 Ω
	Snubber Capacitance	inf
Active Switch	Internal Resistance	0.001 Ω
	Snubber Resistance	1×10^5 Ω
	Snubber Capacitance	inf
IGBT	Internal Resistance	0.01 Ω
	Snubber Resistance	1×10^5 Ω
	Snubber Capacitance	1000 F

The proposed topology outperforms the previous techniques and provides an excellent boosting capacity at a very low ST duty ratio with the high modulation index. Table 4 shows the values of the boosting factor against the different values of the ST duty ratio and modulation index.

Detailed simulation results are presented from Figures 8–13. The simulation model is designed for $D = 0.2855291$ and $M = 0.825$, and it offers a boost factor of 8.96, which is the same as the theoretical analysis (see (33)). The proposed inverter provides the 343 V pole voltages against the 40 V DC input voltage. The pole voltages are shown in Figure 8.

Table 4. Boosting ability against different values of M and D.

ST Duty Ratio (D)	Modulation Index (M)	Boost Factor (B)
0.11	1.0277	1.6567
0.13	1.0046	1.8525
0.15	0.9815	2.0909
0.17	0.9584	2.3878
0.19	0.9353	2.7674
0.21	0.9122	3.2703
0.23	0.8891	3.9677
0.25	0.866	5.0
0.27	0.8429	6.6842
0.2855	0.825	8.96
0.29	0.8198	9.9231
0.31	0.7967	18.7143
0.32	0.7852	33
0.33	0.7736	133

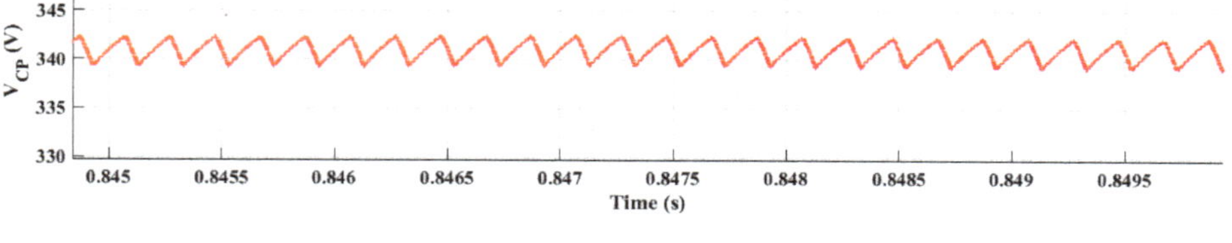

Figure 8. Pole voltage V_{RO}, V_{YO} and V_{BO}.

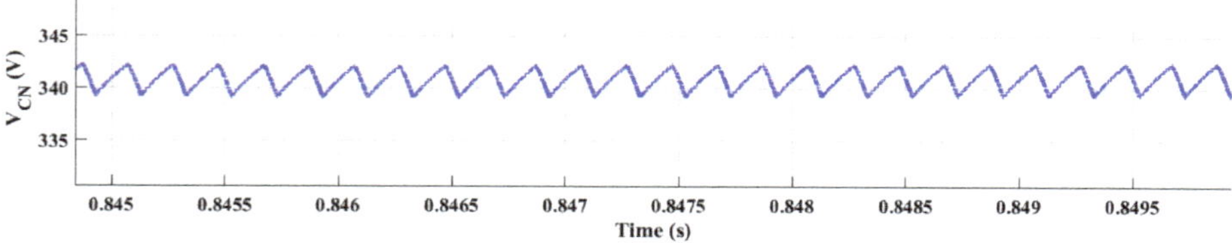

Figure 9. Voltage across capacitors V_{CP} and V_{CN}.

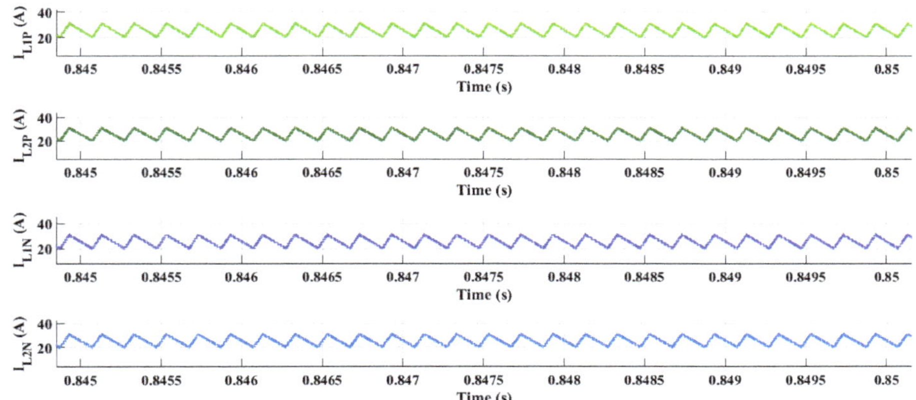

Figure 10. Inductor currents I_{L1P}, I_{L2P}, I_{L1N} and I_{L2N}.

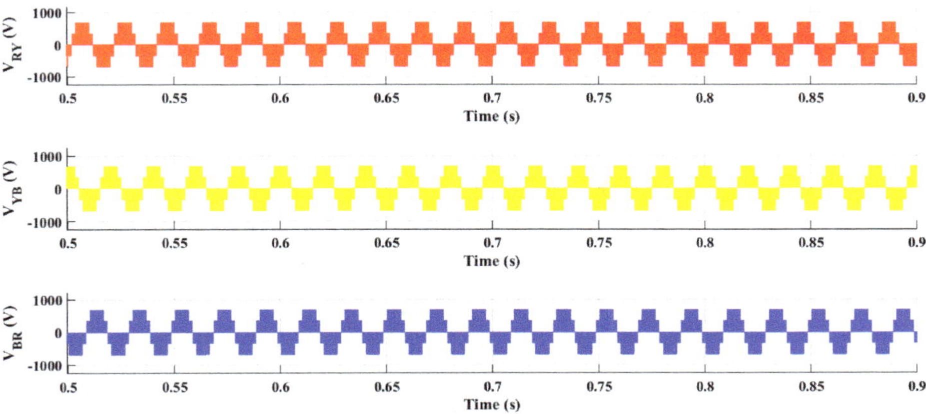

Figure 11. Line voltage V_{RY}, V_{YB}, and V_{BR}.

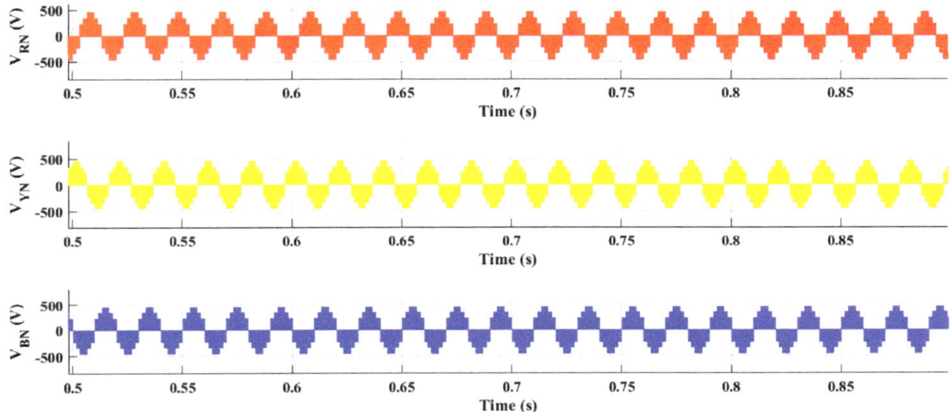

Figure 12. Phase voltage V_{RN}, V_{YN}, and V_{BN}.

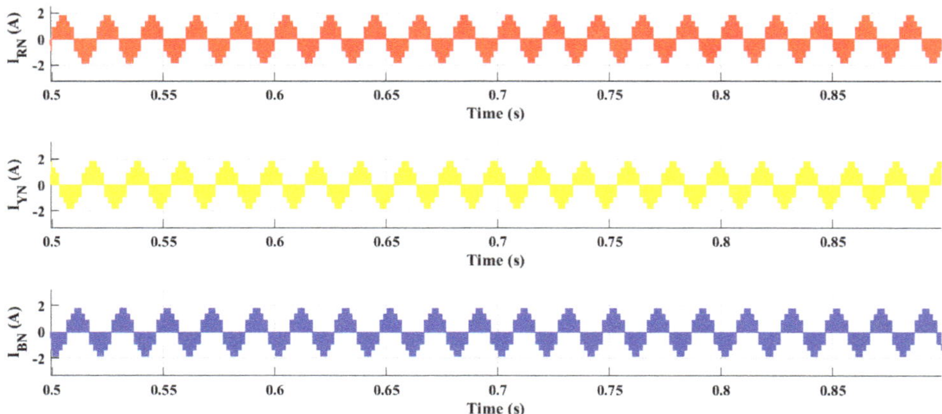

Figure 13. Load currents I_{RN}, I_{YN}, and I_{BN}.

From (21) and (24), the voltage across capacitors V_{CP} and V_{CN} are the same as the pole voltages, having values of 343 V, and the same results are depicted in Figure 9. This ensures the agreement between the mathematically-calculated and simulation results. The voltages across both capacitors are well balanced.

The waveforms for the inductor currents I_{L1P}, I_{L2P}, I_{L1N}, and I_{L2N} are shown in Figure 10.

Line voltage and phase voltage are shown in Figures 11 and 12, respectively. Line voltages are the difference of the pole voltages having a value of 687 V as depicted in the simulation results. Line voltages and phase voltages (the voltages between the phase and neutral points of star-connected load) are interrelated as:

$$\begin{cases} V_{RY} = V_{R0} - V_{Y0} \\ V_{YB} = V_{Y0} - V_{B0} \\ V_{BR} = V_{B0} - V_{R0} \end{cases} \tag{39}$$

$$\begin{cases} V_{RN} = \frac{V_{RY} - V_{YB}}{3} \\ V_{BN} = \frac{V_{BR} - V_{RY}}{3} \\ V_{YN} = \frac{V_{YB} - V_{BR}}{3} \end{cases} \tag{40}$$

A star-connected resistive load having a resistance of 250 Ω per phase is deployed at the output of the inverter. The waveforms of phase currents are shown in Figure 13, that are in-phase with the voltage, thus improving the power quality.

All the simulation results are identical to the theoretical analysis performed for the inverter.

6. Comparison with Previous Topologies

Different parameters, i.e., boosting ability, modulation index, duration of ST duty ratio, and voltage stress across switches, are considered for the comparative analysis purposes to show the effectiveness of the proposed topology. A lot of improvements in the aforementioned parameters offered by the proposed topology are found over the previous topologies.

Figure 14 compares the boost factor versus modulation index for the proposed and the previous topologies, which indicates that the boost factor of the proposed topology is much higher than that of the previous topologies for the same values of modulation index. The proposed inverter is capable of exhibiting higher boosting ability even at larger values of modulation index.

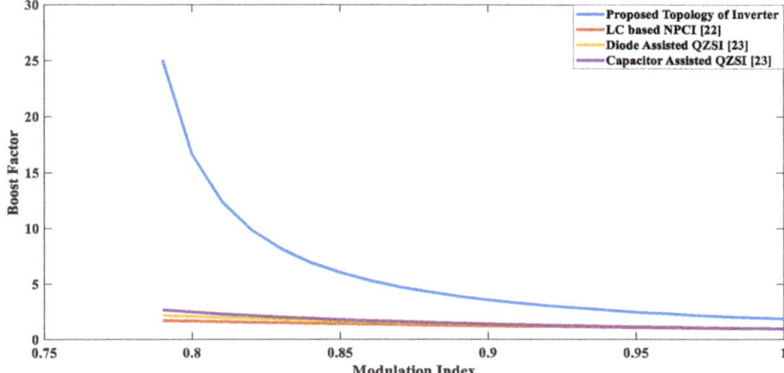

Figure 14. Boost factor versus modulation index.

In addition, the proposed topology is most appropriate to achieve a higher boost factor by utilizing a smaller value of D with a wide range of modulation index. Figure 15 shows a relationship of boost factor versus the ST duty ratio for the proposed and the previous topologies. It can be seen that the proposed topology offers better results, even with smaller values of the ST duty ratio. Thus, this topology can be deployed in applications where higher boost is required with a smaller ST duty ratio.

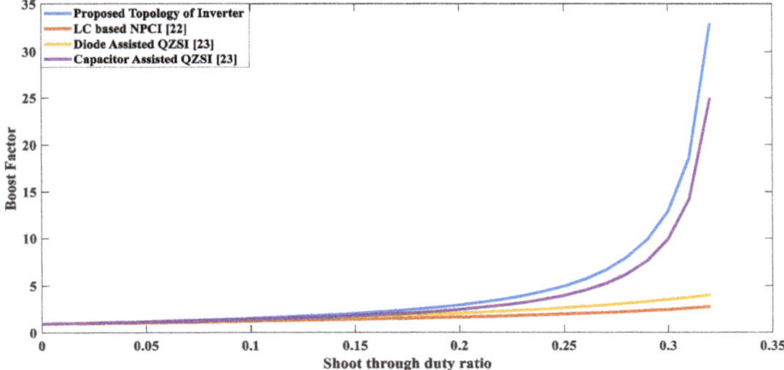

Figure 15. Boost factor versus ST duty ratio.

The proposed topology also shows a remarkable boosting ability with lower voltage stress across the switches. Figure 16 shows a graph between stress across switches and voltage gain, for the proposed topology and the previous topologies, indicating that the proposed topology offers much better results. It enables the availability of higher boost without increasing the stress much. Lowering in switching voltage stresses leads to the reduction of the rating of switches and the size of inverter.

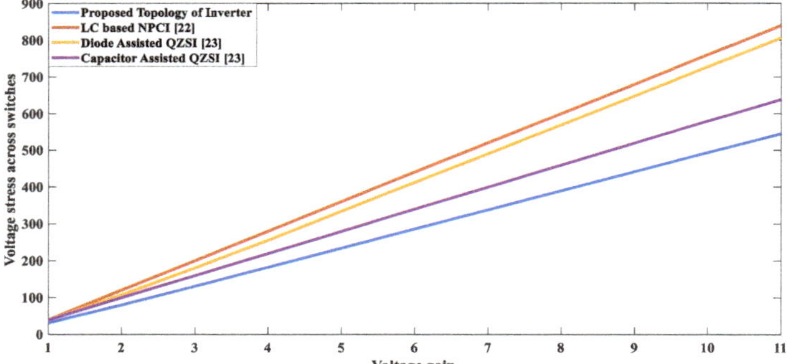

Figure 16. Voltage stress versus gain.

Figure 17 depicts the voltage gain versus the ST duty ratio, which indicates the superiority of proposed topology over the previous topologies in terms of higher voltage gain with lower values of shoot-through duty ratio.

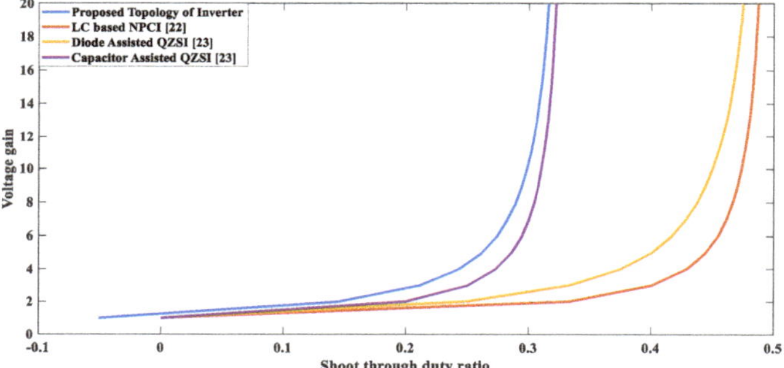

Figure 17. ST duty ratio versus gain.

Figure 18 demonstrates the relationship between boost factor and voltage gain and exhibits that the proposed topology offers the higher voltage gain against the appropriate value of boost factor due to the availability of higher modulation index, whereas, with previous topologies, significantly lower voltage gain can be achieved from the given boost factor. It can be seen from the graph that with the proposed topology, the values of voltage gain are near the corresponding values of the boost factor.

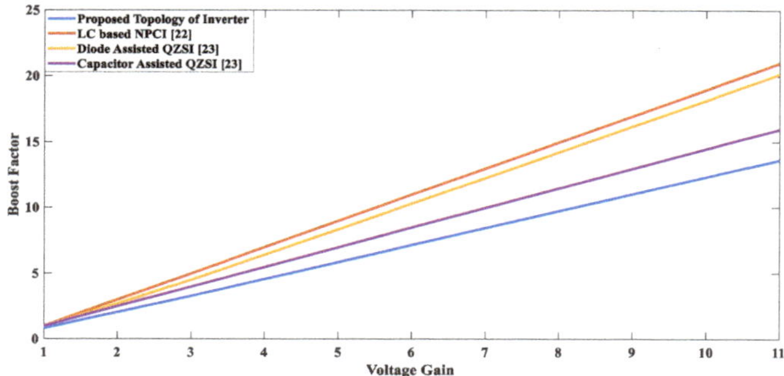

Figure 18. Gain versus boost factor.

In Figure 19, a graph is plotted between BV_{in}/GV_{in} versus the voltage gain, showing clearly that the proposed topology offers better results as compared to previous topologies.

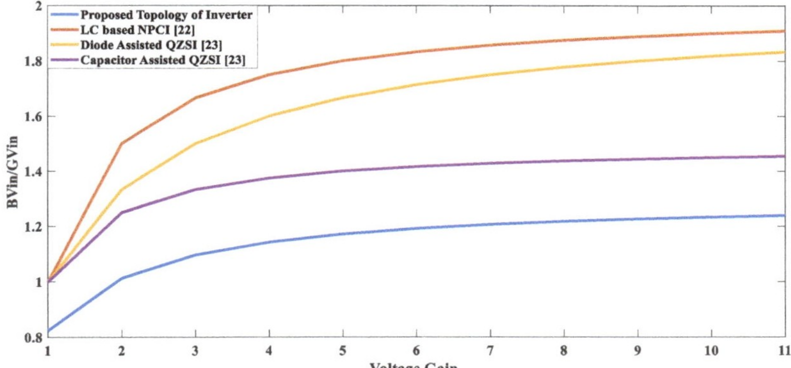

Figure 19. BV_{in}/GV_{in} versus gain.

The simulation results ensure the superiority of the proposed inverter topology over the existing ones. These simulation results can easily be translated into mathematical relations.

In a nutshell, mathematically speaking, a detailed comparison of all such parameters is shown in Table 5. On taking any combination of the input parameters, the output parameters (column 1 of Table 5) are found improved for the proposed topology over the existing ones, just like the simulation results.

Table 5. Comparison with previous topologies.

Parameter	Proposed Inverter Topology	LC-based NPCI [22]	Diode Assisted [23] QZSI	Capacitor Assisted QZSI [23]
Boost Factor	$B = \frac{1+D}{1-3D}$ $B = \frac{4-\sqrt{3}M}{3\sqrt{3}M-4}$ $B = \frac{2\sqrt{3}G}{4-3\sqrt{3}G+\sqrt{27G^2-8\sqrt{3}G+16}}$	$B = \frac{1}{1-2D}$ $B = \frac{1}{2M-1}$ $B = 2G-1$	$B = \frac{1}{1+2D^2-3D}$ $B = \frac{1}{2M^2-M}$ $B = \frac{2G^2}{1+G}$	$B = \frac{1}{1-3D}$ $B = \frac{1}{3M-2}$ $B = \frac{3G-1}{2}$
Voltage Gain	$G = \frac{M(4-\sqrt{3}M)}{3\sqrt{3}M-4}$	$G = \frac{M}{2M-1}$	$G = \frac{1}{2M-1}$	$G = \frac{M}{3M-2}$
Modulation Index	$M = \frac{4-3\sqrt{3}G+\sqrt{27G^2-8\sqrt{3}G+16}}{2\sqrt{3}}$	$M = \frac{G}{2G-1}$	$M = \frac{1+G}{2G}$	$M = \frac{2G}{3G-1}$
Shoot- Through Duty Ratio	$D = \frac{3\sqrt{3}G-\sqrt{27G^2-8\sqrt{3}G+16}}{4}$	$D = \frac{G-1}{2G-1}$	$D = \frac{G-1}{2G}$	$D = \frac{G-1}{3G-1}$
Stress Across Switches	$V_s = \frac{(4-\sqrt{3}M)V_{in}}{3\sqrt{3}M-4}$ $V_s = \left(\frac{2\sqrt{3}G}{4-3\sqrt{3}G+\sqrt{27G^2-8\sqrt{3}G+16}}\right)V_{in}$	$V_s = \frac{V_{in}}{2M-1}$ $V_s = (2G-1)V_{in}$	$V_s = \left(\frac{1}{2M^2-M}\right)V_{in}$ $V_s = \left(\frac{2G^2}{1+G}\right)V_{in}$	$V_s = \left(\frac{1}{3M-2}\right)V_{in}$ $V_s = \left(\frac{3G-1}{2}\right)V_{in}$

Efficiency analysis of the proposed inverter topology is also performed and compared with the previous topologies. Table 6 shows the values of components/parameters used for the analysis. For comparison purposes, they are assumed to be the same for all the topologies. For the simplicity of efficiency analysis, the power losses across the inductors, capacitors, diodes, and active switches (where applicable) due to parasitic resistance of inductors and capacitors, the forward voltage drop of diodes, and on-resistance of active switches are considered.

Table 6. Detail of components/parameters used in the efficiency analysis.

Component/Parameter	Symbol	Value
ESR of Capacitor	R_C	0.08 Ω
DCR of Inductor	R_L	0.07 Ω
On-Resistance of Active Switch	R_S	0.001 Ω
Load Resistance	R_l	250 Ω
Applied DC Voltage	V_{in}	40 V
Voltage Drop Across Diode	V_F	0.7 V

Table 7 shows the equivalent circuits during the NST and ST state of the topologies to be compared. Correspondingly, the power losses that occur across the inductors, capacitors, diodes, and active switches (where applicable) during NST and ST were calculated by utilizing the technique given in [33,34], Expressions of U series (U_1, U_2, U_3, and U_4), V series (V_1, V_2, V_3, and V_4), W series (W_1, W_2, W_3 and W_4), and Z series (Z_3, Z_4) in Table 7 show the power losses across the inductors, capacitors, diodes, and active switches during NST and ST state, respectively.

The efficiency of each topology against the overall voltage gain is computed. As can be observed from Figure 20, efficiency curves of all considered topologies are comparable with one another with a minor difference. The proposed topology offers more than 90% efficiency with a voltage gain of up to 10. With this level of efficiency, it offers several advantages in the form of higher boosting ability, lesser voltage stresses across the switches, lower shoot-through duty ratio, availability of higher modulation index, and improved quality of output waveform by reducing the THD. The reduction in voltage stresses across the switches ensures the utilization of lower rating components/devices even for a higher voltage gain, whereas in the previous topologies, the stresses across the devices drastically increase as the voltage gain increases, as depicted in Figure 16. This situation demands an enormous increase in the rating of components/devices used in previous topologies of the inverter, which leads to a tremendous increase in the cost of components/devices and the size of the inverter, which makes it bulky.

To extend our discussion further, by utilizing Table 5 and Figures 16–20, a detailed comparison of the proposed inverter topology with the previous inverter topologies in terms of overall efficiency, stresses across switches, rating of components/devices, cost, and size is performed and depicted in Figure 21, for a voltage gain of 10 (this voltage is taken as a sample, although the analysis for other values is also true).

It is clear from the comparison that the proposed topology is more feasible for the practical applications as compared to the previous topologies, especially when more voltage gain and boosting ability are required and when cost, size, and lower rating of components are the main concerns. Since the proposed topology belongs to the ZSI family, it finds its applications where all other ZSI are applicable, such as in variable speed drive systems, grid-connected photovoltaic systems, distributed generation systems, hybrid electric vehicles, laminators, conveyor belts, and so on [35–37].

Table 7. Efficiency analysis of various topologies.

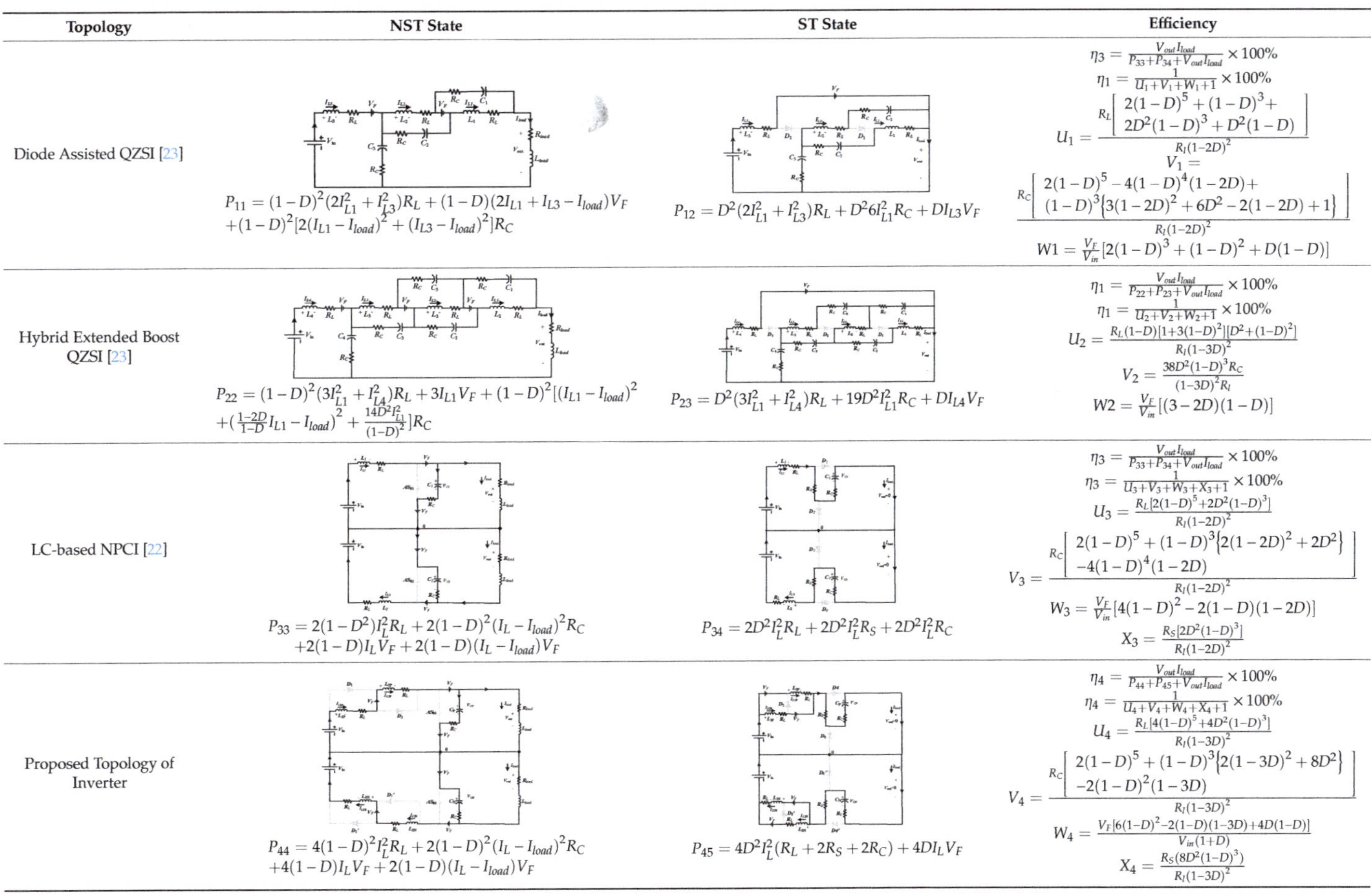

Topology	NST State	ST State	Efficiency
Diode Assisted QZSI [23]	$P_{11} = (1-D)^2(2I_{L1}^2 + I_{L3}^2)R_L + (1-D)(2I_{L1} + I_{L3} - I_{load})V_F$ $+ (1-D)^2[2(I_{L1} - I_{load})^2 + (I_{L3} - I_{load})^2]R_C$	$P_{12} = D^2(2I_{L1}^2 + I_{L3}^2)R_L + D^2 6I_{L1}^2 R_C + DI_{L3}V_F$	$\eta_3 = \frac{V_{out}I_{load}}{P_{33}+P_{34}+V_{out}I_{load}} \times 100\%$ $\eta_1 = \frac{1}{U_1+V_1+W_1+1} \times 100\%$ $U_1 = \frac{R_L\left[\begin{array}{l}2(1-D)^5 + (1-D)^3 + \\ 2D^2(1-D)^3 + D^2(1-D)\end{array}\right]}{R_l(1-2D)^2}$ $V_1 = \frac{R_C\left[\begin{array}{l}2(1-D)^5 - 4(1-D)^4(1-2D) + \\ (1-D)^3\{3(1-2D)^2 + 6D^2 - 2(1-2D) + 1\}\end{array}\right]}{R_l(1-2D)^2}$ $W1 = \frac{V_F}{V_{in}}[2(1-D)^3 + (1-D)^2 + D(1-D)]$
Hybrid Extended Boost QZSI [23]	$P_{22} = (1-D)^2(3I_{L1}^2 + I_{L4}^2)R_L + 3I_{L1}V_F + (1-D)^2[(I_{L1} - I_{load})^2$ $+ (\frac{1-2D}{1-D}I_{L1} - I_{load})^2 + \frac{14D^2 I_{L1}^2}{(1-D)^2}]R_C$	$P_{23} = D^2(3I_{L1}^2 + I_{L4}^2)R_L + 19D^2 I_{L1}^2 R_C + DI_{L4}V_F$	$\eta_1 = \frac{V_{out}I_{load}}{P_{22}+P_{23}+V_{out}I_{load}} \times 100\%$ $\eta_1 = \frac{1}{U_2+V_2+W_2+1} \times 100\%$ $U_2 = \frac{R_L(1-D)[1+3(1-D)^2][D^2+(1-D)^2]}{R_l(1-3D)^2}$ $V_2 = \frac{38D^2(1-D)^3 R_C}{(1-3D)^2 R_l}$ $W2 = \frac{V_F}{V_{in}}[(3-2D)(1-D)]$
LC-based NPCI [22]	$P_{33} = 2(1-D^2)I_L^2 R_L + 2(1-D)^2(I_L - I_{load})^2 R_C$ $+ 2(1-D)I_L V_F + 2(1-D)(I_L - I_{load})V_F$	$P_{34} = 2D^2 I_L^2 R_L + 2D^2 I_L^2 R_S + 2D^2 I_L^2 R_C$	$\eta_3 = \frac{V_{out}I_{load}}{P_{33}+P_{34}+V_{out}I_{load}} \times 100\%$ $\eta_3 = \frac{1}{U_3+V_3+W_3+X_3+1} \times 100\%$ $U_3 = \frac{R_L[2(1-D)^5 + 2D^2(1-D)^3]}{R_l(1-2D)^2}$ $V_3 = \frac{R_C\left[\begin{array}{l}2(1-D)^5 + (1-D)^3\{2(1-2D)^2 + 2D^2\} \\ -4(1-D)^4(1-2D)\end{array}\right]}{R_l(1-2D)^2}$ $W_3 = \frac{V_F}{V_{in}}[4(1-D)^2 - 2(1-D)(1-2D)]$ $X_3 = \frac{R_S[2D^2(1-D)^3]}{R_l(1-2D)^2}$
Proposed Topology of Inverter	$P_{44} = 4(1-D)^2 I_L^2 R_L + 2(1-D)^2(I_L - I_{load})^2 R_C$ $+ 4(1-D)I_L V_F + 2(1-D)(I_L - I_{load})V_F$	$P_{45} = 4D^2 I_L^2(R_L + 2R_S + 2R_C) + 4DI_L V_F$	$\eta_4 = \frac{V_{out}I_{load}}{P_{44}+P_{45}+V_{out}I_{load}} \times 100\%$ $\eta_4 = \frac{1}{U_4+V_4+W_4+X_4+1} \times 100\%$ $U_4 = \frac{R_L[4(1-D)^5 + 4D^2(1-D)^3]}{R_l(1-3D)^2}$ $V_4 = \frac{R_C\left[\begin{array}{l}2(1-D)^5 + (1-D)^3\{2(1-3D)^2 + 8D^2\} \\ -2(1-D)^2(1-3D)\end{array}\right]}{R_l(1-3D)^2}$ $W_4 = \frac{V_F[6(1-D)^2 - 2(1-D)(1-3D) + 4D(1-D)]}{V_{in}(1+D)}$ $X_4 = \frac{R_S(8D^2(1-D)^3)}{R_l(1-3D)^2}$

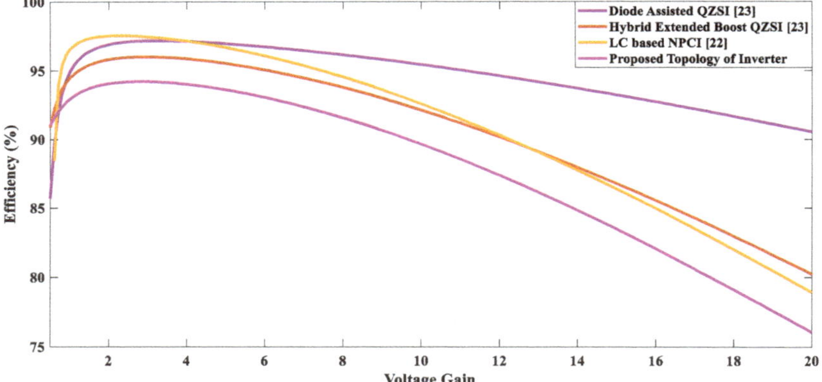

Figure 20. Efficiency versus voltage gain.

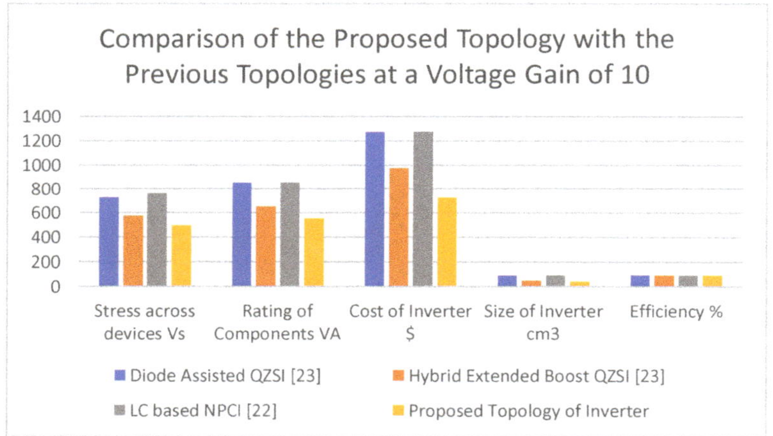

Figure 21. A detailed comparison of the proposed topology with the previous topologies.

7. Conclusions

This research work focused on the development of the three-level high voltage gain NPCI topology to boost up the DC voltage at the desired level and offers the three-level AC output in a single stage. It also detained all the merits of previous topologies of three-level NPCZSI/QZSI, such as continuity in input current and voltage balance across the capacitors. The results validated that the proposed topology ensures the remarkable boosting ability by utilizing the smaller duration of ST state and higher range of modulation index, which enables it to keep the lower stresses across the devices even at higher values of voltage gain; that is the most desirable feature for low voltage applications. The proposed topology has a slightly lower efficiency compared to other topologies; however, it reduces cost and size by utilizing the low rating components at higher voltage gain operations. This unique feature makes it more feasible for practical applications as compared to other topologies that oblige the higher rating components for their operation; this drawback associated with previous topologies not only affects the cost of the inverter but also makes the size of inverter bulky and voluminous.

Author Contributions: Conceptualization, G.A. and I.K.; methodology, M.A.A. and R.M.; software, A.B.A.; validation, U.F. and S.S.K.; formal analysis, S.S.K.; investigation, M.A.A.; resources, G.A.; data curation, M.A.A.; writing—original draft preparation, G.A.; writing—review and editing, I.K.; visualization, U.F. and R.M.;

supervision, A.B.A.; project administration, G.A.; funding acquisition, I.K. All authors have read and agreed to the published version of the manuscript.

Funding: This research received no external funding.

Conflicts of Interest: The authors declare no conflict of interest.

References

1. Tang, Y.; Xie, S.; Zhang, C. An improved Z-source inverter. *IEEE Trans. Power Electron.* **2010**, *26*, 3865–3868. [CrossRef]
2. Ding, X.; Qian, Z.; Yang, S.; Cui, B.; Peng, F. A high-performance Z-source inverter operating with small inductor at wide-range load. In Proceedings of the APEC 07-Twenty-Second Annual IEEE Applied Power Electronics Conference and Exposition, Anaheim, CA, USA, 25 Feburary–1 March 2007; pp. 615–620.
3. Tang, Y.; Xie, S.; Zhang, C.; Xu, Z. Improved Z-source inverter with reduced Z-source capacitor voltage stress and soft-start capability. *IEEE Trans. Power Electron.* **2009**, *24*, 409–415. [CrossRef]
4. Vinnikov, D.; Roasto, I.; Jalakas, T. An improved high-power DC/DC Converter for distributed power generation. In Proceedings of the 2009 10th International Conference on Electrical Power Quality and Utilisation, Lodz, Poland, 15–17 September 2009; pp. 1–6.
5. Anderson, J.; Peng, F.Z. A class of quasi-Z-source inverters. In Proceedings of the 2008 IEEE Industry Applications Society Annual Meeting, Edmonton, AB, Canada, 5–9 October 2008; pp. 1–7.
6. Anderson, J.; Peng, F.Z. Four quasi-Z-source inverters. In Proceedings of the 2008 IEEE Power Electronics Specialists Conference, Rhodes, Greece, 15–19 June 2008; pp. 2743–2749.
7. Peng, F.Z.; Yuan, X.; Fang, X.; Qian, Z. Z-source inverter for adjustable speed drives. *IEEE Power Electron. Lett.* **2003**, *1*, 33–35. [CrossRef]
8. Li, Y.; Anderson, J.; Peng, F.Z.; Liu, D. Quasi-Z-source inverter for photovoltaic power generation systems. In Proceedings of the 2009 Twenty-Fourth Annual IEEE Applied Power Electronics Conference and Exposition, Washington, DC, USA, 15–19 February 2009; pp. 918–924.
9. Ismeil, M.A.; Kouzou, A.; Kennel, R.; Abu-Rub, H.; Orabi, M. A new switched-inductor quasi-Z-source inverter topology. In Proceedings of the 2012 15th International Power Electronics and Motion Control Conference (EPE/PEMC), Novi Sad, Serbia, 4–6 September 2012; pp. DS3d.2-1–DS3d.2-6.
10. Mosa, M.; Abu-Rub, H.; Rodríguez, J. High performance predictive control applied to three phase grid connected Quasi-Z-Source Inverter. In Proceedings of the IECON 2013-39th Annual Conference of the IEEE Industrial Electronics Society, Vienna, Austria, 10–13 November 2013; pp. 5812–5817.
11. Loh, P.C.; Gao, F.; Blaabjerg, F.; Goh, A.L. Buck-boost impedance networks. In Proceedings of the 2007 European Conference on Power Electronics and Applications, Aalborg, Denmark, 2–5 September 2007; pp. 1–10.
12. Loh, P.C.; Gao, F.; Blaabjerg, F. Embedded EZ-source inverters. *IEEE Trans. Ind. Appl.* **2009**, *46*, 256–267.
13. Loh, P.C.; Gao, F.; Blaabjerg, F.; Feng, S.Y.C.; Soon, K.N.J. Pulsewidth-Modulated $ Z $-Source Neutral-Point-Clamped Inverter. *IEEE Trans. Ind. Appl.* **2007**, *43*, 1295–1308. [CrossRef]
14. Loh, P.C.; Blaabjerg, F.; Wong, C.P. Comparative evaluation of pulse-width modulation strategies for Z-source neutral-point-clamped inverter. In Proceedings of the 2006 37th IEEE Power Electronics Specialists Conference, Jeju, Korea, 18–22 June 2006; pp. 1–7.
15. Loh, P.C.; Lim, S.W.; Gao, F.; Blaabjerg, F. Three-level Z-source inverters using a single LC impedance network. *IEEE Trans. Power Electron.* **2007**, *22*, 706–711. [CrossRef]
16. Loh, P.C.; Gao, F.; Blaabjerg, F.; Lim, S.W. Operational analysis and modulation control of three-level Z-source inverters with enhanced output waveform quality. *IEEE Trans. Power Electron.* **2009**, *24*, 1767–1775. [CrossRef]
17. Maheshwari, R.; Munk-Nielsen, S.; Busquets-Monge, S. Design of neutral-point voltage controller of a three-level NPC inverter with small DC-link capacitors. *IEEE Trans. Ind. Electron.* **2012**, *60*, 1861–1871. [CrossRef]
18. Mo, W.; Loh, P.C.; Blaabjerg, F.; Wang, P. Trans-Z-source and Γ-Z-source neutral-point-clamped inverters. *IET Power Electron.* **2014**, *8*, 371–377. [CrossRef]
19. Wang, X.; Liu, H.; Li, Y. A novel coupled inductor Z-source three-level inverter. *IEICE Electron. Express* **2017**, *14*. [CrossRef]

20. Battiston, A.; Miliani, E.-H.; Pierfederici, S.; Meibody-Tabar, F. A novel quasi-Z-source inverter topology with special coupled inductors for input current ripples cancellation. *IEEE Trans. Power Electron.* **2015**, *31*, 2409–2416. [CrossRef]

21. Yahya, A.; Waqar, T.; Husain, N.; Ali, S.M.U. Hybrid Multilevel Topology Based High Power Quality Inverter. In Proceedings of the APEC'98 Thirteenth Annual Applied Power Electronics Conference and Exposition, Anaheim, CA, USA, 15–19 February 1998.

22. Sahoo, M.; Keerthipati, S. A three-level LC-switching-based voltage boost NPC inverter. *IEEE Trans. Ind. Electron.* **2016**, *64*, 2876–2883. [CrossRef]

23. Gajanayake, C.J.; Luo, F.L.; Gooi, H.B.; So, P.L.; Siow, L.K. Extended-boost $ Z $-source inverters. *IEEE Trans. Power Electron.* **2010**, *25*, 2642–2652. [CrossRef]

24. Peng, F.Z. Z-Source Inverter. *IEEE Trans. Ind. Appl.* **2003**, *39*. [CrossRef]

25. Nabae, A.; Takahashi, I.; Akagi, H. A new neutral-point-clamped PWM inverter. *IEEE Trans. Ind. Appl.* **1981**, *5*, 518–523. [CrossRef]

26. McGrath, B.P.; Holmes, D.G. Multicarrier PWM strategies for multilevel inverters. *IEEE Trans. Ind. Electron.* **2002**, *49*, 858–867. [CrossRef]

27. Rathore, S.; Kirar, M.K.; Bhardwaj, S.K. Simulation of cascaded H-bridge multilevel inverter using PD, POD, APOD techniques. *Electr. Comput. Eng. Int. J. ECIJ* **2015**, *4*, 27–41. [CrossRef]

28. Haskar Reddy, V.N.; Babu, C.S.; Suresh, K. Advanced Modulating Techniques for Diode Clamped Multilevel Inverter Fed Induction Motor. *ARPN J. Eng. Appl. Sci.* **2011**, *6*, 90–99.

29. Colak, I.; Kabalci, E.; Bayindir, R. Review of multilevel voltage source inverter topologies and control schemes. *Energy Convers. Manag.* **2011**, *52*, 1114–1128. [CrossRef]

30. McGrath, B.P.; Holmes, D.G. A comparison of multicarrier PWM strategies for cascaded and neutral point clamped multilevel inverters. In Proceedings of the 2000 IEEE 31st Annual Power Electronics Specialists Conference. Conference Proceedings (Cat. No. 00CH37018), Galway, Ireland, 23 June 2000; Volume 2, pp. 674–679.

31. Peng, F.Z.; Shen, M.; Qian, Z. Maximum boost control of the Z-source inverter. *IEEE Trans. Power Electron.* **2005**, *20*, 833–838. [CrossRef]

32. Shen, M.; Wang, J.; Joseph, A.; Peng, F.Z.; Tolbert, L.M.; Adams, D.J. Constant boost control of the Z-source inverter to minimize current ripple and voltage stress. *IEEE Trans. Ind. Appl.* **2006**, *42*, 770–778. [CrossRef]

33. Pan, L. LZ-source inverter. *IEEE Trans. Power Electron.* **2014**, *29*, 6534–6543. [CrossRef]

34. Zhu, X.; Zhang, B.; Qiu, D. Enhanced boost quasi-Z-source inverters with active switched-inductor boost network. *IET Power Electron.* **2018**, *11*, 1774–1787. [CrossRef]

35. Rodriguez, J.; Lai, J.-S.; Peng, F.Z. Multilevel inverters: A survey of topologies, controls, and applications. *IEEE Trans. Ind. Electron.* **2002**, *49*, 724–738. [CrossRef]

36. Peng, F.Z.; Joseph, A.; Wang, J.; Shen, M.; Chen, L.; Pan, Z.; Ortiz-Rivera, E.; Huang, Y. Z-source inverter for motor drives. *IEEE Trans. Power Electron.* **2005**, *20*, 857–863. [CrossRef]

37. Khlebnikov, A.S.; Kharitonov, S.A. Application of the Z-source converter for aircraft power generation systems. In Proceedings of the 2008 9th International Workshop and Tutorials on Electron Devices and Materials, Novosibirsk, Russia, 1–5 July 2008; pp. 211–215.

Article

A New Efficient Step-Up Boost Converter with CLD Cell for Electric Vehicle and New Energy Systems

Muhammad Zeeshan Malik [1,*], Haoyong Chen [2,*], Muhammad Shahzad Nazir [1], Irfan Ahmad Khan [3], Ahmed N. Abdalla [4], Amjad Ali [5] and Wan Chen [1]

[1] Faculty of Automation, Huaiyin Institute of Technology, Huai'an 223003, China; msn_bhutta88@yahoo.com (M.S.N.); calvinchenw1979@163.com (W.C.)

[2] Department of Electrical Engineering, South China University of Technology Guangzhou, Guangdong 510641, China

[3] Marine Engineering Technology Department in a Joint Appointment with the Electrical and Computer Engineering Department, Texas A&M University, Galveston, TX 77554, USA; irfankhan@tamu.edu

[4] Faculty of Electronic Information Engineering, Huaiyin Institute of Technology, Huai'an 223003, China; dramaidecn@gmail.com

[5] Center of Research Excellence in Renewable Energy, King Fahad University of Petroleum and Minerals, Dhahran 31261, Saudi Arabia; amjad.ali@kfupm.edu.sa

* Correspondence: malik4one@yahoo.com (M.Z.M.); eehychen@scut.edu.cn (H.C.); Tel.: +86-131-1525-6515 (M.Z.M.)

Received: 13 January 2020; Accepted: 4 April 2020; Published: 8 April 2020

Abstract: An increase in demand for renewable energy resources, energy storage technologies, and electric vehicles requires high-power level DC-DC converters. The DC-DC converter that is suitable for high-power conversion applications (i.e., resonant, full-bridge or the dual-active bridge) requires magnetic transformer coupling between input and output stage. However, transformer design in these converters remains a challenging problem, with several non-linear scaling issues that need to be simultaneously optimized to reduce losses and maintain acceptable performance. In this paper, a new transformer-less high step-up boost converter with a charge pump capacitorand capacitor-inductor-diode CLD cell is proposed using dynamic modeling. The experimental and simulation results of the proposed converter are carried out in a laboratory and through Matlab Simulink, where 10 V is given as an input voltage, and at the output, 100 V achieved in the proposed converter. A comparative analysis of the proposed converter has also been done with a conventional quadratic converter that has similar parameters. The results suggest that the proposed converter can obtain high voltage gain without operating at the maximum duty cycle and is more efficient than the conventional converter.

Keywords: dynamic modeling; DC-DC converter; electric vehicle (EV); charge pump capacitor

1. Introduction

In recent years, due to the shortage of fossil fuels, research on renewable energy sources such as photovoltaic PV, wind, and fuel cells (FC), has gained immense popularity [1,2]. The intermittency associated with PV systems and low voltages at load end and electric vehicle (EV) or hybrid electric vehicle (HEV) charging needs a boost converter as depicted in Figures 1 and 2 [3–5]. Generally, EVs are powered by fuel cell stacks, supercapacitors, and battery systems. Thus, there is a need to step up the voltages [6–8].

A review of related literature depicts that various boost converters have been presented to overcome high voltage gain at the output. The traditional boost converters are preferred for their simple structure and low cost; however, they usually produce high input current ripples. For high voltage gain, this converter needs to work at a high duty ratio cycle, which causes switching problems [9–13].

The traditional solution to implement a DC-DC converter, with a high voltage transformation ratio, typically involves a magnetic transformer with a high turn's ratio, Due to the high cost of magnetic transformer components, several transformer-less topologies have been proposed that can achieve a high step-up ratio e at lower cost and size. The most common transformer-coupled topologies are derived from isolated versions of the basic converter types, i.e., buck, boost, buck-boost, cúk, single-ended primary-inductor converter (SEPIC) [14,15]. The forward converter is based on the buck topology; the flyback converter is based on the buck-boost topology; Full-bridge as well as Half-bridge converters can be both buck and boost derived. Moreover, converter types can be classified based on transformer core utilization [16]. Forward and flyback topologies have a net DC current in the transformer winding, which means the flux in the core must be reset at the end of each switching cycle. Full bridge, half-bridge and push-pull topologies provide bidirectional excitation to the transformer core; the net current in the winding is AC. This means that the flux in the core is reset automatically at the end of each switching cycle. The concept of using switching capacitor (SC) is obtained from [17] in which switching converters with wide DC conversion range are discussed by including a hybrid cell that multiplies current or voltage, as described in [18–20]. The high-frequency rectifier stage can be passive (half-wave, center-tapped half wave, full wave) or active, with switching devices. Some designs eliminate the high-frequency transformer and replace it with a parallel capacitor [21]. The shortcomings of wide frequency operation are EMI issues [22,23], higher switching loss at low power levels, difficulty in the design of magnetic components and the circulating currents independent of power level. Furthermore, phase-controlled converters operate at fixed frequencies and adjust power flow by varying phase shifts between the switching legs of a full-bridge inverter [24–26]. Similarly, modular DC–DC converters with input- and output-side series/parallel configurations are typically used for conditions where power processing exceeds the capacity of any single converter. However, these converters have control-related issues that do not provide balanced power-share between the constituent converter modules for different load conditions. Quadratic converters are DC–DC converters with a voltage conversion ratio that has a quadratic dependence on-duty ratio [27,28]. They are synthesized by cascading two converters in series and then eliminating redundant switches and controllers [29,30].

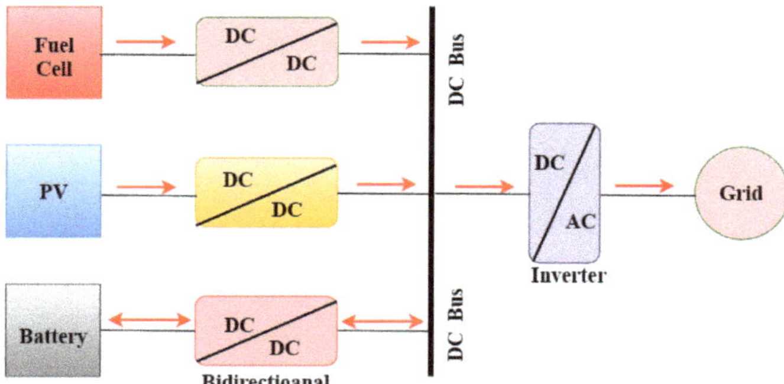

Figure 1. New energy technologies.

One of the main advantages of these converters is their higher voltage transformation ratio with a reduced number of components. However, the limitations of these converters arise from their cascaded nature. In addition, their combined efficiency is lower than the individual converter stages; and the voltage stresses on the boost diode and switch of the final stage are equal to the output voltage. Such problems can be solved by adding a switch to form a three-level boost converter [31]. Many other structures can also be integrated with quadratic converters like a capacitor-diode combination of the voltage multiplier or single/three-phase transformer windings [32]. There are different variations of boost converters, for instance, the quadratic boost zeta converter, quadratic tapped-boost converter

ZVS/ZCS quadratic converters, and quadratic converters based on non-cascading structures [33,34]. (SC) converters are a class of DC–DC converters, where the power stage consists of a network of only capacitors and switches. The capacitors are charged and discharged in sequence via input voltage to achieve the required voltage conversion ratio at the output [35]. The most significant advantage of SC converter is the absence of inductive storage devices. This makes the topology appropriate for monolithic integration and high-power density as both switches and capacitors can be fabricated on the same substrate using standard semiconductor manufacturing processes. One of its major shortcomings, which has been observed, is that the process of charge transfer results in impulse currents due to capacitors acting as charge pumps, which can lead to an increase in device stress and EMI issues. Another drawback is that at a fixed voltage transformation ratio, the efficiency of the converter drops quickly as the conversion ratio moves away from the designed operating mode. Furthermore, an additional limitation is that the voltage stress across all active switches is not equal [36].

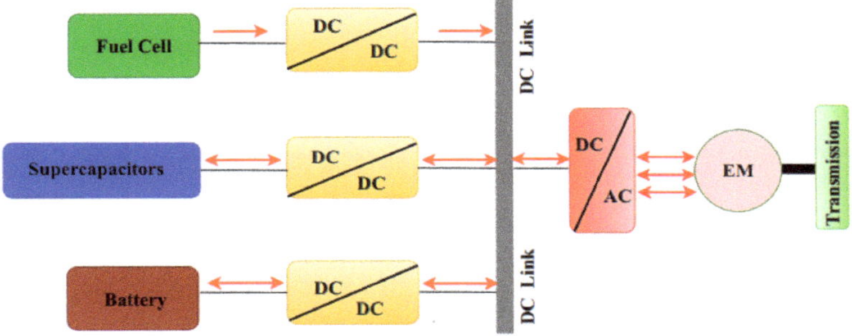

Figure 2. Block diagrams of the electric vehicle.

2. Related Work

The Cockroft–Walton (CW) circuit or voltage multiplier-based hybrid DC–DC Converters were introduced in 1932. This circuit operates as a voltage multiplier, which is fed by a pulsating low voltage waveform to generate a high voltage DC on its output terminal. Its main application is to generate high voltage DC required for insulation testing generators and particle accelerators. As the multiplication ratio increases, the CW circuit suffers from poor load regulation resulting in a significant voltage drop at its output terminal. For this reason, the CW circuit needs another converter to provide output voltage regulation. In [18,19], an isolated hybrid DC–DC converter composed of series-resonant and a CW circuit is proposed for a medical-use high-voltage X-ray power generator. In [37], a classical transformer-less multilevel boost converter with a hybrid connection of CW is introduced. Both converters in [37,38] can provide a considerable DC voltage step-up ratio at high efficiency. However, these circuits cannot be used for high-power applications as they share a similar high current-spike problem, which results from charging and discharging capacitors connected in parallel. This boost circuit is necessary to provide a pulsating input voltage waveform for the CW circuit and to produce output voltage regulation through use of Pulse Width Modulation (PWM). The presence of this boost circuit is inappropriate for high-power applications. Moreover, voltage multiplier hybrid DC–DC converters cannot achieve DC voltage step-down operation.

In this article, a new modified boost converter with a charge–pump capacitor and CLD cell is proposed (see Figure 3). The proposed converter can achieve very high voltage gain at the output without intensive duty ratio, and with lower switching stress across the switch. As the proposed converter consists of a charge pump capacitor which can control input current ripples, CLD cell can support high voltage gain at the output. Figure 3 represents the block diagram of proposed converter where V_s is an input voltage, the four capacitors are V_{c1}, V_{c2}, V_{c3}, and V_{c0}, S indicates metal–oxide–semiconductor field-effect transistor (MOSFET) switch three inductors are L_1, L_2, L_3

with same duty ratio. For proposed converter, the following assumptions have been made: (1) The proposed converter works with continuous conduction mode (CCM), (2) All components including input power are taken as ideal (no voltage drop and resistance),and (3) All the capacitors are large enough with constant voltages.

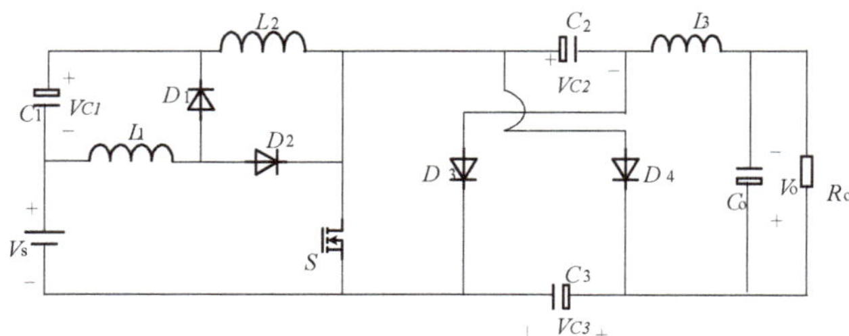

Figure 3. Diagram of the proposed converter.

3. Working Method of the Proposed Converter

The switching frequency of the proposed converter is fixed and has two switching states with one PWM signal. For one state ($t_0 - t_1$), MOSFET is switched ON; for other state ($t_1 - t_2$), MOSFET is switched OFF as illustrated in Figures 4 and 5. Furthermore, the characteristics of steady-state waveform, voltages, and currents of the proposed converter are shown in Figure 6.

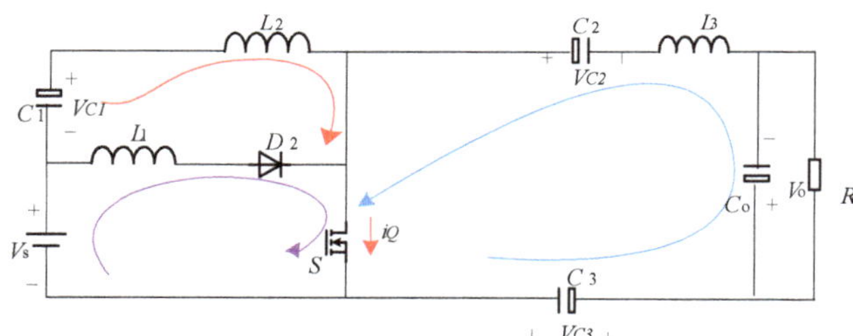

Figure 4. State-I.

3.1. MOSFET Switch ON State ($t_0 - t_1$)

In this state, when switch "S" is ON, diode D_2 will be forward biased while other diodes (D_1, D_3, and D_4) will work in reverse biased mode. This configuration of DC–DC converter is presented in Figure 4 with arrows indicating the direction of the current flow. The input voltage V_S/L_1 developed across inductor L_1, linearly increases the (iL1) inductor current and rise in $(V_S + V_{C1})/L_2$ across inductor L_2, increases the inductor current (iL2). In this state, capacitors C_2 and C_3 are in series, and the voltage output of these two capacitors is given as $V_{C2} + V_{C3} = 2V_{C2}$. And the inductor L_3 increases by $(2V_{C2} - V_0)/L_3$. The equation of this state is derived in Equations (1)–(5).

$$i_{L1} = i_{L1-t_0} + \frac{V_s}{L_1}t \tag{1}$$

$$ui_{L2} = i_{L2-t_0} + \frac{V_s + V_{c1}}{L_2}t \tag{2}$$

$$i_{L3} = i_{L3-t_0} + \frac{2V_{c2} - V_0}{L_3}t \tag{3}$$

$$i_Q = i_{L1} + i_{L2} + i_{L3} \tag{4}$$

$$i_{in} = i_{L1} + i_{L2} \tag{5}$$

3.2. MOSFET Switch OFF State (t_1–t_2)

In this state, when semiconductor switch S is turned OFF, diode D_2 will be in reverse biased mode, as shown in Figure 5. As the voltage across the inductor L_1 becomes negative of V_{C1}/L_1, current changes its path from switch to diode D_1 and decreases linearly. At the same time, the voltage across inductor L_2 is $(V_S + V_{C1} - V_{C2})/L_2$, and inductor L_3 is $(V_{C2} - V_0)/L_3$, respectively. All the inductors current decreases linearly during this state. Equations for this state are (6)–(10) derived using Figure 5, as given below.

$$i_{L1} = i_{L1-t_1} - \frac{V_{c1}}{L_1}t \tag{6}$$

$$i_{L2} = i_{L2-t_1} - \frac{V_s + V_{c1} - V_{c2}}{L_2}t \tag{7}$$

$$i_{L3} = i_{L3-t_1} - \frac{V_{c2} - V_0}{L_3}t \tag{8}$$

$$i_Q = 0 \tag{9}$$

$$i_{in} = i_{L2} \tag{10}$$

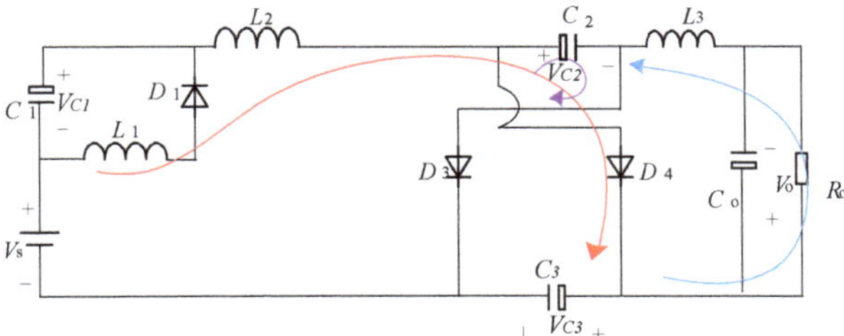

Figure 5. State-II.

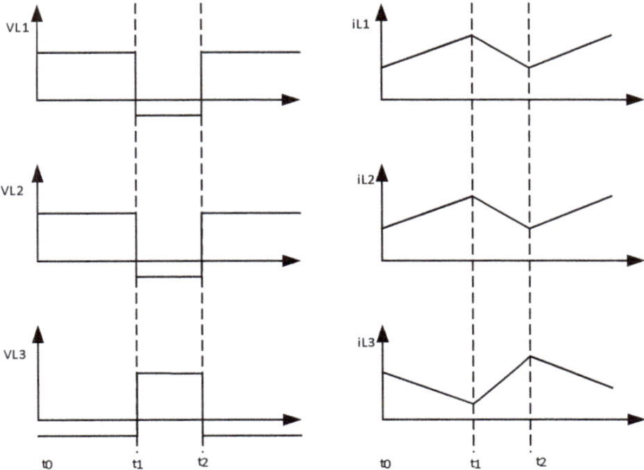

Figure 6. Steady-state behavior.

4. Steady-State Analysis of the Proposed Converter

The steady-state analysis of the proposed converter is discussed below.

4.1. DC Conversion Ratio

The relationship between input and output voltage can be obtained through Equations (11)–(17) after applying inductor volt second balance (VSB) at inductor L_1, L_2, and L_3 for each state. Equations (11)–(13) describe state-I while Equations (14)–(16) refer to state-II.

$$V_{L1} = V_s \tag{11}$$

$$V_{L2} = V_s + V_{c1} \tag{12}$$

$$V_{L3} = 2V_{c2} - V_0 \tag{13}$$

$$V_{L1} = -V_{c1} \tag{14}$$

$$V_{L2} = V_s + V_{c1} - V_{c2} \tag{15}$$

$$V_{L3} = V_{c2} - V_0 \tag{16}$$

$$G = \frac{V_0}{V_s} = \frac{1+D}{(1-D)^2} \tag{17}$$

The voltage and current stress across semiconductor components are determined through Equations (18)–(25)

$$V_S = V_{c2} \tag{18}$$

$$V_{D1} = V_S + V_{c1} \tag{19}$$

$$V_{D2} = V_{c2} - V_S - V_{c1} \tag{20}$$

$$V_{D3,4} = V_{c2} \tag{21}$$

$$i_{D1,2} = \frac{V_s}{(1-D)^5 R_L} \tag{22}$$

$$i_{D3} = \frac{V_s}{(1-D)^2 R_L} \tag{23}$$

$$i_{D4} = \frac{V_s}{(1-D)^4 R_L} \tag{24}$$

$$i_Q = \frac{V_s}{(D) R_L} \tag{25}$$

4.2. Dynamic Modeling of the Proposed Converter

In this section, we assume that inductor current AC ripples and capacitor voltage ripples are monomers [39]. Figure 7 is derived from Figure 4 where,

V_s: input voltage,

$i_{L1,2,3}$: current across the inductor

$V_{c1,2,3,4}$: voltage across the capacitor.

When the switch "S" is ON, current across the diode $D_1 = i_{L1}$ and current across the switch $S = i_{L1} + i_{L2} + i_{L3}$. During this stage, diode $D_{1,\,2,\,3}$ are reversed biased, and the voltage across these diodes is equal to $(V_s + V_{c1})$, (V_{c2}), (V_{c3}), as shown in Figure 2.

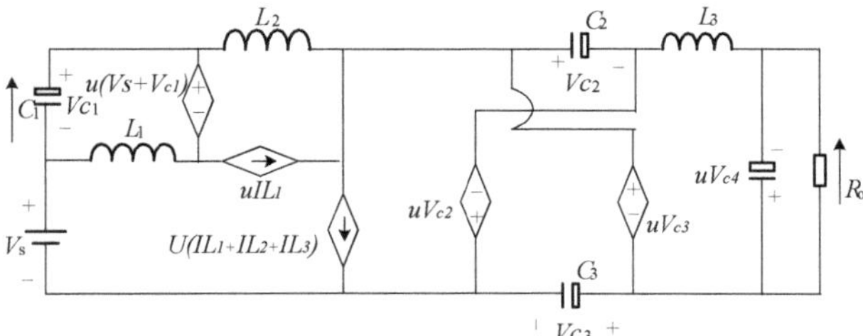

Figure 7. Average equivalent circuit of the proposed converter.

Figure 7 shows the equivalent circuit of the proposed converter where diode D_2 and switch are replaced by a current source, and diode D_1, D_3, and D_4 are replaced by the voltage source. In the equivalent circuit, after applying circuit theory, the equation derived is (26),

$$
\begin{bmatrix} \dot{i}_{L1} \\ \dot{i}_{L2} \\ \dot{i}_{L3} \\ \dot{V}_{c1} \\ \dot{V}_{c2} \\ \dot{V}_{c3} \\ \dot{V}_{c4} \end{bmatrix} =
\begin{bmatrix}
0 & 0 & 0 & -(1-u)/L_1 & 0 & 0 & 0 \\
0 & 0 & L_3/L_2 & 1/L_2 & -1/L_2 & -1/L_2 & -1/L_2 \\
0 & 0 & L_2/L_3 & 0 & -1/L_3 & 1/L_3 & 1/L_3 \\
1-u/C_1 & -1/C_1 & 1/C_1 & 0 & 0 & 0 & 0 \\
0 & 1/C_2 & -1/C_2 & 0 & 0 & 0 & 0 \\
0 & 1/C_3 & -1/C_3 & 0 & 0 & 0 & 0 \\
0 & 1/C_4 & -1/C_4 & 0 & 0 & 0 & -1/R_LC_4
\end{bmatrix}
\begin{bmatrix} i_{L1} \\ i_{L2} \\ i_{L3} \\ V_{c1} \\ V_{c2} \\ V_{c3} \\ V_{c4} \end{bmatrix}
$$
$$
+ \begin{bmatrix} \frac{u}{L_1} \\ \frac{1}{L_2} \\ -\frac{1}{L_3} \\ 0 \\ 0 \\ 0 \\ 0 \end{bmatrix} [e(t)]
\tag{26}
$$

Without considering losses, the above equation can be written as,

$$
\dot{x}(t) = A(u)x(t) + B(u)e(t)
\tag{27}
$$

where $e(t)$ is the input vector, $x(t) = [i_{L1}, i_{L2}, i_{L3}, V_{c1}, V_{c2}, V_{c3}, V_{c4}]^T R^7$ is the average value of state vector, $A(u)$ is a matrix in $R^{7 \times 7}$ and $B(u)$ is a vector in R^7; $V_s \in R$ = input voltage; R_L is the resistive load and u is a function of switch S with e binary value [0, 1]. Subsequently, [0] demonstrates that the switch is OFF, whereas [1] indicates that the switch is ON. Equation (27) is based on non-linear, whereas matrix $A(u)$, $B(u)$ is dependent on the control signal of $u(t) \in R$. The behavior of the proposed converter obtained after the linearization process results in small perturbations around an operating point. Therefore, the nominal steady-state operating condition of the proposed converter can be written by setting (27) as $AX + BE = 0$ and and is shown in Equations (28)–(34),

The voltage across the capacitors are given in Equations (28)–(31),

$$
V_{c1} = \frac{D}{1-D} Vs
\tag{28}
$$

$$
V_{c2} = \frac{1}{(1-D)^2} Vs
\tag{29}
$$

$$
V_{c3} = \frac{1}{(1-D)^2} Vs
\tag{30}
$$

$$V_{c4} = \frac{1+D}{(1-D)^2} Vs \tag{31}$$

Current across the inductor is given in Equations (32)–(34),

$$i_{L1} = \frac{V_s}{(1-D)^5 R_L} \tag{32}$$

$$i_{L2} = \frac{V_s}{(1-D)^4 R_L} \tag{33}$$

$$i_{L3} = \frac{V_s}{(1-D)^2 R_L} \tag{34}$$

where D is the duty cycle of the switch S,

$$e(t) = E + \widetilde{e} \tag{35}$$

$$u(t) = U + \widetilde{u} \tag{36}$$

where $\widetilde{u}$ is the small-signal perturbations of the nominal dusty cycle D, and $\widetilde{e}$ is the nominal input voltage V_s. Thus, the relationship between voltage and duty cycle above is given as (35)–(36), Where, $\widetilde{e} << E$ and $\widetilde{u} << D$, which can also be written as (37)–(38),

$$X(t) = X + \widetilde{x} \tag{37}$$

$$V_0(t) = V_0 + \widetilde{v}_0 \tag{38}$$

After substituting Equations (28)–(34) and (35)–(36) into Equation (26), the linear mode can be derived by assuming fewer perturbations, and avoiding non-linear terms as mentioned in Equation (39),

$$
\begin{bmatrix} \dot{\widetilde{i}}_{L1} \\ \dot{\widetilde{i}}_{L2} \\ \dot{\widetilde{i}}_{L3} \\ \dot{\widetilde{V}}_{c1} \\ \dot{\widetilde{V}}_{c2} \\ \dot{\widetilde{V}}_{c3} \\ \dot{\widetilde{V}}_{c4} \end{bmatrix}
=
\begin{bmatrix}
0 & 0 & 0 & -\frac{1-u}{L_1} & 0 & 0 & 0 \\
0 & 0 & \frac{L_3}{L_2} & \frac{1}{L_2} & -\frac{1}{L_2} & -\frac{1}{L_2} & -\frac{1}{L_2} \\
0 & 0 & \frac{L_2}{L_3} & 0 & -\frac{1}{L_3} & \frac{1}{L_3} & \frac{1}{L_3} \\
1-\frac{u}{C_1} & -\frac{1}{C_1} & \frac{1}{C_1} & 0 & 0 & 0 & 0 \\
0 & \frac{1}{C_2} & -\frac{1}{C_2} & 0 & 0 & 0 & 0 \\
0 & \frac{1}{C_3} & -\frac{1}{C_3} & 0 & 0 & 0 & 0 \\
0 & \frac{1}{C_4} & -\frac{1}{C_4} & 0 & 0 & 0 & -\frac{1}{R_L C_4}
\end{bmatrix}
\begin{bmatrix} \widetilde{i}_{L1} \\ \widetilde{i}_{L2} \\ \widetilde{i}_{L3} \\ \widetilde{V}_{c1} \\ \widetilde{V}_{c2} \\ \widetilde{V}_{c3} \\ \widetilde{V}_{c4} \end{bmatrix}
+
\begin{bmatrix}
\frac{D}{(1-D)L_1} Vs & \frac{u}{L_1} \\
\frac{1}{(1-D)^2 L_2} Vs & \frac{1}{L_2} \\
\frac{1}{(1-D)^2 L_2} Vs & -\frac{1}{L_3} \\
\frac{1+D}{(1-D)^2 L_3} Vs & 0 \\
\frac{V_s}{(1-D)^5 R_{Lc1}} & 0 \\
\frac{V_s}{(1-D)^4 R_{Lc2}} & 0 \\
\frac{V_s}{(1-D)^2 R_{Lc3}} & 0
\end{bmatrix}
\begin{bmatrix} \widetilde{u} \\ \widetilde{e} \end{bmatrix}
\tag{39}
$$

Furthermore, the above equation can be written as (40),

$$\dot{x}(t) = Ax(t) + Bv(t) \tag{40}$$

where, $x(t) \in R^7$ is the state vector, $A \in R^{7 \times 7}$ and B are the constant matrix in $R^{7 \times 2}$, R^2 is vector of input voltage and $V_{(t)} = \begin{bmatrix} \widetilde{u} & \widetilde{e} \end{bmatrix}$. When the perturbation input voltage is neglected $\widetilde{e} = 0$, matrix B is removed and small signal state-space model can be written as (41).

$$
\begin{bmatrix} \widetilde{\dot{i}_{L1}} \\ \widetilde{\dot{i}_{L2}} \\ \widetilde{\dot{i}_{L3}} \\ \widetilde{\dot{V}_{c1}} \\ \widetilde{\dot{V}_{c2}} \\ \widetilde{\dot{V}_{c3}} \\ \widetilde{\dot{V}_{c4}} \end{bmatrix} =
\begin{bmatrix}
0 & 0 & 0 & -(1-u)/L_1 & 0 & 0 & 0 \\
0 & 0 & L_3/L_2 & 1/L_2 & -1/L_2 & -1/L_2 & -1/L_2 \\
0 & 0 & L_2/L_3 & 0 & -1/L_3 & 1/L_3 & 1/L_3 \\
1-u/C_1 & -1/C_1 & 1/C_1 & 0 & 0 & 0 & 0 \\
0 & 1/C_2 & -1/C_2 & 0 & 0 & 0 & 0 \\
0 & 1/C_3 & -1/C_3 & 0 & 0 & 0 & 0 \\
0 & 1/C_4 & -1/C_4 & 0 & 0 & 0 & 1/R_LC_4
\end{bmatrix}
\begin{bmatrix} \widetilde{i_{L1}} \\ \widetilde{i_{L2}} \\ \widetilde{i_{L3}} \\ \widetilde{V_{c1}} \\ \widetilde{V_{c2}} \\ \widetilde{V_{c3}} \\ \widetilde{V_{c4}} \end{bmatrix}
$$

$$
+ \begin{bmatrix}
\dfrac{D}{(1-D)L_1}V_S \\[4pt]
\dfrac{1}{(1-D)^2 L_2}V_S \\[4pt]
\dfrac{1}{(1-D)^2 L_2}V_S \\[4pt]
\dfrac{1+D}{(1-D)^2 L_3}V_S \\[4pt]
\dfrac{V_s}{(1-D)^5 R_{Lc1}} \\[4pt]
\dfrac{V_s}{(1-D)^4 R_{Lc2}} \\[4pt]
\dfrac{V_s}{(1-D)^2 R_{Lc3}}
\end{bmatrix} [\widetilde{u}]
\tag{41}
$$

5. Simulation Results & Discussion

The simulation results of the proposed converter, which are performed in the Matlab Simulink environment, are shown in Figure 8. Figure 8a describes the switching signal of the MOSFET switch S; Figure 8b represents 10 V as an input voltage; in Figure 8c 100 V output voltages are obtained at the duty cycle of D = 0.6; and Figure 8d,e is the output voltage of capacitors C_1 and C_2, which are 62.50 V. In Figure 8f voltage stress across the switch S is 62.50 V. Thus, the output voltage of the proposed converter is higher than voltage stress. Furthermore, a comparison of results demonstrates that voltage gain of the conventional quadratic converter is significantly less than the proposed converter, whereas voltage stress across the switch is equal to output voltages [11]. Figure 8g–i shows the current across the inductor L_1, L_2, and L_3.

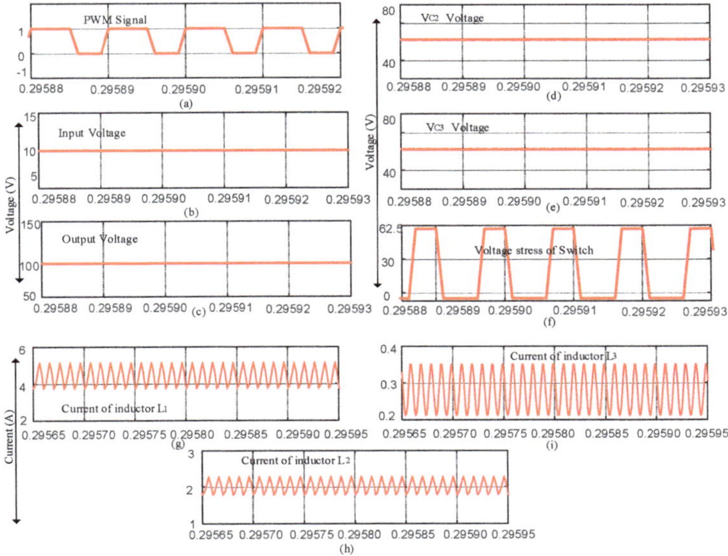

Figure 8. Simulation results of the proposed converter.

6. Experimental Results & Discussion

To validate results and effectiveness of the proposed converter, an experimental setup is prepared in the research laboratory with circuit parameters as given in Table 1. Experimental waveforms of the proposed converter are depicted in Figure 9 and a prototype of this converter is presented in Figure 10. The duty cycle of the proposed converter is assumed as D = 0.6. Figure 9a shows the switching signals of the proposed converter, and Figure 9b illustrates waveforms of input voltage, which is 10 V. Figure 9c shows that the output voltage of the proposed converter is 98 V, which is in accordance with voltage gain determined through Equation (17). The output voltage indicates that the proposed converter can achieve high gain without operating at a maximum duty cycle. The voltage conversion ratio of the proposed converter is ten times higher than the input voltage, which is quite high when compared to the traditional quadratic converter.

Table 1. Parameters for prototype.

Components	Symbol	Parameters
Output power	P_0	50 (W)
Input voltage	V_{in}	10 (V)
Output voltage	V_0	100 (V)
Load resistance	R	200 (Ω)
Frequency	F_S	100 (KHz)
Filter inductor	L_{1-3}	220 (μH)
Capacitor	C_{1-4}	440 (μF)
Diodes	D	MUR860
MOSFET Switch	S	IRFZ46N

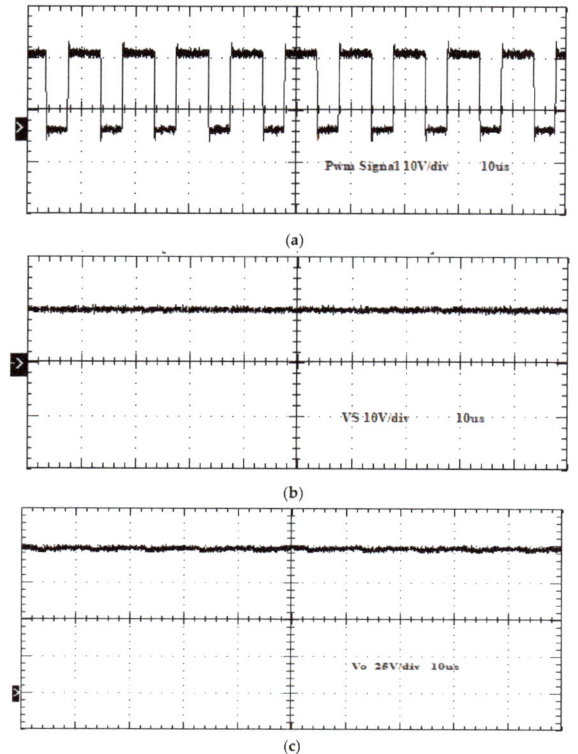

Figure 9. *Cont.*

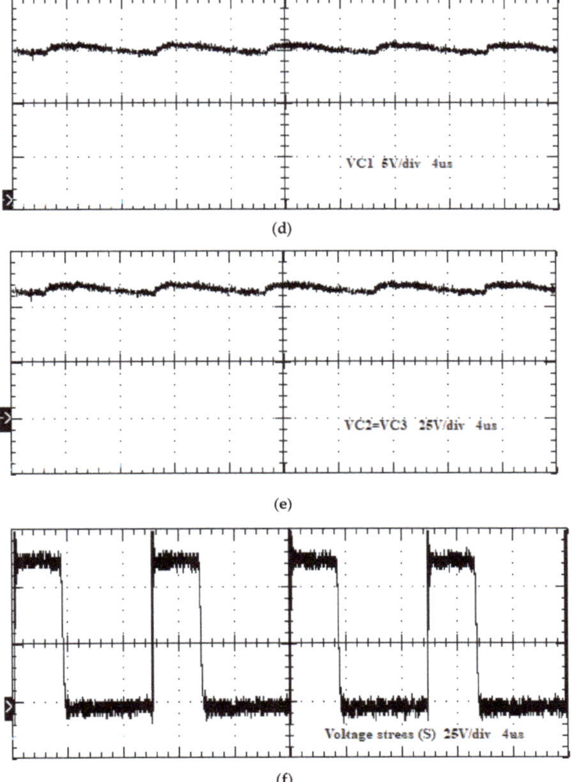

(d)

(e)

(f)

Figure 9. Experimental results of the proposed converter.

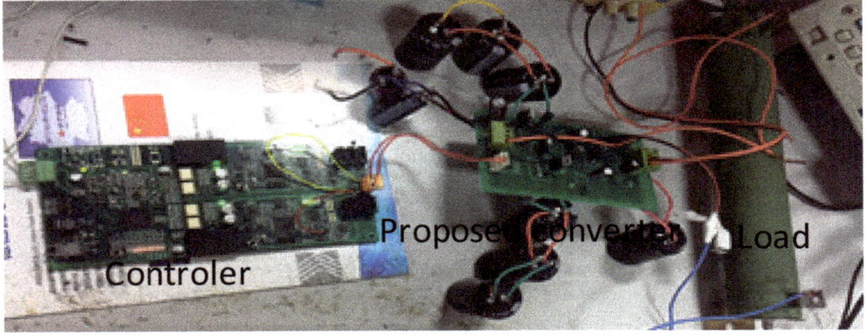

Figure 10. The prototype of the proposed converter.

In Figure 9d,e waveform of capacitor voltages V_{C1}, V_{C2}, and V_{C3} are shown, where the output voltage of capacitor V_{C1} is 15 V and of capacitor V_{C2} and V_{C3} is 61 V, that are in close proximity to capacitor Equations (28)–(31). Furthermore, Figure 9f shows the waveforms of switching stress across the MOSFETs switch S, where voltage stress across the switch is 62 V, which is less than the output voltage. The conventional quadratic converter has stress across the switch Vs-stress = V_0. From experimental and theoretical analysis, it is proved that the proposed converter has many advantages over conventional quadratic converter.

In Figure 11, an efficiency graph of the proposed and conventional quadratic converter is given. The efficiency of both converters is calculated for different loads. It is observed that the maximum efficiency of the proposed converter is 95.91% and is achieved at an output power of 195 W. This is due to the fact that higher input voltages allow the converter to operate at higher output power. As maximum power is limited by thermal limit of the switches and lower current, it results in lower losses in switches, and attainment of minimum efficiency (86.74%) at an output power of 33 W. As compared to the conventional quadratic converter at the same load, maximum efficiency achieved by the proposed converter is 93.97%, and the minimum is 83.29%, which indicate that it has excellent efficiency at high voltage gain.

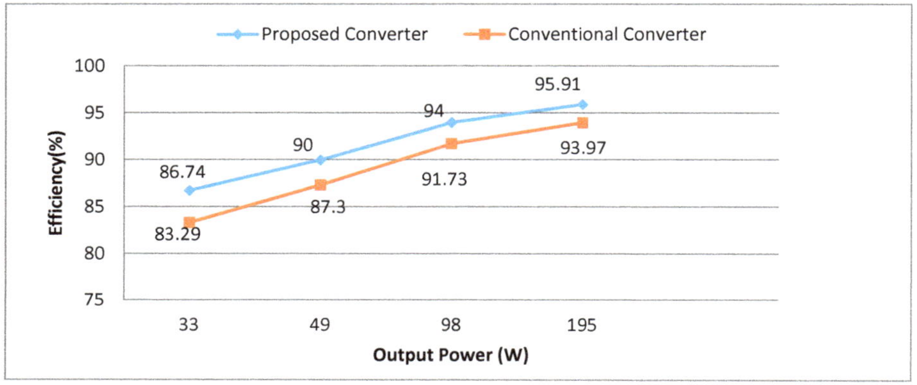

Figure 11. Efficiency graph.

7. Conclusions

In this paper, a DC–DC boost converter with charge pump–capacitor and CLD cell is introduced with its static and dynamic working principle. To investigate operational principles of the proposed converter, the simulation was performed in MATLAB/SIMULINK, and to validate the results, an experimental setup was developed in the laboratory. It is evident from results (Simulation and experimental) that the proposed converter holds certain advantages over traditional converters, such as efficient high voltage gain and low losses across the switch than the output voltage. The presented converter is suitable for converting low voltage to high voltage in various applications such as photovoltaic systems, where the wires from the PV modules are isolated from the earth, and there would not be a problem. If the input and output are (galvanically) separated and at different potentials. However, it may have some shortcomings for high voltage floating ground applications, for instance, where input source and output load share a common ground (some cases in automotive). In the floating ground system, it may have some safety concerns, because there is no low impedance path to ground. However, this type of grounding can facilitate in isolating a system from interference problems caused by ground loops.

Author Contributions: Conceptualization, M.Z.M. and W.C.; methodology, A.N.A.; software, M.S.N.; validation, A.A. and W.C.; formal analysis, I.A.K.; investigation, H.C.; resources, A.A.; data curation, A.N.A.; writing—original draft preparation, M.Z.M.; writing—review and editing, I.A.K.; visualization, A.A.; supervision, H.C.; project administration, I.A.K.; funding acquisition, H.C. All authors have read and agreed to the published version of the manuscript.

Funding: This work is supported by the National Natural Science Foundation of China (51937005).

Conflicts of Interest: The authors declare that there is no conflict of interest regarding the publication of this paper.

References

1. Ahmad, B.; Kyyra, J.; Martinez, W. Efficiency optimisation of an interleaved high step-up converter. *J. Eng.* **2019**, *2019*, 4167–4172. [CrossRef]
2. Cornea, O.; Andreescu, G.-D.; Muntean, N.; Hulea, D. Bidirectional power flow control in a DC microgrid through a switched-capacitor cell hybrid DC–DC converter. *IEEE Trans. Ind. Electron.* **2016**, *64*, 3012–3022. [CrossRef]
3. Pop-Calimanu, I.-M.; Lica, S.; Popescu, S.; Lascu, D.; Lie, I.; Mirsu, R. A New Hybrid Inductor-Based Boost DC-DC Converter Suitable for Applications in Photovoltaic Systems. *Energies* **2019**, *12*, 252. [CrossRef]
4. Khan, I.; Li, Z.; Xu, Y.; Gu, W. Distributed control algorithm for optimal reactive power control in power grids. *Int. J. Electr. Power Energy Syst.* **2016**, *83*, 505–513. [CrossRef]
5. Shahzad Nazir, M.; Wu, Q.; Li, M. Symmetrical Short-Circuit Parameters Comparison of DFIG–WT. *Int. J. Electr. Comput. Eng. Syst.* **2017**, *8*, 77–83. [CrossRef]
6. Itani, K.; De Bernardinis, A.; Khatir, Z.; Jammal, A.; Oueidat, M. Regenerative braking modeling, control, and simulation of a hybrid energy storage system for an electric vehicle in extreme conditions. *IEEE Trans. Transp. Electrif.* **2016**, *2*, 465–479. [CrossRef]
7. Kouchachvili, L.; Yaïci, W.; Entchev, E. Hybrid battery/supercapacitor energy storage system for the electric vehicles. *J. Power Sources* **2018**, *374*, 237–248. [CrossRef]
8. Akar, F.; Tavlasoglu, Y.; Ugur, E.; Vural, B.; Aksoy, I. A bidirectional nonisolated multi-input DC–DC converter for hybrid energy storage systems in electric vehicles. *IEEE Trans. Veh. Technol.* **2015**, *65*, 7944–7955. [CrossRef]
9. Ren, G.; Ma, G.; Cong, N. Review of electrical energy storage system for vehicular applications. *Renew. Sustain. Energy Rev.* **2015**, *41*, 225–236. [CrossRef]
10. Kim, T.; Feng, D.; Jang, M.; Agelidis, V.G. Common mode noise analysis for cascaded boost converter with silicon carbide devices. *IEEE Trans. Power Electron.* **2016**, *32*, 1917–1926. [CrossRef]
11. Malik, M.Z.; Xu, Q.; Farooq, A.; Chen, G. A new modified quadratic boost converter with high voltage gain. *IEICE Electron. Express* **2016**, *14*, 20161176. [CrossRef]
12. Ellahi, M.; Abbas, G.; Khan, I.; Koola, P.M.; Nasir, M.; Raza, A.; Farooq, U. Recent Approaches of Forecasting and Optimal Economic Dispatch to Overcome Intermittency of Wind and Photovoltaic (PV) Systems: A Review. *Energies* **2019**, *12*, 4392. [CrossRef]
13. Khan, I.; Xu, Y.; Sun, H.; Bhattacharjee, V. Distributed optimal reactive power control of power systems. *IEEE Access* **2017**, *6*, 7100–7111. [CrossRef]
14. García-Sánchez, J.R.; Hernández-Márquez, E.; Ramírez-Morales, J.; Marciano-Melchor, M.; Marcelino-Aranda, M.; Taud, H.; Silva-Ortigoza, R. A Robust Differential Flatness-Based Tracking Control for the "MIMO DC/DC Boost Converter–Inverter–DC Motor" System: Experimental Results. *IEEE Access* **2019**, *7*, 84497–84505. [CrossRef]
15. Farooq, A.; Malik, Z.; Sun, Z.; Chen, G. A Review of non-isolated high step-down Dc-Dc Converters. *Int. J. Smart Home* **2015**, *9*, 133–150. [CrossRef]
16. Thenathayalan, D.; Lee, C.; Park, J.-H. High-order resonant converter topology with extremely low-coupling contactless transformers. *IEEE Trans. Power Electron.* **2015**, *31*, 2347–2361. [CrossRef]
17. Maksimovic, D.; Cuk, S. Switching converters with wide DC conversion range. *IEEE Trans. Power Electron.* **1991**, *6*, 151–157. [CrossRef]
18. Axelrod, B.; Berkovich, Y.; Ioinovici, A. Switched-capacitor/switched-inductor structures for getting transformerless hybrid DC–DC PWM converters. *IEEE Trans. Circuits Syst. I Regul. Pap.* **2008**, *55*, 687–696. [CrossRef]
19. Chen, M.; Li, K.; Hu, J.; Ioinovici, A. Hybrid switched-capacitor quadratic boost converters with very high DC gain and low voltage stress on their semiconductor devices. In Proceedings of the 2016 IEEE Energy Conversion Congress and Exposition (ECCE), Milwaukee, WI, USA, 18–22 September 2016; pp. 1–8.
20. Lin, B.-R. Investigation of a resonant dc–dc converter for light rail transportation applications. *Energies* **2018**, *11*, 1078. [CrossRef]
21. Lee, S.-W.; Do, H.-L. High step-up coupled-inductor cascade boost DC–DC converter with lossless passive snubber. *IEEE Trans. Ind. Electron.* **2018**, *65*, 7753–7761. [CrossRef]

22. Zhang, G.; Iu, H.H.-C.; Zhang, B.; Li, Z.; Fernando, T.; Chen, S.-Z.; Zhang, Y. An impedance network boost converter with a high-voltage gain. *IEEE Trans. Power Electron.* **2017**, *32*, 6661–6665. [CrossRef]

23. Naderi, A.; Abbaszadeh, K. High step-up DC–DC converter with input current ripple cancellation. *IET Power Electron.* **2016**, *9*, 2394–2403. [CrossRef]

24. Malik, M.Z.; Ali, A.; Xu, Q.; Chen, G. A New Quadratic Boost Converter with Voltage Multiplier Cell: An Analysis and Assessment. *Int. J. Smart Home* **2016**, *10*, 281–294. [CrossRef]

25. Richelli, A.; Colalongo, L.; Tonoli, S.; Kovacs-Vajna, Z.M. A 0.2$-\hbox {1.2} $ V DC/DC Boost Converter for Power Harvesting Applications. *IEEE Trans. Power Electron.* **2009**, *24*, 1541–1546. [CrossRef]

26. Bhaskar, M.S.; Padmanaban, S.; Kulkarni, R.; Blaabjerg, F.; Seshagiri, S.; Hajizadeh, A. Novel LY converter topologies for high gain transfer ratio—A new breed of XY family. In Proceedings of the 4th IET Clean Energy and Technology Conference (CEAT 2016), Kuala Lumpur, Malaysia, 14–15 November 2016.

27. Maalandish, M.; Hosseini, S.H.; Ghasemzadeh, S.; Babaei, E.; Alishah, R.S.; Jalilzadeh, T. Six-phase interleaved boost dc/dc converter with high-voltage gain and reduced voltage stress. *IET Power Electron.* **2017**, *10*, 1904–1914. [CrossRef]

28. Wang, H.; Li, Z. A PWM LLC type resonant converter adapted to wide output range in PEV charging applications. *IEEE Trans. Power Electron.* **2017**, *33*, 3791–3801. [CrossRef]

29. Leyva-Ramos, J.; Mota-Varona, R.; Ortiz-Lopez, M.G.; Diaz-Saldierna, L.H.; Langarica-Cordoba, D. Control strategy of a quadratic boost converter with voltage multiplier cell for high-voltage gain. *IEEE J. Emerg. Sel. Top. Power Electron.* **2017**, *5*, 1761–1770. [CrossRef]

30. Wu, H.; Zhan, X.; Xing, Y. Interleaved LLC resonant converter with hybrid rectifier and variable-frequency plus phase-shift control for wide output voltage range applications. *IEEE Trans. Power Electron.* **2016**, *32*, 4246–4257. [CrossRef]

31. Malik, M.Z.; Farooq, A.; Ali, A.; Chen, G. A DC-DC boost converter with extended voltage gain. In Proceedings of the MATEC Web of Conferences Singapore. *EDP Sci.* **2016**, *40*, 7001.

32. Nejad, M.L.; Poorali, B.; Adib, E.; Birjandi, A.A.M. New cascade boost converter with reduced losses. *IET Power Electron.* **2016**, *9*, 1213–1219. [CrossRef]

33. Tomaszuk, A.; Krupa, A. Step-up DC/DC converters for photovoltaic applications–theory and performance. *Electr. Rev.* **2013**, *89*, 51–57.

34. Farooq, A.; Malik, Z.; Qu, D.; Sun, Z.; Chen, G. A three-phase interleaved floating output boost converter. *Adv. Mater. Sci. Eng.* **2015**, *2015*, 409674. [CrossRef]

35. Malek, S. A new nonlinear controller for DC-DC boost converter fed DC motor. *Int. J. Power Electron.* **2015**, *7*, 54–71. [CrossRef]

36. Freitas, A.A.A.; Tofoli, F.L.; Júnior, E.M.S.; Daher, S.; Antunes, F.L.M. High-voltage gain dc–dc boost converter with coupled inductors for photovoltaic systems. *IET Power Electron.* **2015**, *8*, 1885–1892. [CrossRef]

37. Rosas-Caro, J.C.; Ramirez, J.M.; Peng, F.Z.; Valderrabano, A. A DC–DC multilevel boost converter. *IET Power Electron.* **2010**, *3*, 129–137. [CrossRef]

38. Sun, J.; Ding, X.; Nakaoka, M.; Takano, H. Series resonant ZCS–PFM DC–DC converter with multistage rectified voltage multiplier and dual-mode PFM control scheme for medical-use high-voltage X-ray power generator. *IEE Proc. Electr. Power Appl.* **2000**, *147*, 527–534. [CrossRef]

39. Davoudi, A.; Jatskevich, J.; Chapman, P.L. Simple method of including conduction losses for average modelling of switched-inductor cells. *Electron. Lett.* **2006**, *42*, 1246–1248. [CrossRef]

Article

A Novel Algorithm for MPPT of an Isolated PV System Using Push Pull Converter with Fuzzy Logic Controller

Tehzeeb-ul Hassan [1], Rabeh Abbassi [2], Houssem Jerbi [3], Kashif Mehmood [4,5,*], Muhammad Faizan Tahir [6], Khalid Mehmood Cheema [4], Rajvikram Madurai Elavarasan [7], Farman Ali [8] and Irfan Ahmad Khan [9,*]

[1] Department of Electrical Engineering, Punjab University, Lahore 54000, Pakistan; tehzibulhasan@gmail.com
[2] Department of Electrical Engineering, College of Engineering, University of Ha'il, Hail 1234, Saudi Arabia; r.abbassi@uoh.edu.sa
[3] Department of Industrial Engineering, College of Engineering, University of Ha'il, Hail 1234, Saudi Arabia; h.jerbi@uoh.edu.sa
[4] School of Electrical Engineering, Southeast University, Nanjing 210000, China; kmcheema@seu.edu.cn
[5] Department of Electrical Engineering, The University of Lahore, Lahore 54000, Pakistan
[6] School of Electric Power, South China University of Technology, Guangzhou 510000, China; epfaizantahir_2k7@mail.scut.edu.cn
[7] Electrical and Automotive Parts Manufacturing Unit, AA Industries, Chennai 600123, India; rajvikram787@gmail.com
[8] Lahore University of Management Sciences, Lahore 54000, Pakistan; farman.ali@lums.edu.pk
[9] Marine Engineering Technology Department in a Joint Appointment with Electrical and Computer Engineering, Texas A&M University, Galveston, TX 77553, USA
*   Correspondence: kashif_mehmood@seu.edu.cn or kashif.mehmood@ee.uol.edu.pk (K.M.); irfankhan@tamu.edu (I.A.K.)

Received: 22 June 2020; Accepted: 30 July 2020; Published: 3 August 2020

Abstract: Photovoltaic (PV) is a highly promising energy source because of its environment friendly property. However, there is an uncertainty present in the modeling of PV modules owing to varying irradiance and temperature. To solve such uncertainty, the fuzzy logic control-based intelligent maximum power point tracking (MPPT) method is observed to be more suitable as compared with conventional algorithms in PV systems. In this paper, an isolated PV system using a push pull converter with the fuzzy logic-based MPPT algorithm is presented. The proposed methodology optimizes the output power of PV modules and achieves isolation with high DC gain. The DC gain is inverted into a single phase AC through a closed loop fuzzy logic inverter with a low pass filter to reduce the total harmonic distortion (THD). Dynamic simulations are developed in Matlab/Simulink by MathWorks under linear loads. The results show that the fuzzy logic algorithms of the proposed system efficiently track the MPPT and present reduced THD.

Keywords: fuzzy logic control; maximum power point tracking; photovoltaic; push pull converter; off-grid voltage source inverter

---

1. Introduction

Energy is the need of the modern world, but its conventional sources are depleting with each passing day, which includes thermal, nuclear, and natural gas [1–3]. These sources are insufficient and not environment friendly [4,5]. Therefore, it is becoming imperative to find alternate sources of energy that can meet the requirements of energy in the future and should be environmentally friendly [6]. High prices of electrical energy from thermal power plants and intermittency of renewable energy sources [2] move extra emphasis towards hydro-thermal scheduling [7]. Additionally, owing to the continuous diminishing of conventional sources [1,2], the active power generation and the control and compensation of reactive power are becoming the focus for the economic dispatch [8–12], and researchers have also found the way to optimally generate, shed, and forecast the electrical power [13–15]. Recently, solar energy has become one of the free, clean, and reliable sources of energy [3,16].

Neeraj et al. employ hybrid neural network and fuzzy-logic control for maximum power point tracking (MPPT) of photovoltaic (PV) [17], while Gul et al. use a fuzzy controller that depends on a distinct combination of inputs and outputs to track MPPT [18]. Ref. [19], ant colony optimization (ACO) algorithm with MPPT is used in case of the hybrid PV-wind system to produce power for rural areas. Moreover, an advanced MPPT method considering temperature variability for a PV system to attain maximum tracking performance is designed by [20]. Therefore, with the passage of time, various MPPT techniques such as incremental conductance (IC), perturb and observe (P&O), and artificial neural networks (ANN) are developed to ensure maximum gain from solar modules [21–23]. Furthermore, in Ref. [24], the center of inertia technique is used to evaluate the performance of an electronic inverter-based PV power system. However, these techniques have shortcomings including an inability to track continuous power and oscillations near the maximum point. Furthermore, conventional techniques are not able to detect the maximum power point accurately when weather conditions change rapidly [25] and computational time is relatively long to calculate the maximum power point. Usually, single switch buck-boost converters are used with these conventional techniques [26]. These converters do not provide the isolation between input and output sides and a high voltage conversion ratio. From the system point of view and utilization of AC loads, the conventional converters produce DC voltages from the PV panels and then invert it into AC through the open loop inverters. These inverters present high total harmonic distortion (THD) [27–30]. Therefore, an active, computationally fast, resilient, and efficient MPPT algorithm is required.

In this paper, a fuzzy logic controller is employed for MPPT to overcome the aforementioned shortcomings by tracking the maximum power point (MPP) in real time. Fuzzy logic based control offers an advantage in that it does not oscillate near the MPP [31,32]. This kind of control is unique for push pull current-fed boost converters where the high frequency transformer is used to provide the galvanic isolation between the input and output side along with a high conversion ratio [33,34]. DC voltage is inverted to AC through a voltage source inverter (VSI) with a fuzzy logic closed loop controller, which improves the power quality of the AC voltage and provides very low THD. In this work, 5% THD is considered, which is tolerable according to the Institute of Electrical and Electronics Engineers (IEEE) standard [35].

The proposed work covers many of the shortcomings mentioned in the introduction section. However, to distinguish the proposed research from the previously published literature, the main contributions of this proposed work are summarized as follows:

1. Design of fuzzy logic based MPPT, which can track the continuous power without oscillations and noise near the maximum point.
2. Implementation of a push pull current-fed boost chopper in which high frequency transformer is used to provide the galvanic isolation between input and output, along with a high conversion ratio.
3. Implementation of a voltage source inverter (VSI) with a fuzzy logic closed loop controller, which improves the power quality of the AC voltage and provides very low THD.
4. Applications of two fuzzy logic controllers (FLCs) are employed in the proposed system and each has its unique fuzzy rule. The first one tracks the MPPT, and the second is used in VSI with a proper designed low pass filter to reduce the THD value.

Furthermore, a comparative section is added at the end, which is based on the literature on the fuzzy logic principle. The comparison includes the methodology used with fuzzy logic, implementation complexity, generalization in terms of symmetrical and asymmetrical membership function, inputs to the membership functions, hardware implementation, noise and oscillations near MPP, THD analysis, and the proper filter design. This comparative analysis also distinguishes the proposed research from the previously published literature. As a result, the proposed work has advantages in term of simple, accurate, and faster convergence to the operating point with minimum noise and THD levels.

This paper is organized as follows. An equivalent circuit for an individual PV cell based on a single diode model is presented in Section 2 that serves as a basis for MPPT and defines the proposed methodology. The push pull converter and its design aspects are explained in Section 3, while the fuzzy logic based MPPT algorithm is described in Section 4. Furthermore, the push pull converter, VSI, and low pass filter design are explained in Section 4. Simulation results and discussions are in Section 5 and, finally, the conclusions are drawn in Section 6.

2. Proposed Work Methodology

Before discussing the proposed work topology and fuzzy logic control, it is important to elaborate on the equivalent solar circuit.

2.1. Solar Cell Equivalent Circuit

The equivalent circuit of the solar cell is presented in Figure 1 and indicates a current source is connected with parallel diode. Here, R_{se} is a series resistance, R_p is connected in parallel, while R_L is the load resistance. The reverse saturation current of the diode is I_s. The resistance R_P is very high compared with R_{se}. The diode anode current and V_{PV} can be obtained as [36] by applying Kirchhoff current law (KCL) to the solar cell equivalent circuit:

$$I_{Source} - I_d - \frac{V_d}{R_p} - I_{pv} = 0 \tag{1}$$

$$I_d = I_s \left(e^{\frac{qV_d}{NKT}} - 1 \right) \tag{2}$$

$$V_{PV} = V_d - R_{se}I_{PV} \tag{3}$$

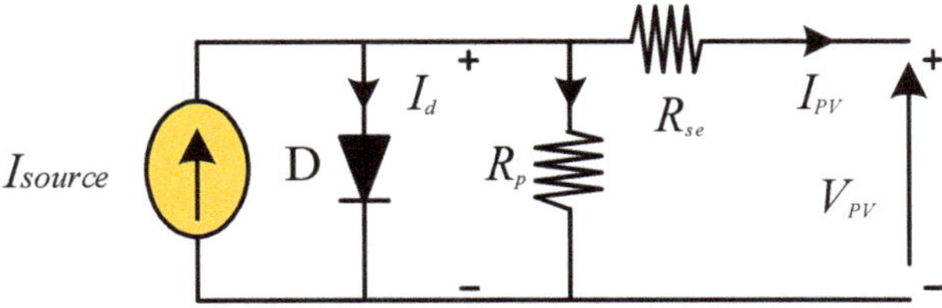

Figure 1. Equivalent circuit of practical single diode model solar cell.

2.2. Proposed Topology

An isolated photovoltaic system is designed for a 500 W solar panel with a fuzzy logic MPPT algorithm. Fuzzy logic closed loop voltage source inverter and low pass filter are employed. Figure 2 shows the proposed system components. Both the voltage and current of the solar PV array are measured to calculate the power. On the basis of the present and previous values of power and current, error and change in error are computed for the fuzzy logic controller that gives the fuzzy rules. On the basis of these fuzzy rules, the fuzzy logic controller sets the duty ratio for pulse width modulation (PWM), 40 kHz triangular wave is compared with the fuzzy logic controller output. The PWM generator produces the switching signal for push pull boost converter switches (which are insulated-gate bipolar transistors (IGBTs)). The push pull boost converter boosts the 60 V DC to 340 V DC. Then, the DC voltages are inverted into 220 V AC through voltage source inverter (VSI). The low pass filter removes the high frequency harmonic content. The switching of VSI is performed through the unipolar sinusoidal pulse width modulation (USPWM) generator and the unipolar switching technique is employed to mitigate the low order harmonic contents.

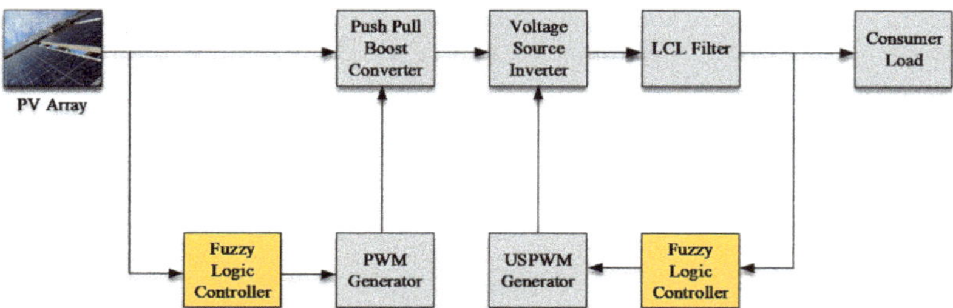

Figure 2. Block diagram of the proposed methodology. USPWM, unipolar sinusoidal pulse width modulation; PV, photovoltaic.

2.3. Fuzzy Logic Control

In the fuzzy logic MPPT algorithm, voltage and current at each instant k are sensed to calculate the active power [31]. The active power is then compared with the power at last instant $k-1$ to obtain the change in power ($\Delta P(k)$). Similarly, the current at instant k is compared with the current at instant $k-1$ to achieve the current error ($\Delta I(k)$). Afterwards, the power error is divided by the current error to achieve the error (e), which is compared with the previous error to calculate the change in error ($\Delta e(k)$) as in Equations (4) and (5), respectively. In this way, error $e(k)$ and $\Delta e(k)$ become the crisp inputs of the fuzzy logic controllers. The flow chart for fuzzy logic MPPT is shown in Figure 3.

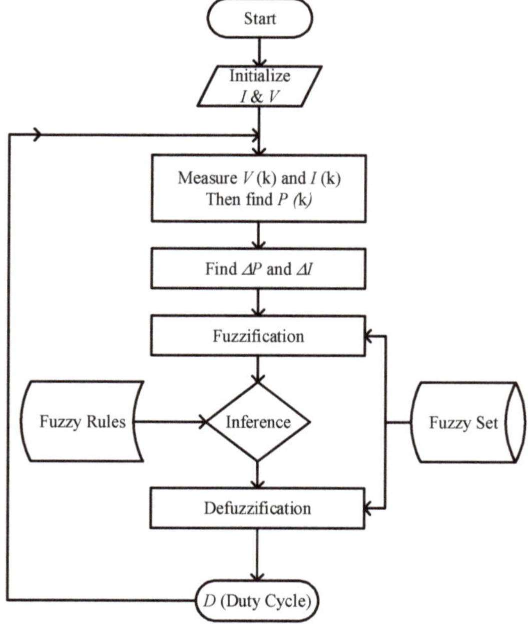

Figure 3. Fuzzy logic maximum power point tracking (MPPT) flow chart.

In this work, Mamdani inference technique, A-type membership functions, and 49-element rule base were used for the fuzzy logic control because of the fact that Mamdani inference technique is efficient and straightforward in defining the fuzzy output sets and is more popular among researchers than other inference techniques [37]. The A-type or triangular type membership function is used because it has fewer complexities when splitting values (low, med, and high (Membership Function) MF) comparing other membership functions. Moreover, it was observed that the triangle membership function gives the faster response and less overshoot than others [38]. The 49-element rule base was employed because it exhibits good performance [39,40].

$$e(k) = \frac{\Delta P(k)}{\Delta I(k)} = \frac{P(k) - P(k-1)}{I(k) - I(k-1)} \tag{4}$$

$$\Delta e(k) = e(k) - e(k-1) \tag{5}$$

Generally, a fuzzy logic controller consist of three components: (i) fuzzifier, (ii) inference, and (iii) defuzzifier [41]. Each component is individually described below.

2.3.1. Fuzzifier

This component of the fuzzy logic controller receives the data from the input and analyzes them according to the user user-defined chart called membership function. Fuzzifier receives the data in the non-linear form and assigns them grade from 0 to 1. Membership functions have different shapes. These shapes depend on the type of data, and but the common shapes are S, π, A, and Z [30]. 'A' shape has been used in this work for fuzzification operation.

2.3.2. Interference

The inference system consists of a fuzzy rule plays an important role in representing the expert control or modeling knowledge between the input and output side. In the literature, different techniques are used in the inference system. Mamdani inference technique with the fuzzy rule is employed in this work. If-then else statements are used in the system for fuzzy inference [42]. For example, we consider a simple two-input one-output example that has three fuzzy rules.

Rule (1)   IF X is A_2 OR Y is B_1 Then Z is C_1
Rule (2)   IF X is A_2 AND Y is B_2 Then Z is C_3
Rule (3)   IF X is A_1 Then Z is C_3

The fuzzy logic membership functions designer, fuzzy logic rule editor, fuzzy logic rules, fuzzy logic member ship function input error, change in error, and membership function output are shown below in Figures 4–7.

The following are the fuzzy rules in Table 1, which are used for desired MPP of push pull converter PWM.

Table 1. Fuzzy logic rules for the push pull converter. NB, negative big; NM, negative medium; NS, negative small; ZE, zero; PS, positive small; PM, positive medium; PB, positive big.

Input		E						
		NB	**NM**	**NS**	**ZE**	**PS**	**PM**	**PB**
	NB	ZE	ZE	ZE	NB	NB	NB	NM
	NM	ZE	ZE	ZE	NS	NM	NM	NM
	NS	NS	ZE	ZE	ZE	NS	NS	NS
ΔE	**ZE**	NM	NS	ZE	ZE	ZE	PS	PM
	PS	PS	PM	PM	PS	ZE	ZE	ZE
	PM	PM	PM	PM	ZE	ZE	ZE	ZE
	PB	PB	PB	PB	ZE	ZE	ZE	ZE

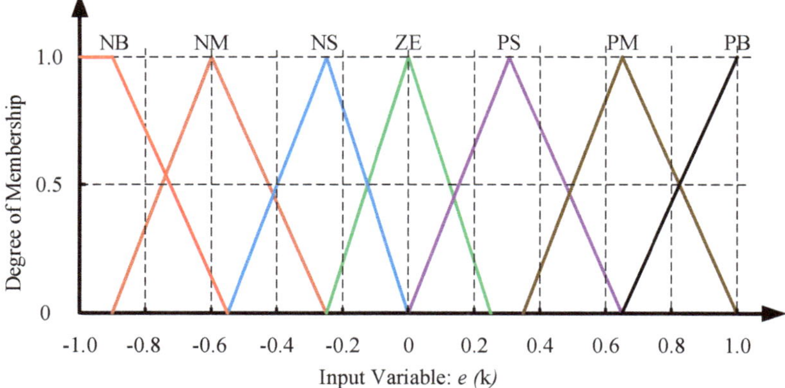

Figure 4. Membership function input error. NB, negative big; NM, negative medium; NS, negative small; ZE, zero; PS, positive small; PM, positive medium; PB, positive big.

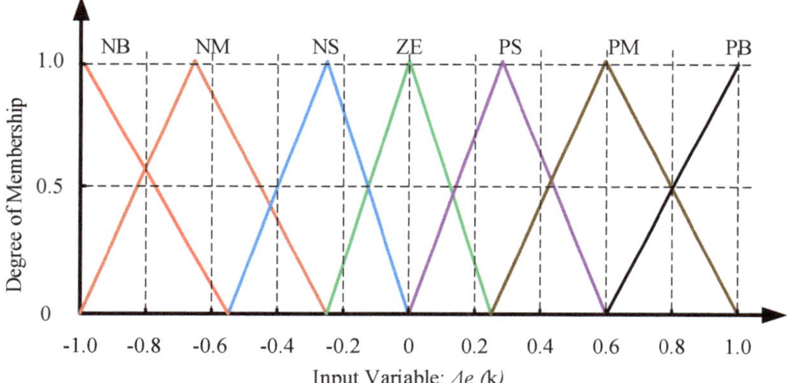

Figure 5. Membership function change in error.

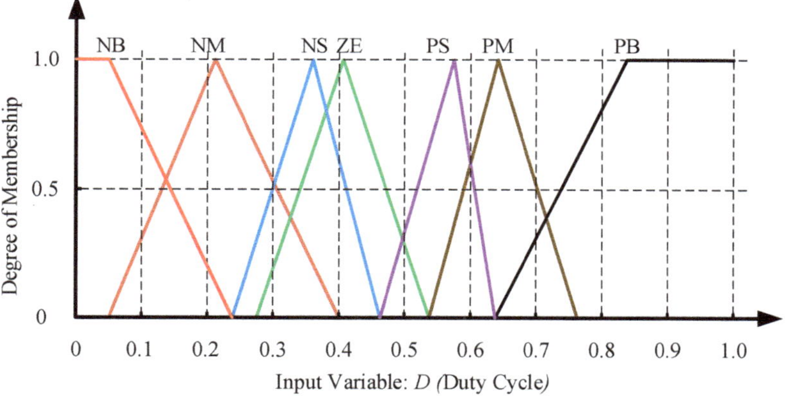

Figure 6. Membership function output.

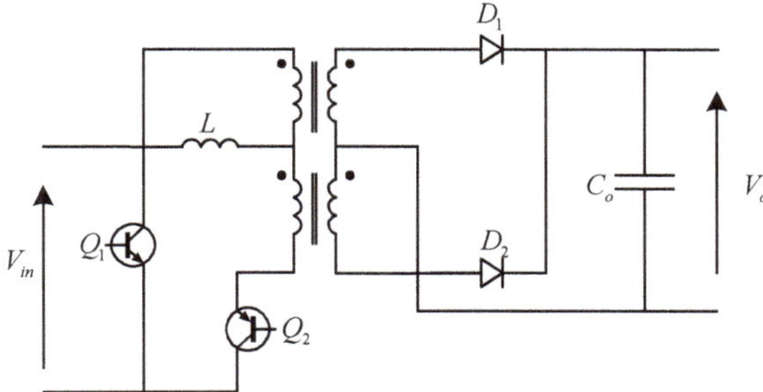

Figure 7. Circuit diagram for the current fed push pull boost converter.

2.3.3. Defuzzification

In the defuzzification process, fuzzy logic controllers use the fuzzy rules to obtain the output value. This output value of the fuzzy logic controller depends upon the method of defuzzification. Hence, it is the gained value of a fuzzy logic controller with respect to the label value in the fuzzy logic membership function. Seven fuzzy membership functions were used in this research, as enlisted in Table 1. These seven functions are negative big (NB), negative medium (NM), negative small (NS), zero (ZE), positive small (PS), positive medium (PM), and positive big (PB). During the defuzzification, the FLC converts the fuzzy logic value into a data value. Numerous methods are available for the defuzzification process such as the average weight (AW) method, center of gravity (COG), mean of maximum (MOM), and smallest of maximum (SOM) [30]. The COG method is used for MPPT in which all fuzzy values are converged at one point [26]. The fuzzy logic rules are used in the push pull boost converter design for MPPT, which is given in Table 1 [42].

3. Push Pull Converter and Its Design Aspects

The push pull converter consists of a centrally tapped transformer, two push pull switches Q_1 and Q_2, series inductor L, two rectifier diodes D_1 and D_2, and a parallel capacitor C_0, as shown in Figure 7. Push pull converter can operate in four states. By applying PWM, the designed inductor and specifications of load ensure that the converter always operates in the continuous conduction mode (CCM) [36,42].

A designed current fed DC–DC push pull boost converter is shown in Figure 7 and its operating characteristics are given in Table 2.

Table 2. Design parameters of the push pull converter. PWM, pulse width modulation.

Parameters	Values
Power	500 W
Input voltage	60 V
Output voltage	340 V
Turn ratio (n)	1:6
Switching frequency	40 kHz
Duty cycle	0.0523
Inductor	518 µH
Output capacitor	100 µF
Input Inductor current	9.38 A
PWM switching frequency	10 KHz
Input DC voltage	340 V
Output voltage	V rms
Resistive load	100 Ω

This circuit consists of a center tap transformer, two push pull switches Q_1 and Q_2, series inductor L, two rectifier diodes D_1 and D_2, and a parallel capacitor with the output load.

The turn ratio of the transformer is calculated as follows:

$$n = \frac{2V_o(1-D)}{V_{in}} \tag{6}$$

The switches' on and off time are selected as follows:

$$t_{on} = \frac{T}{2} - \left(D - \frac{1}{2}\right) = (1-D)T \tag{7}$$

$$t_{off} = \left(D - \frac{1}{2}\right)T \tag{8}$$

When both switches are in an off state, the voltage across these switches is double. This voltage stress is compensated by selecting a transformer tapping voltage as follows:

$$V_o \cong 1.05 \times V_{in,\text{max}} \tag{9}$$

During the dead time, when both switches are in the off position, the inductor increases in a linear mode of operation.

$$V_{in} = \frac{2L\Delta I}{t_{off}} \tag{10}$$

When only one switch is on, the energy is transferred to the secondary side of the transformer, then

$$V_o = V_{in} + \frac{2L\Delta I}{t_{on}} \tag{11}$$

Hence,

$$V_o = V_{in} + \frac{V_{in} \times t_{off}}{t_{on}} = V_{in} \times \frac{1}{2(1-D)} \tag{12}$$

The input current of the inductor with efficiency η is given by the following:

$$I_i = \frac{p_o}{\eta \times V_d} \tag{13}$$

The inductor size is selected carefully, a very value inductor may cause the converter to operate in the discontinuous mode and very high-value inductor may cause an increase in the size and weight of the converter.

$$\Delta I = x I_i \tag{14}$$

where $0.05 \leq x \leq 0.3$ for optimal operation.

$$V_{in} = V_o \times 2(1-D) = \frac{2L\Delta I}{t_{\;off}} \tag{15}$$

Rearranging the equation

$$L\Delta I = V_o \times 2(1-D) \times \left(D - \frac{1}{2}\right)T \tag{16}$$

Or

$$(\Delta I)_{\text{max}} = \frac{V_{ct}}{16Lf_s} \tag{17}$$

The proposed converter operates in CCM; therefore, minimum inductance is required at the output side, which is calculated as represented in Equation (18).

$$L = \frac{V_o}{16 \times f_s (\Delta I)_{max}}$$ (18)

To operate the converter in CCM, the value of the output side capacitor is calculated as shown in Equation (19).

$$C_0 = \frac{P_o(2D - 1)}{4V_r \times V_o^2 \times f_s}$$ (19)

where V_r is DC ripple voltage, which is 3% allowable.

The proposed converter topology offers the benefit of isolation between the input and output side, maximum efficiency, constant input current, high voltage conversion, the minimum number of switches, simplicity of configuration, and thus low conduction losses. Furthermore, there is no need for a filter capacitor at the input side that makes the system simple and compact [34].

4. VSI and Low Pass Filter Design

In the proposed system, a single-phase full bridge inverter is used to feed the consumer load, which inverts the 340 V DC into 220 V AC at 50 Hz frequency. The unipolar sinusoidal pulse width modulation (USPWM) technique is used to turn on/off the inverter switches. This technique reduces THD and power losses during switching [26]. In USPWM, two control signals are used: a sinusoidal wave and its 180° out of phase version at 50 Hz. These control signals are compared with high frequency triangular carrier signal of 10 kHz. Control signal 1 is compared with the carrier signal, resulting in a logic signal that generates the output voltage between 0 and $+V_{dc}$. Control signal 2 is compared with the carrier signal, resulting in a logic signal that produces the output voltage between 0 and $+V_{dc}$. In every inverter, a filter is necessary for improving power quality. Therefore, a low pass filter is used for smoothing the output current from VSI. The LCL filter is shown in Figure 8.

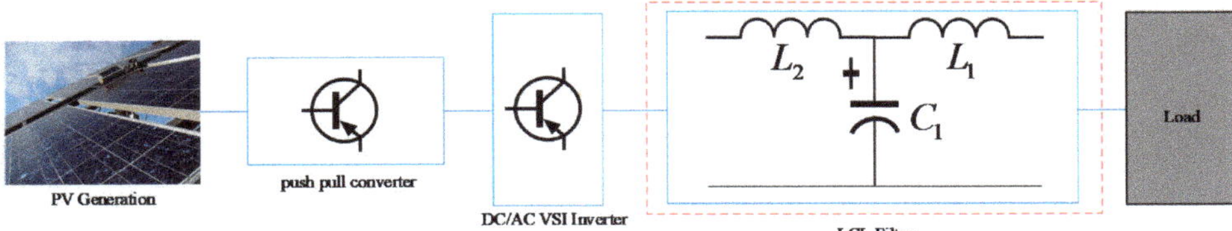

Figure 8. Application of LCL filter in PV system.

The components of the filter are obtained as represented in Equations (20)–(25).

The filter value refers to base impedence

$$Z_b = \frac{E\,n^2}{P_n} = \frac{V_0^2}{P_n}$$ (20)

where Z_b is the base impedance and V_0 and P_n are the output voltage and power of the inverter, respectively. The maximum power variation is considered as 5% and the base impedance is adjusted as computed in Equation (21):

$$C_f = 0.05\,C_b$$ (21)

where L_1 is inductance on inverter side. For 10% ripple, L_1 is calculated by considering the rated current of the inductor.

$$\Delta I_{Lmax} = 0.1\,I_{max}$$ (22)

Further, $I_{\max}$ is calculated as shown in Equation (23):

$$I_{\max} = \frac{P_n}{V_o} \tag{23}$$

L_1 is calculated as represented in Equation (24):

$$L_1 = \frac{V_{dc}}{16\, f_s\, \Delta I_{L\max}} \tag{24}$$

where f_s is switching frequency and L_2 is calculated as follows:

$$L_2 = \frac{\sqrt{\frac{1}{K_a^2} + 1}}{C_f\, \omega_s^2} \tag{25}$$

where K_a is the attenuation factor and is taken as $K_a = 0.2$, while $\omega_s = 2\pi f_s$ is angular switching frequency [41,43].

For LCL filter, derived parameters are $L_1 = 9.3$ mH, $L_2 = 37.5$ mH, and $C = 1.6$ µF with the unity power factor. The LCL filter designing algorithm is shown in Figure 9. After the filtration, (root mean square) RMS output voltages are sensed for the fuzzy logic controller input. The output of the controller is compared with the sinusoidal AC voltages at the fundamental frequency, and then PWM set the duty of VSI.

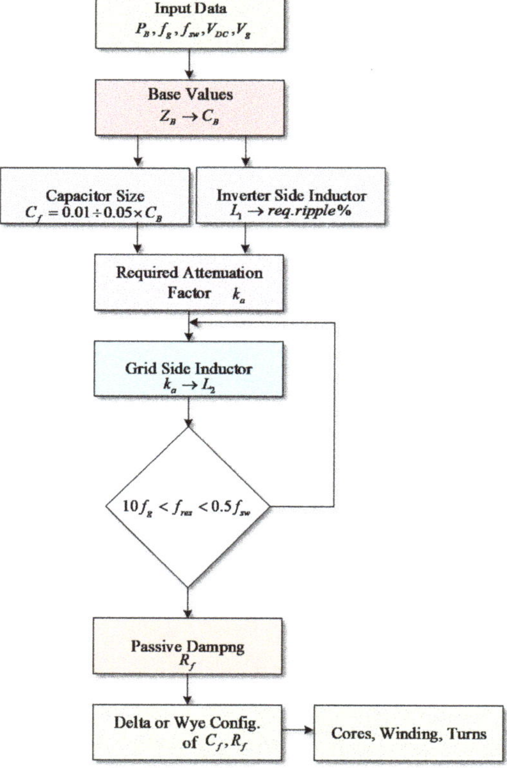

Figure 9. LCL filter design algorithm.

The fuzzy logic structure of VSI is the same as the MPP push pull converter; however, the fuzzy rules used in FLC of VSI are listed in Table 3.

Table 3. Fuzzy logic rules for the voltage source inverter (VSI).

Input		E						
		NB	NM	NS	ZE	PS	PM	PB
	NB	NB	NB	NB	NB	NM	NS	ZE
	NM	NB	NB	NB	NM	NS	ZE	PS
	NS	NB	NB	NM	NS	ZE	PS	PM
ΔE	ZE	NB	NM	NS	ZE	PS	PM	PB
	PS	NM	NS	ZE	PS	PM	PB	PB
	PM	NS	ZE	PS	PM	PB	PB	PB
	PB	ZE	PS	PM	PB	PB	PB	PB

5. Results and Discussion

The proposed isolated photovoltaic system with a fuzzy logic controller, the current fed push pull DC–DC boost converter, is presented here. The DC–DC boost converter operates in continuous conduction mode, and the voltage source inverter with fuzzy logic closed loop and low pass filter are simulated in Matlab/Simulink. The parameters of the employed PV array (Canadian solar CS5P 250-M) are given in Table 4. The performance of the developed system is tested at different irradiance intensity at 25 °C and a linear load of 200 Ω, as shown in Figure 10. Voltage and current are sensed to calculate the power.

Table 4. Parameters of photovoltaic (PV) array.

PV Array	Parameters
No. of Cells and Connections	96
Open Circuit Voltage	59.4 V
Maximum Power Voltage	48.7 V
Short Circuit Current	5.49 A
Maximum Power Current	5.14 A
Maximum Power	250.318 W
Diode saturation current	2.9177×10^{-11}
Diode ideality factor	0.93246
Shunt resistance	428.442 Ohm

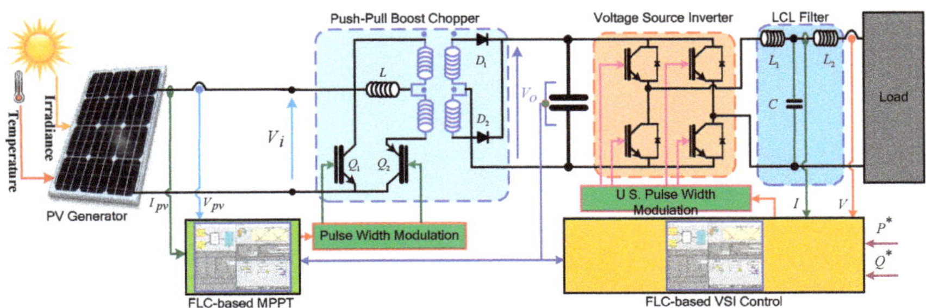

Figure 10. Schematic of the proposed control architecture of the studied system. VSI, voltage source inverter; FLC, fuzzy logic controller.

In the first case, the system is simulated at a constant temperature of 25 °C and constant irradiance 1000 W/m². It tracks the maximum power 250 W in a very small amount of time, approximately 0.005 s, as shown in Figure 11.

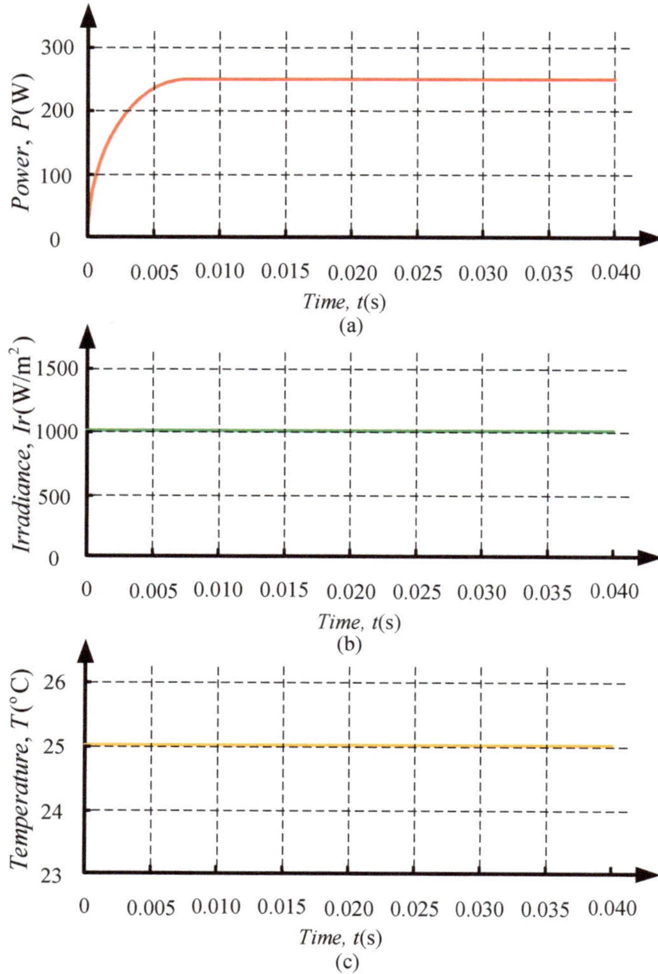

Figure 11. Output: (**a**) Power verses time at; (**b**) constant irradiance; (**c**) constant temperature.

In the second case, the system is simulated for various irradiance levels as follows: 800 W/m² for 0.015 s, 600 W/m² for 0.03 s, and 1000 W/m² for the rest of the time. In this scenario, it again tracks the maximum power point within the same designed spam of time (0.005 s) and gives the power of 200 W, 150 W, and 250 W, respectively, as shown in Figure 12. This power is tracked through the fuzzy logic controller, where fuzzy rules are employed as given in Table 1.

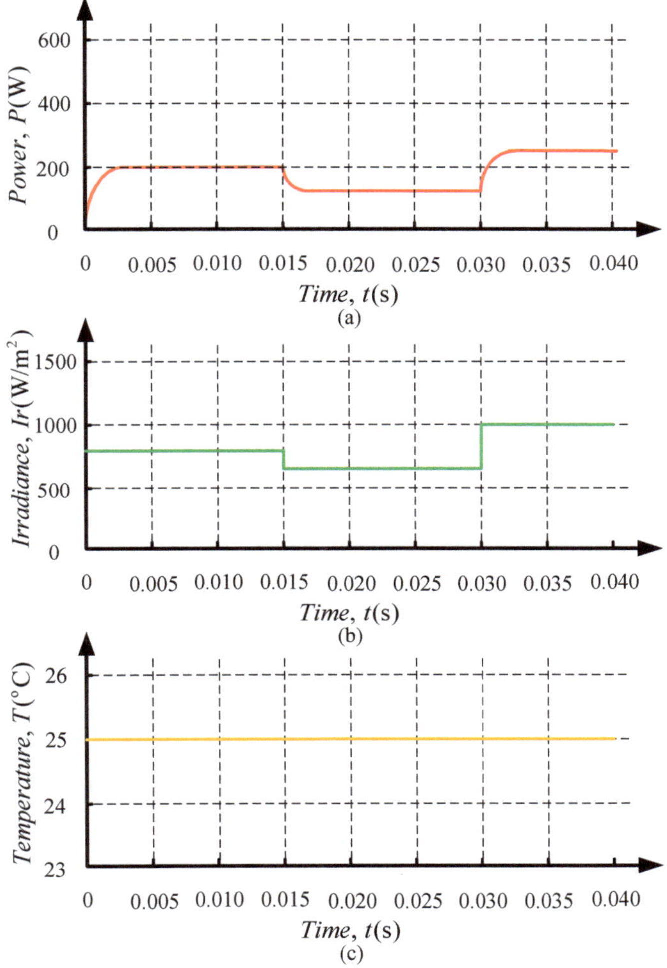

Figure 12. Output: (**a**) Power verses time at; (**b**) different irradiance and; (**c**) constant temperature.

On the basis of the designed fuzzy rules, the fuzzy logic controllers generate the fuzzy logic PWM and decide what would be the duty ratio of the push pull boost converter switches shown in Figure 13a. The push pull boost converter operates in four states. In state 1, when the switch Q_1 is on, then the inductor would discharge, and output voltage would be positive. In state 2, switches Q_1 and Q_2 are ON simultaneously. During state 2, the inductor is charged, as shown in Figure 13c, that is, 4.3 A current flows, and the output voltage would be zero because the flux generated in both windings cancels each other out. In this state, the output capacitor provides the voltage to the load, which means that the capacitor would be discharged. In state 3, when the switch Q_2 is ON, then the voltage would be negative. Similarly, in state 4, both switches are ON simultaneously for zero output voltages, as shown in Figure 13b, that is, modified sine wave voltages, which are converted into 340 V DC through the rectifier diodes D_1 and D_2, as shown in Figure 13d.

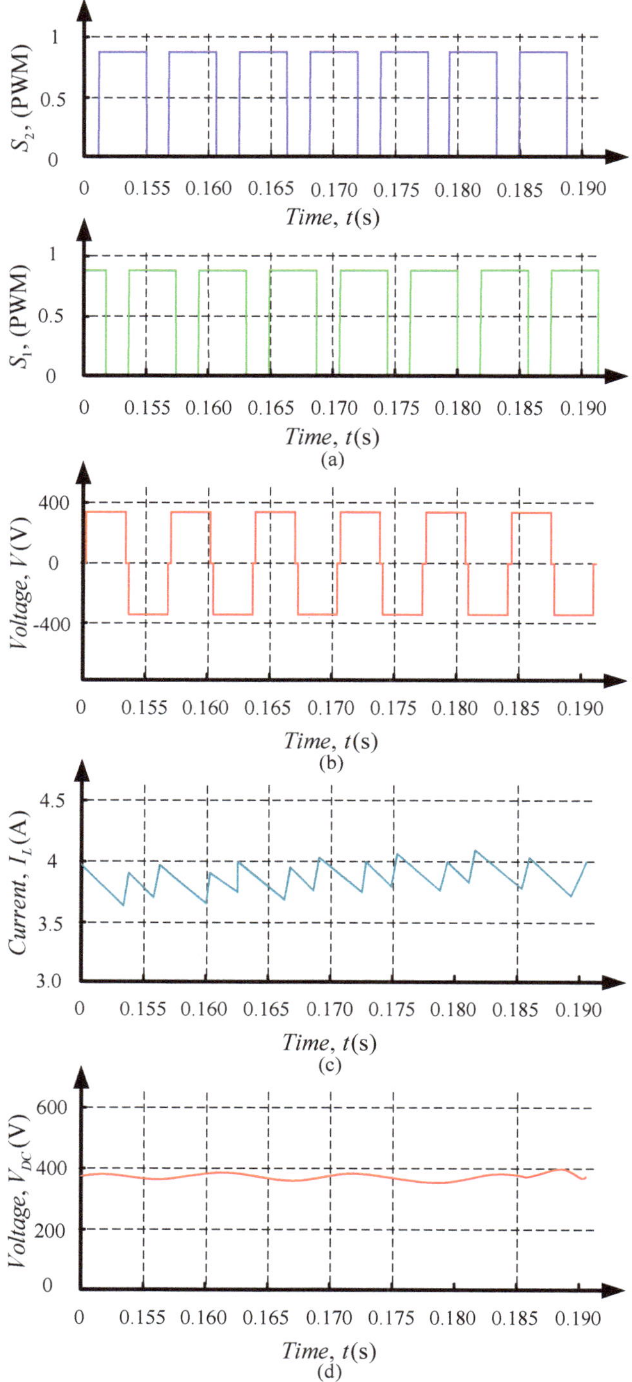

Figure 13. Output: (**a**) Fuzzy logic PWM; (**b**) transformer output; (**c**) inductor current; (**d**) DC output voltage.

The DC voltages are inverted into 220 V AC through the VSI. These voltages are sensed through the voltage sensor and compared with the sinusoidal AC voltages; error and change in error are calculated for fuzzy FLC. This controller generates the reference signal for the PWM generator, which operates the inverter switches and is operated according to the fuzzy logic controller. The PWM of the inverter is shown in Figure 14a. FLC generates the reference signal by using the fuzzy rules of Table 3, settled down by removing the lower order harmonic content in AC voltages. However, these voltages still have the higher-order harmonic content shown in Figure 14b. The higher-order harmonic contents are removed through the low pass filter. The output voltages and current of inverter after removing the higher-order harmonic content are shown in Figure 15, where Figure 15a presents the AC voltages and Figure 15b shows the AC current, which is 1.5 A.

To check the quality of output voltages and current, the fast Fourier transform (FFT) analysis is also carried out and obtained THD are shown in Figures 16 and 17. The FFT analysis of the proposed algorithm provides only 1.41% THD for output voltage and current at 50 Hz.

To prove the validity of the conducted research, a comparison between the results of the fuzzy logic-based MPPT algorithm is compared with P&O and incremental conductance algorithms available in the literature, in the time domain function at irradiance 1000 W/m^2 and at 25 °C, as listed in Table 5.

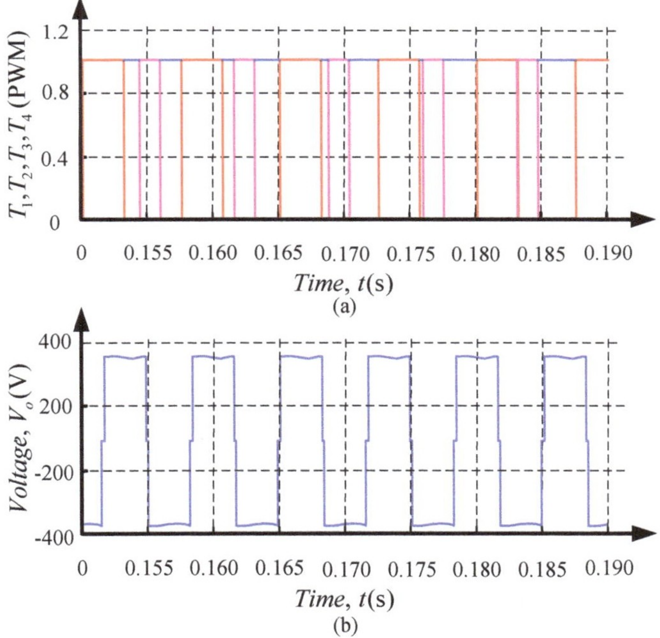

Figure 14. Output: (**a**) Inverter fuzzy logic PWM; (**b**) output voltage before filter.

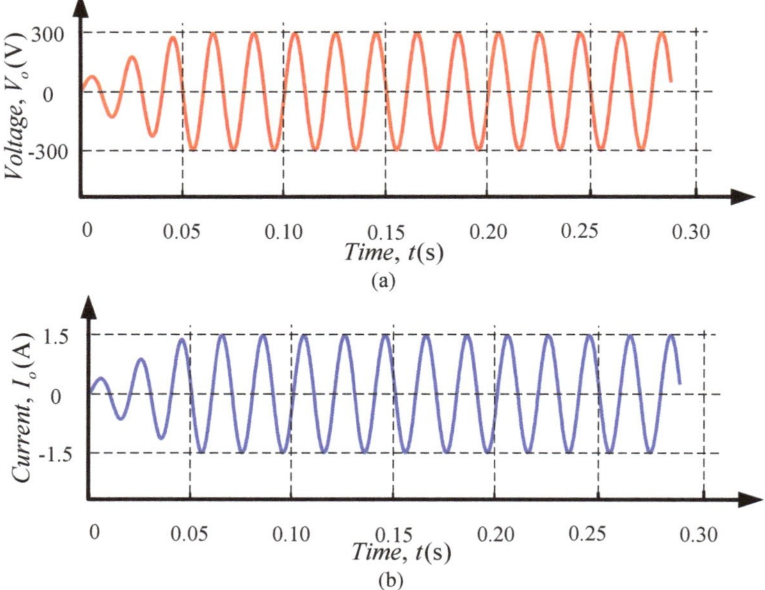

Figure 15. Output: (a) Inverter output voltage; (b) output current.

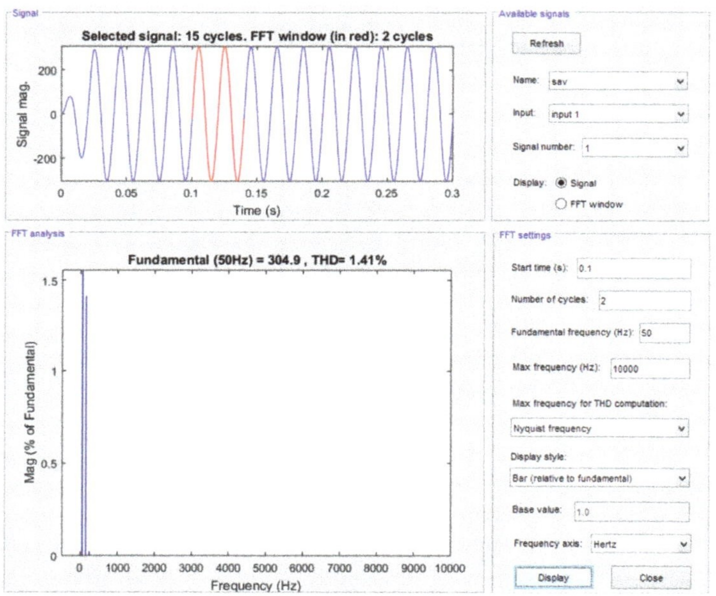

Figure 16. Fast Fourier transform (FFT) analysis of output voltage. THD, total harmonic distortion.

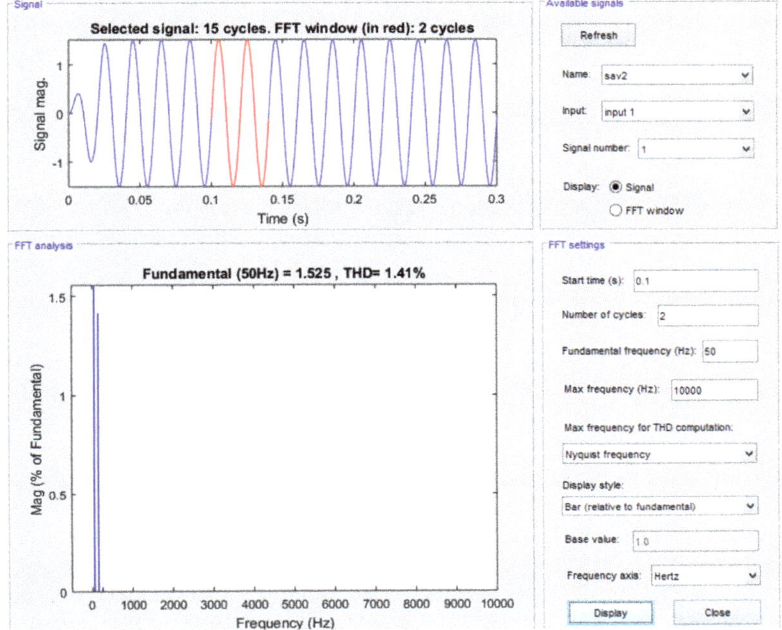

Figure 17. FFT analysis of output current.

Table 5. Tracking time performance comparison [44]. P&O, perturb and observe; INC, incremental conductance.

MPPT Algorithms	Tracking Time of PV Power
P&O	0.300 s
INC	0.250 s
Fuzzy Logic	0.005 s

Comparative Analysis

In this section, a comparative analysis is performed, which is represented in Table 6. The comparative analysis is based on the literature on the fuzzy logic principle. The comparison includes the methodology used with fuzzy logic, implementation complexity, generalization in terms of symmetrical and asymmetrical membership function, inputs to the membership functions, hardware implementation, noise and oscillations near MPP, THD analysis, and the proper filter design. This comparative analysis also distinguishes the proposed research from the previous published literature. As a result, the proposed work has advantages in terms of simple, accurate, and faster convergence to the operating point with minimum noise and THD.

Table 6. Comparative analysis of fuzzy logic based MPPT.

Ref.	Methodology	Implementation	Generalization	Input MFs	Hardware	Noise and Oscillations Near MPP	THD	Proper Filter Design
Proposed	Dual FL based MPPT with PPC	Simple	Sym./Asym. membership	$\Delta P/\Delta I$	✗	✗	✓	✓
[45]	AFL based MPPT	Simple	Sym./Asym. membership	$\Delta(\Delta P/\Delta I)$	✓	✓	✗	✗
[46]	FL based MPPT	Complex	Asym. membership	$\Delta P/\Delta V$, $\Delta(\Delta P/\Delta V)$	✓	✓	✗	✗
[32]	FL based MPPT with PSO	Complex	Asym. membership	$\Delta P, \Delta V$	✓	✓	✗	✗
[47]	FL based MPPT	Complex	Sym. membership	$\Delta P, \Delta V$	✓	✓	✗	✗
[48]	FL based MPPT	Simple	Sym./Asym. membership	$\Delta P/\Delta t$, $\Delta V/\Delta t$	✗	✓	✗	✗
[49]	FL based MPPT	Simple	Asym. membership	$\Delta P/\Delta V$, $\Delta(\Delta P/\Delta V)$	✓	✓	✗	✗
[50]	FL based MPPT	Simple	Sym. membership	$V, \Delta V$	✗	✓	✗	✗
[51]	FL + P&O MPPT	Complex	Sym. membership	$\Delta P, \Delta I$	✓	✓	✗	✗
[52]	FL + HC MPPT	Complex	Asym. membership	$\Delta P/\Delta V$, $\Delta(\Delta P/\Delta V)$	✓	✓	✗	✗
[53]	FL + FO MPPT	Complex	Sym. membership	$\Delta P, \Delta I$	✓	✓	✗	✗

6. Conclusions

In this paper, an off-grid photovoltaic system with a fuzzy logic MPPT-controlled push pull boost converter is designed. The proposed system is simulated in Matlab/Simulink and tested for various weather conditions. The results proved the efficiency of the fuzzy logic algorithm, which ouperforms the conventional algorithms in terms of MPPT accuracy and minimization of fluctuations, regardless of irradiance rapid changes. In addition, the fuzzy logic-based controller designed for the VSI and the adopted LCL filter allow to achieve high performance. The proposed interfacing system between the PV generator and the load is very effective because the provided total harmonic distortion (THD) is 1.41%, which is in agreement with the IEEE standard at the operating frequency.

Author Contributions: Conceptualization, T.-u.H., and F.A.; methodology, F.A.; software, F.A. and K.M.C.; validation, F.A., formal analysis, M.F.T.; investigation, M.F.T.; resources, M.F.T. and K.M.C.; data curation, F.T.; writing—original draft preparation, K.M.; writing—review and editing, K.M., R.A. and H.J.; visualization, I.A.K. and R.M.E.; supervision, T.-u.H.; project administration, T.-u.H.; Funding acquisition, R.A. and H.J. All authors have read and agreed to the published version of the manuscript.

Funding: This research received no external funding.

Conflicts of Interest: The authors declare no conflict of interest.

References

1. Tahir, M.F.; Haoyong, C.; Mehmood, K.; Ali, N.; Bhutto, J.A. Integrated energy system modeling of China for 2020 by incorporating demand response, heat pump and thermal storage. *IEEE Access* **2019**, *7*, 40095–40108. [CrossRef]

2. Tahir, M.F.; Haoyong, C.; Khan, A.; Javed, M.S.; Laraik, N.A.; Mehmood, K. Optimizing size of variable renewable energy sources by incorporating energy storage and demand response. *IEEE Access* **2019**, *7*, 103115–103126. [CrossRef]

3. Cheema, K.M. A comprehensive review of virtual synchronous generator. *Int. J. Electr. Power Energy Syst.* **2020**, *120*, 106006. [CrossRef]

4. Tahmasebi, H. *Boost Integrated High Frequency Isolated Half-Bridge DC-DC Converter: Analysis, Design, Simulation and Experimental Results*; University of Victoria Libraries: Victoria, BC, Canada, 2015.

5. Dolara, A.; Faranda, R.; Leva, S. Energy comparison of seven MPPT techniques for PV systems. *J. Electromagn. Anal. Appl.* **2009**, *2009*, 725. [CrossRef]

6. Tahir, M.F.; Chen, H.; Javed, M.S.; Jameel, I.; Khan, A.; Adnan, S. Integration of different individual heating scenarios and energy storages into hybrid energy system model of China for 2030. *Energies* **2019**, *12*, 2083. [CrossRef]

7. Alquthami, T.; Butt, S.E.; Tahir, M.F.; Mehmood, K. Short-term optimal scheduling of hydro-thermal power plants using artificial bee colony algorithm. *Energy Rep.* **2020**, *6*, 984–992.

8. Khan, I.; Li, Z.; Xu, Y.; Gu, W. Distributed control algorithm for optimal reactive power control in power grids. *Int. J. Electr. Power Energy Syst.* **2016**, *83*, 505–513. [CrossRef]

9. Bhattacharjee, V.; Khan, I. A non-linear convex cost model for economic dispatch in microgrids. *Appl. Energy* **2018**, *222*, 637–648. [CrossRef]

10. Khan, I.; Xu, Y.; Sun, H.; Bhattacharjee, V. Distributed optimal reactive power control of power systems. *IEEE Access* **2017**, *6*, 7100–7111. [CrossRef]

11. Khan, I.; Xu, Y.; Kar, S.; Sun, H. Compressive sensing-based optimal reactive power control of a multi-area power system. *IEEE Access* **2017**, *5*, 23576–23588. [CrossRef]

12. Li, Z.; Mehmood, K.; Zhan, R.; Yang, X.; Qin, Y. Voltage-current double loop control strategy for magnetically controllable reactor based reactive power compensation. In Proceedings of the 2019 IEEE Sustainable Power and Energy Conference (iSPEC), Beijing, China, 20–24 November 2019; pp. 825–830.

13. Mehmood, K.; Hassan, H.T.U.; Raza, A.; Altalbe, A.; Farooq, H. Optimal Power generation in energy-deficient scenarios using bagging ensembles. *IEEE Access* **2019**, *7*, 155917–155929. [CrossRef]

14. FaizanTahir, M. Optimal load shedding using an ensemble of artificial neural networks. *Int. J. Electr. Comput. Eng. Syst.* **2016**, *7*, 39–46.

15. Muhammad Faizan, T.; Chen, H.; Kashif, M.; Noman Ali, L.; Asad, K.; Muhammad Sufyan, J. Short term load forecasting using bootstrap aggregating based ensemble artificial neural network. *Recent Adv. Electr. Electron. Eng.* **2019**, *12*, 1–11. [CrossRef]

16. Li, J.; Wang, H. Maximum power point tracking of photovoltaic generation based on the fuzzy control method. In Proceedings of the 2009 International Conference on Sustainable Power Generation and Supply, Nanjing, China, 6–7 April 2009; pp. 1–6.

17. Priyadarshi, N.; Padmanaban, S.; Mihet-Popa, L.; Blaabjerg, F.; Azam, F. Maximum power point tracking for brushless DC motor-driven photovoltaic pumping systems using a hybrid ANFIS-FLOWER pollination optimization algorithm. *Energies* **2018**, *11*, 1067. [CrossRef]

18. Kebir, T.; Filiz, G.; Larbes, C.; Ilinca, A.; Obeidi, T.; Tchoketch Kebir, S. Study of the intelligent behavior of a maximum photovoltaic energy tracking fuzzy controller. *Energies* **2018**, *11*, 3263. [CrossRef]

19. Priyadarshi, N.; Ramachandaramurthy, V.K.; Padmanaban, S.; Azam, F. An ant colony optimized MPPT for standalone hybrid PV-wind power system with single Cuk converter. *Energies* **2019**, *12*, 167. [CrossRef]

20. Wang, Y.; Yang, Y.; Fang, G.; Zhang, B.; Wen, H.; Tang, H.; Fu, L.; Chen, X. An advanced maximum power point tracking method for photovoltaic systems by using variable universe fuzzy logic control considering temperature variability. *Electronics* **2018**, *7*, 355. [CrossRef]

21. Ma, S.; Chen, M.; Wu, J.; Huo, W.; Huang, L. Augmented nonlinear controller for maximum power-point tracking with artificial neural network in grid-connected photovoltaic systems. *Energies* **2016**, *9*, 1005. [CrossRef]

22. Li, C.; Chen, Y.; Zhou, D.; Liu, J.; Zeng, J. A high-performance adaptive incremental conductance MPPT algorithm for photovoltaic systems. *Energies* **2016**, *9*, 288. [CrossRef]

23. Piegari, L.; Rizzo, R.; Spina, I.; Tricoli, P. Optimized adaptive perturb and observe maximum power point tracking control for photovoltaic generation. *Energies* **2015**, *8*, 3418–3436. [CrossRef]

24. Cheema, K.M.; Mehmood, K. Improved virtual synchronous generator control to analyse and enhance the transient stability of microgrid. *IET Renew. Power Gener.* **2020**, *14*, 495–505. [CrossRef]

25. Yau, H.-T.; Wu, C.-H. Comparison of extremum-seeking control techniques for maximum power point tracking in photovoltaic systems. *Energies* **2011**, *4*, 2180–2195. [CrossRef]

26. Salas, V.; Olias, E.; Barrado, A.; Lazaro, A. Review of the maximum power point tracking algorithms for stand-alone photovoltaic systems. *Sol. Energy Mater. Sol. Cells* **2006**, *90*, 1555–1578. [CrossRef]

27. Srndovic, M.; Familiant, Y.L.; Grandi, G.; Ruderman, A. Time-domain minimization of voltage and current total harmonic distortion for a single-phase multilevel inverter with a staircase modulation. *Energies* **2016**, *9*, 815. [CrossRef]

28. Dolara, A.; Leva, S. Power quality and harmonic analysis of end user devices. *Energies* **2012**, *5*, 5453–5466. [CrossRef]

29. Trubitsyn, A. *High Efficiency DC/AC Power Converter for Photovoltaic Applications*; Massachusetts Institute of Technology: Cambridge, MA, USA, 2010.

30. Sharaf, F.Y. Designing Power Inverter with Minimum Harmonic Distortion Using Fuzzy Logic Control. 2014. Available online: https://195.189.210.17/handle/20.500.12358/18855 (accessed on 15 June 2020).
31. Li, J.; Wang, H. A novel stand-alone PV generation system based on variable step size INC MPPT and SVPWM control. In Proceedings of the 2009 IEEE 6th International Power Electronics and Motion Control Conference, Wuhan, China, 17–20 May 2009; pp. 2155–2160.
32. Cheng, P.-C.; Peng, B.-R.; Liu, Y.-H.; Cheng, Y.-S.; Huang, J.-W. Optimization of a fuzzy-logic-control-based MPPT algorithm using the particle swarm optimization technique. *Energies* **2015**, *8*, 5338–5360. [CrossRef]
33. Maiti, D.; Mondal, N.; Biswas, S. *Design Procedure of a Push Pull Current-Fed DC-DC Converter*; Jadavpur University: Kolkata, India, 2010.
34. Blooming, T.M.; Carnovale, D.J. Application of IEEE Std 519-1992 harmonic limits. In Proceedings of the Conference Record of 2006 Annual Pulp and Paper Industry Technical Conference, Appleton, WI, USA, 18–22 June 2006; pp. 1–9.
35. Adhikari, N.; Singh, B.; Vyas, A.L.; Chandra, A. Analysis and design of isolated solar-PV energy generating system. In Proceedings of the 2011 IEEE Industry Applications Society Annual Meeting, Orlando, FL, USA, 9–13 October 2011; pp. 1–6.
36. Singh, S.; Mathew, L.; Shimi, S. Design and simulation of intelligent control MPPT technique for PV module using MATLAB/SIMSCAPE. *Int. J. Adv. Res. Electr. Electron. Instrum. Eng.* **2013**, *2*, 4554–4566.
37. Wang, C. A Study of Membership Functions on Mamdani-Type Fuzzy Inference System for Industrial Decision-Making. Master's Thesis, Lehigh University, Bethlehem, PA, USA, 2015.
38. Usta, M.A.; Akyazi, Ö.; Altaş, İ.H. Design and performance of solar tracking system with fuzzy logic controller used different membership functions. In Proceedings of the 2011 7th International Conference on Electrical and Electronics Engineering (ELECO), Bursa, Turkey, 1–4 December 2011; pp. II-381–II-385.
39. Mudi, R.K.; Pal, N.R. A robust self-tuning scheme for PI-and PD-type fuzzy controllers. *IEEE Trans. Fuzzy Syst.* **1999**, *7*, 2–16. [CrossRef]
40. Shehata, A.; Metered, H.; Oraby, W.A. Vibration control of active vehicle suspension system using fuzzy logic controller. In *Vibration Engineering and Technology of Machinery*; Springer: Berlin/Heidelberg, Germany, 2015; pp. 389–399.
41. Reznik, A.; Simões, M.G.; Al-Durra, A.; Muyeen, S. $ LCL $ filter design and performance analysis for grid-interconnected systems. *IEEE Trans. Ind. Appl.* **2013**, *50*, 1225–1232. [CrossRef]
42. Mendel, J.M. Fuzzy logic systems for engineering: A tutorial. *Proc. IEEE* **1995**, *83*, 345–377. [CrossRef]
43. Kahlane, A.; Hassaine, L.; Kherchi, M. LCL filter design for photovoltaic grid connected systems. *J. Renew. Energies* **2014**, *2014*, 227–232.
44. Suwannatrai, P.; Liutanakul, P.; Wipasuramonton, P. Maximum power point tracking by incremental conductance method for photovoltaic systems with phase shifted full-bridge dc-dc converter. In Proceedings of the 8th Electrical Engineering/Electronics, Computer, Telecommunications and Information Technology (ECTI) Association of Thailand-Conference 2011, Khon Kaen, Thailand, 17–19 May 2011; pp. 637–640.
45. Rezk, H.; Aly, M.; Al-Dhaifallah, M.; Shoyama, M. Design and hardware implementation of new adaptive fuzzy logic-based MPPT control method for photovoltaic applications. *IEEE Access* **2019**, *7*, 106427–106438. [CrossRef]
46. Liu, C.-L.; Chen, J.-H.; Liu, Y.-H.; Yang, Z.-Z. An asymmetrical fuzzy-logic-control-based MPPT algorithm for photovoltaic systems. *Energies* **2014**, *7*, 2177–2193. [CrossRef]
47. Ozdemir, S.; Altin, N.; Sefa, I. Fuzzy logic based MPPT controller for high conversion ratio quadratic boost converter. *Int. J. Hydrogen Energy* **2017**, *42*, 17748–17759. [CrossRef]
48. Robles Algarín, C.; Taborda Giraldo, J.; Rodríguez Álvarez, O. Fuzzy logic based MPPT controller for a PV system. *Energies* **2017**, *10*, 2036. [CrossRef]
49. El Khateb, A.; Abd Rahim, N.; Selvaraj, J.; Uddin, M.N. Fuzzy-logic-controller-based SEPIC converter for maximum power point tracking. *IEEE Trans. Ind. Appl.* **2014**, *50*, 2349–2358. [CrossRef]
50. Kottas, T.L.; Boutalis, Y.S.; Karlis, A.D. New maximum power point tracker for PV arrays using fuzzy controller in close cooperation with fuzzy cognitive networks. *IEEE Trans. Energy Convers.* **2006**, *21*, 793–803. [CrossRef]
51. Zainuri, M.A.A.M.; Radzi, M.A.M.; Soh, A.C.; Abd Rahim, N. Development of adaptive perturb and observe-fuzzy control maximum power point tracking for photovoltaic boost dc–dc converter. *IET Renew. Power Gener.* **2013**, *8*, 183–194. [CrossRef]

52. Alajmi, B.N.; Ahmed, K.H.; Finney, S.J.; Williams, B.W. Fuzzy-logic-control approach of a modified hill-climbing method for maximum power point in microgrid standalone photovoltaic system. *IEEE Trans. Power Electron.* **2010**, *26*, 1022–1030. [CrossRef]

53. Tang, S.; Sun, Y.; Chen, Y.; Zhao, Y.; Yang, Y.; Szeto, W. An enhanced MPPT method combining fractional-order and fuzzy logic control. *IEEE J. Photovolt.* **2017**, *7*, 640–650. [CrossRef]

Article

Distribution Network Hierarchically Partitioned Optimization Considering Electric Vehicle Orderly Charging with Isolated Bidirectional DC-DC Converter Optimal Efficiency Model

Qiushi Zhang [1], Jian Zhao [1,*], Xiaoyu Wang [1], Li Tong [2], Hang Jiang [2] and Jinhui Zhou [2]

[1] College of Electric Power Engineering, Shanghai University of Electric Power, Shanghai 200090, China; zhangnicper@163.com (Q.Z.); xiaoyuw@ieee.org (X.W.)
[2] State Grid Zhejiang Electric Power Research Institute, Hangzhou 310006, China; tonyhust@126.com (L.T.); jhshiep@163.com (H.J.); zhoujinhui_hz@163.com (J.Z.)
* Correspondence: zhaojian@shiep.educ.cn; Tel.: +86-199-4560-8525

Abstract: The access of large-scale electric vehicles (EVs) will increase the network loss of medium voltage distribution network, which can be alleviated by adjusting the network structure and orderly charging for EVs. However, it is difficult to accurately evaluate the charging efficiency in the orderly charging of electric vehicle (EV), which will cause the scheduling model to be insufficiently accurate. Therefore, this paper proposes an EV double-layer scheduling model based on the isolated bidirectional DC–DC (IBDC) converter optimal efficiency model, and establishes the hierarchical and partitioned optimization model with feeder–branch–load layer. Firstly, based on the actual topology of medium voltage distribution network, a dynamic reconfiguration model between switching stations is established with the goal of load balancing. Secondly, with the goal of minimizing the branch layer network loss, a dynamic reconstruction model under the switch station is established, and the chaotic niche particle swarm optimization is proposed to improve the global search capability and iteration speed. Finally, the power transmission loss model of IBDC converter is established, and the optimal phase shift parameter is determined to formulate the double-layer collaborative optimization operation strategy of electric vehicles. The example verifies that the above model can improve the system load balancing degree and reduce the operation loss of medium voltage distribution network.

Keywords: medium voltage distribution network; switch station; electric vehicle; DC–DC converter; reconfiguration; orderly charging

Citation: Zhang, Q.; Zhao, J.; Wang, X.; Tong, L.; Jiang, H.; Zhou, J. Distribution Network Hierarchically Partitioned Optimization Considering Electric Vehicle Orderly Charging with Isolated Bidirectional DC-DC Converter Optimal Efficiency Model. *Energies* **2021**, *14*, 1614. https://doi.org/10.3390/en14061614

Academic Editor: Irfan Ahmad Khan

Received: 4 February 2021
Accepted: 11 March 2021
Published: 14 March 2021

Publisher's Note: MDPI stays neutral with regard to jurisdictional claims in published maps and institutional affiliations.

1. Introduction

In recent years, due to the large-scale access of distributed new energy sources and electric vehicles (EVs), the economy and reliability of the distribution network have been severely challenged [1,2]; especially with the surge of electric vehicle (EV) users, disorderly charging behavior will aggravate the imbalance of the distribution network load [3]. Therefore, there are usually two solutions to the above problems. The first is distribution network reconstruction [4,5], that is, the topology of the distribution network is adjusted, and then the power flow direction is adjusted by changing the closing and opening of switches. The other is optimal scheduling for controllable load.

Nowadays, relying on the rapid development of information collection, communication, and processing technology, the active distribution network can collect a large amount of data to provide a data basis for the distribution network reconfiguration plan. Thus, by changing the switch combination state, load balance can be achieved, system loss can be reduced, and the economic reliability of the distribution network can be improved [6,7]. Taking into account the temporal and spatial characteristics of loads such as EVs, and the uncertainty of the charging time, it is necessary to dynamically adjust the switch state to adapt to the impact of this uncertainty [8]. Some articles have researched dynamic reconstruction [9–11]. Furthermore, in the actual distribution network, the switching station is a facility that connects or cuts the user's electrical equipment selectively through the switching device, which is usually taken as the research object. The topological structure under the switch station is also complicated. There may be a contact relationship between the ring network cabinets under a certain switch station, and there may also be a contact relationship between the ring network cabinets under different switch stations [12]. Few people have established an optimization model under the switch station. If the topology reconstruction of the branch layer under the switch station is not considered, the line loss problem under the switch station is still not solved. Therefore, not only the reconstruction between the switch stations, but also the reconstruction under the switch station must be considered.

On the other hand, after the optimization of topology is completed, orderly charging optimization scheduling for EVs can be performed to further optimize the line loss in the branch layer [13–15]. However, in the process of dispatching EVs, there are two problems that need to be solved urgently. The first one is the optimization of the dispatching method. The traditional method is mostly direct dispatch, that is, the dispatching center directly dispatches all EVs under the transformer area. Considering the increasing number of EVs in the future, the solution dimension and difficulty of this method gradually increase. Therefore, a new scheduling method should be established to reduce the difficulty of solving [16,17]. The second one is the determination of charging efficiency parameters. The charging efficiency is affected by the converter and heat dissipation. The loss of an isolated bidirectional DC–DC (IBDC) converter under dual-phase-shift (DPS) control mainly includes switching loss [18], on-state loss, copper loss, and iron loss caused [19]. Previous studies only roughly estimated the charging efficiency. Modeling and analysis of these main factors are required to achieve the optimal charging efficiency. Once the optimal charging efficiency is determined, an orderly charging optimization model can be established more accurately.

Therefore, this paper establishes a power transmission loss model in the IBDC converter, and the loss is minimized by adjusting parameters. Then, this paper calculates the optimal charging efficiency and applies it to the orderly charging model. Prior to this, this paper proposes a method of topology dynamic reconstruction, which not only optimizes the switch state between switch stations, but also optimizes the specific topology structure under the switch station. The main contributions of this paper can be summarized as follows:

1. Compared with most previous researches on dynamic reconstruction, this paper takes the switch station as the research object. This paper not only establishes a dynamic reconfiguration model between switch stations, but also proposes a dynamic reconfiguration model under a certain switch station. Especially, in the branch layer optimization, the chaotic niche particle swarm optimization (CNPSO) is proposed to speed up the solution convergence speed and prevent falling into the local optimum.
2. In order to reduce the solving difficulty of dispatch, this paper proposes a double-layer distributed optimization scheduling model. Specifically, multiple aggregators are set under a switch station, and the multi-level information interaction mechanism of network–aggregator–vehicle is established to formulate the charging strategy of electric vehicles.
3. This paper innovatively proposes an orderly charging model for electric vehicles considering isolated bidirectional DC–DC converter optimal efficiency model. Specifically, the power loss model of the IBDC converter is established to determine the optimal shift ratio parameter for a given transmission power, and the optimal efficiency is applied to the ordered charging model.

The rest of this paper is organized as follows: In Section 2, medium voltage distribution network dynamic reconfiguration model with EV is established. Specifically, a dynamic reconstruction model between switch stations is established and an internal dynamic reconfiguration model under a switch station is established. Subsequently, not only the principle of dual phase shift control is analyzed, but the transmission power loss model is established in Section 3. Furthermore, the orderly charging model of EV considering IBDC converter optimal efficiency is set up in Section 4. Simulation analysis is implemented in Section 5, and some useful conclusions are finally drawn in Section 6.

2. Distribution Network Dynamic Reconfiguration Model with EV

In Figure 1, A~N represent the switch station. The reconstruction between switch stations described in this paper means to change the state of disconnect switches and tie switches in the network to achieve regional load balance. The lower part of the arrow represents the location information of the specific load node under the switch station. The ring network cabinet adopts the interval power supply mode, and its branches can be directly connected to the distribution transformer, that is, directly supply power to the low voltage transformer area, and can also be connected to the ring network cabinet for external distribution.

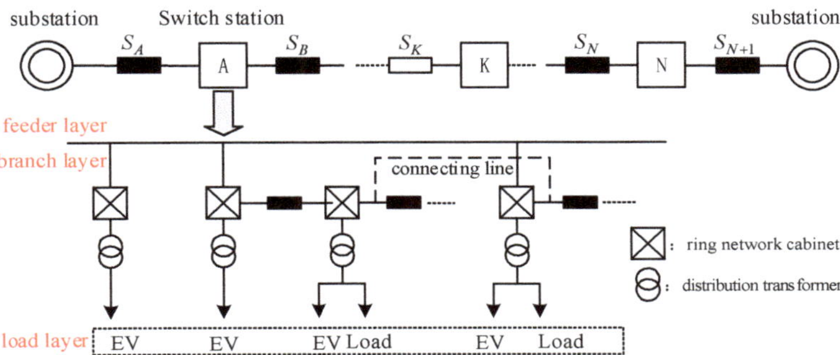

Figure 1. Topological structure of medium-voltage distribution network with EV.

The feeder voltage level in the structure diagram described in Figure 1 is 10 kV. Different numbers of EVs are installed under different distribution transformers. In view of the large number of electric vehicles in the future, this paper gives priority to optimizing the topological structure, and then sets up multiple aggregators under the switch station to guide the charging time of a specific electric vehicle.

In this paper, the DC charging pile is used to charge the electric vehicle. The control panel in the charging pile is used to collect the battery capacity of the electric vehicle and upload the next day's travel demand. The charging module uses IBDC converter to supply power to the high voltage distribution box in the vehicle. The structure of specific load layer is shown in Figure 2.

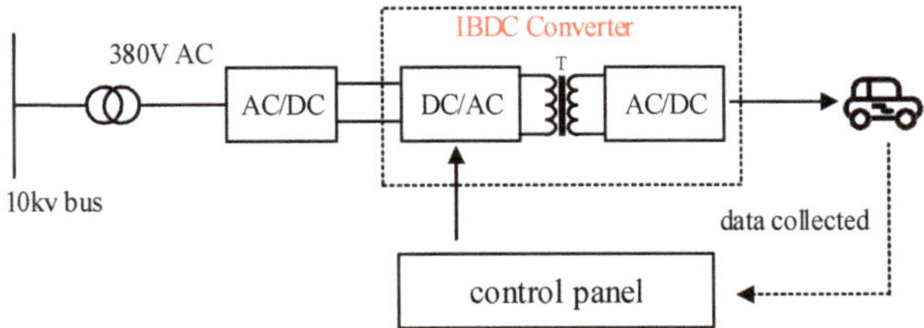

Figure 2. Electric vehicle charging structure diagram.

The driving circuit acts on the power switch to convert the DC voltage after rectifier filtering into AC voltage. Then, the AC voltage is isolated by the high frequency transformer, and the DC pulse is obtained by rectification filtering, thus charging the battery pack.

In this paper, the optimization between switching stations is defined as the feeder layer optimization, the optimization within the switching station is defined as the branch layer optimization, and the charging optimization for electric vehicles is defined as the load layer optimization.

2.1. Dynamic Reconstruction Model between Switch Station Groups

For the optimization of switch stations in feeder layer, it is necessary to analyze the connection mode of 10 kV distribution network, and derive different constraints for different connection modes. The actual typical connection types of 10 kV distribution network are single power supply, series supply, T-type series supply, etc. The feeder layer in Figure 1 is a series connection mode, this paper focuses on the analysis of T-type series supply wiring mode, which is more complicated. The specific connection mode is as follows.

Taking the switch topology of Figure 3 as an example, the power supply demand guarantee constraints are as follows:

$$
\begin{cases}
P_{G,A_1}^t = \sum\limits_{i=1}^{j-1} S_i^t P_{l,i} \\
P_{G,A_2}^t = \sum\limits_{i=k+1}^{n} S_i^t P_{l,i-1} \\
P_{G,A_3}^t = \sum\limits_{i=n+1}^{n+n_1} S_i^t P_{l,i} + \sum\limits_{i=j+1}^{k} S_i^t P_{l,i-1}
\end{cases}
\tag{1}
$$

where P_{G,A_1}^t and P_{G,A_2}^t represent all loads supplied by A_1 and A_2, respectively. $P_{l,i}$ denotes the total load under the switching station after the i-th switch. S_i^t represents the state of the i-th switch. t represents time.

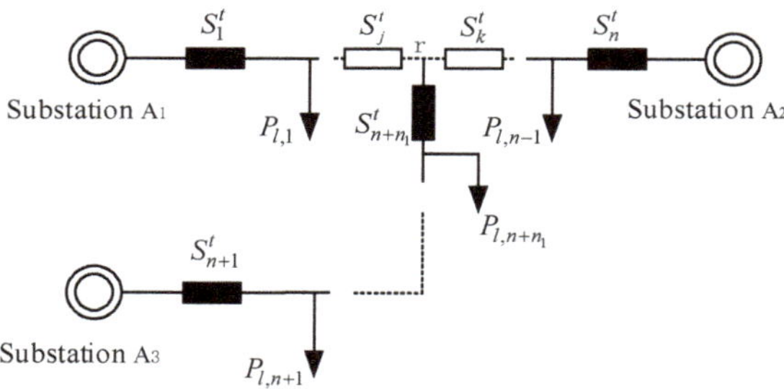

Figure 3. T-type series supply wiring mode.

Three active power equality constraints ensure the power supply demand of each switch station. For the other two topologies, it is only equivalent to the change of power supply position. In the left graph of Figure 4, the original topology can be obtained by exchanging the position of A_2 point and A_3 point. In the right graph of Figure 4, the original topology can be obtained by exchanging the position of A_1 point and A_3 point.

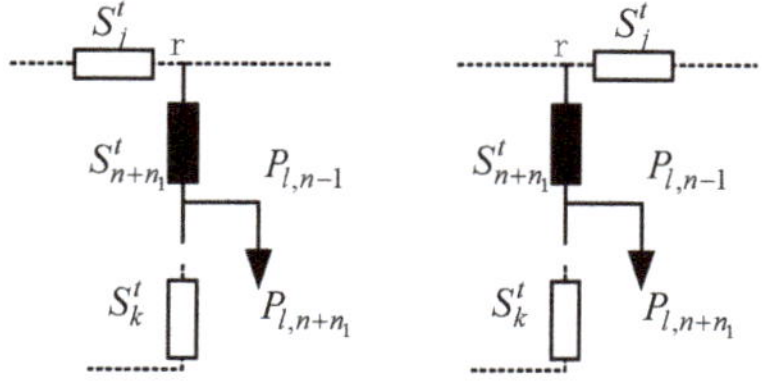

Figure 4. T-type series supply disconnect switch position.

For T-type serial supply structure, topology constraint is established:

$$\begin{cases} \sum\limits_{i=1}^{n+n_1} S_i^t = n + n_1 - 2 \\[2mm] n - 2 \le \sum\limits_{i=1}^{n} S_i^t \le n - 1 \\[2mm] n_1 - 3 + r \le \sum\limits_{i=1}^{r-1} S_i^t + \sum\limits_{i=n+1}^{n+n_1} S_i^t \le n_1 - 2 + r \\[2mm] n_1 + n - r - 1 \le \sum\limits_{i=r}^{n} S_i^t + \sum\limits_{i=n+1}^{n+n_1} S_i^t \le n_1 + n - r \end{cases} \tag{2}$$

Three inequality constraints ensure that there must be one or two disconnected switches between two power points, and equality constraints determine the number of disconnected switches in the T-type series supply structure. The three networks with different switch positions satisfy the constraint condition of Formula (2).

Therefore, according to the typical topology and connection mode of 10 kV main feeder, the total prediction load curve of each switch station is obtained. Taking the minimum load fluctuation level in the region as the goal, the reconstruction scheme between switching stations is established. The objective function is as follows:

$$f = \min \sum_{t=1}^{24} \left[\left(P_{G,A_1}^t - P_{G,A_2}^t \right)^2 + \left(P_{G,A_1}^t - P_{G,A_3}^t \right)^2 + \left(P_{G,A_3}^t - P_{G,A_2}^t \right)^2 \right] \tag{3}$$

Considering the high cost of breaking the switch, this paper needs to calculate the load balancing index to determine whether to perform the switch action.

2.2. Internal Dynamic Reconfiguration Model of Switch Station

2.2.1. Branch Layer Reconstruction Model

After the feeder layer topology optimization is completed, it is necessary to analyze the load in each switch station to realize the branch layer autonomous optimization.

In this paper, the load curve of each ring network cabinet in the switching station is analyzed, and the minimum loss is achieved by changing the switch state between the ring network cabinets. At a single time level, the objective function of minimizing network loss is as follows:

$$f_{\text{loss},i} = \sum_{\text{Br}=1}^{N-1} y_{\text{Br}} r_{\text{Br}} \frac{P_{\text{Br}}^2 + Q_{\text{Br}}^2}{U_{\text{Br}}^2} \tag{4}$$

where y_{Br} is the switch state of branch Br, 0 means open, 1 means close; U_{Br} is the voltage value at the beginning of branch; P_{Br} and Q_{Br} represent the injected active power and reactive power of the first node.

In addition, the power flow equation equality constraints, voltage amplitude constraints and line capacity constraints should be satisfied. The formula is as follows:

$$\begin{cases} P_{\text{sum},i} = U_i \sum\limits_{j=2}^{N} U_j (G_{ij} \cos \theta_{ij} + B_{ij} \sin \theta_{ij}) \\[2mm] Q_{\text{sum},i} = U_i \sum\limits_{j=2}^{N} U_j (G_{ij} \cos \theta_{ij} + B_{ij} \sin \theta_{ij}) \\[2mm] U_{i,\min} \le U_i \le U_{i,\max} \\[1mm] S_{\text{Br}}^2 \le S_{\text{Br},\max}^2 \end{cases} \tag{5}$$

where $S_{\text{Br},\max}$ is the maximum line capacity, $U_{i,\max}$ is the maximum node voltage.

In addition to the above constraints, the model also needs to meet the network topology operation constraints. In this paper, through network coding and simplification, a single-loop matrix is formed, which constitutes a radial constraint and a connected constraint, so as to determine the infeasible optimization solution.

The switches in each loop network are coded, and the single loop matrix is formed by searching the path from each node to the parent node. Each row in the single loop matrix represents a loop. The optimization variable in the reconstruction scheme is a switch in each row of the matrix. In order to make the solution satisfy the constraint of no-island and no-ring network, this paper uses SL correlation matrix and upper node search to determine the feasibility of each solution.

SL correlation matrix is defined as follows:

$$SL = \begin{bmatrix} a_{11} & \cdots & a_{1n} \\ \vdots & \ddots & \vdots \\ a_{n1} & \cdots & a_{nn} \end{bmatrix} \tag{6}$$

where a_{ij} represents the membership relationship between the j-th dimension solution and the i-th loop in the SL correlation matrix. If the solution belongs to this loop, it is 1, otherwise it is 0.

The principle of no-island judgment method is that if two rows are the same in the SL matrix, the solution of the j-dimension belongs to two loops at the same time. At this time, the system has a loop, and the infeasible solution must be eliminated.

The principle of upper node search is that the upper node of each node in turn is found, putting the result into a matrix, and finally it is judged whether there are 0 elements in the matrix except for the first column. If it exists, it means that there are islands in the network, and the infeasible solution must be eliminated.

Considering the high cost of the switch, this paper analyzes the difference between the network loss value before optimization and the optimized network loss value. Once the difference is less than a threshold, the switch optimization action is not performed.

2.2.2. Reconstruction Method Based on CNPSO Algorithm

Aiming at the branch layer dynamic reconstruction model, and in order to prevent particle swarm algorithm from falling into local optimum and accelerate the iterative speed, this paper improves the traditional particle swarm optimization algorithm by introducing logistic chaotic equation and niche elite retention algorithm. Refer to Appendix B for dynamic updates of inertia weights.

To solve the problem of local optimum, by adding a mixed disturbance near the group extremum, the solution space near the optimal solution is searched and the local search is strengthened. The specific steps are as follows:

Step 1: The global optimal solution of the n-th iteration output is mapped to the definition domain of the logistic equation to generate the chaotic variable z_{nj}. The formula is as follows:

$$z_j^n = \frac{x_j^n - x_{\min,j}}{x_{\max,j} - x_{\min,j}}, j = 1, 2, \ldots\ldots d \tag{7}$$

where x_j is the global optimal particle, that is, the disconnected switch combination of loops, n is the number of iterations; $x_{\max,j}$ and $x_{\min,j}$ represent the upper and lower bounds of the j-th dimensional variable, respectively.

Step 2: Using logistic mapping equation to generate d chaotic sequences, and then the chaotic variables are inversed to the original solution space to obtain new optimization variables. The formula is as follows:

$$z_j^{n+1} = \mu z_j^n (1 - z_j^n), j = 1, 2, \ldots\ldots d \tag{8}$$

$$x_j^{n+1} = x_{\min,j} + z_j^{n+1} (x_{\max,j} - x_{\min,j}), j = 1, 2, \ldots\ldots d \tag{9}$$

Step 3: The fitness function value of the new optimization variable is calculated and compared with the fitness value of the original solution x_j. If it is better than the original solution or reaches the maximum number of iterations, the position of the original particle is replaced by the position of the new particle. Otherwise, let $n = n + 1$, and turn to Step 2.

In addition to the local optimal problem, once the population is too large, it will affect the iteration speed. In this paper, the niche technology of sharing fitness mechanism is used to solve, that is, by calculating the sharing degree of individual particles in the group. The greater the degree of sharing, the higher the degree of closeness with other particles.

In this paper, the Euclidean distance between each particle is calculated, and the niche radius of the current population is calculated. The sharing degree of each individual is sorted, and the whole population is pruned. The individuals with higher sharing degree are deleted to ensure the uniform distribution of the solution. The number of all particles in the initial population is N_{sh}, and the shared fitness function is defined as follows:

$$F_i = \frac{1}{\sum\limits_{j=1}^{N_{sh}} 1 - \left(\frac{D_{ij}}{\sigma_{\text{share}}}\right)^{\alpha}}, \quad i, j \in \Omega_{N_{sh}} \tag{10}$$

where $\Omega_{N_{sh}}$ represents the whole group space, σ_{share} represents the niche radius, D_{ij} represents the Euclidean distance between individual x_i and x_j, and d represents dimension of decision variables.

$$D_{ij} = \|X_i - X_j\| = \sqrt{\sum_{d=1}^{m}\left(x_{id} - x_{jd}\right)^2} \quad i, j \in \Omega_{N_{sh}} \tag{11}$$

$$\sigma_{\text{share}} = \frac{1}{N_{sh}}\sum_{i=1}^{N_{sh}} \min\left(D_{ij}\right) \quad i \neq j \ , i, j \in \Omega_{N_{sh}} \tag{12}$$

3. Isolated Bidirectional DC–DC Converter Optimal Efficiency Model

According to the topological structure of the medium voltage distribution network group, after determining the topological structure of the feeder layer and the branch layer, the orderly charging modeling of electric vehicles in each switch station is carried out. However, the charging efficiency of electric vehicles is affected by the transmission loss of the DC–DC converter, which further affects the accuracy of formulating the charging timing strategy for electric vehicles. It is assumed that the charging power is 12 kW and the battery capacity is 80 kWh. If the charging efficiency is increased by 5%, an additional 3 kW can be charged. In some scenarios, the next-day travel demand of EVs can be met in advance, and the charging load at that time at the branch layer can be reduced.

Therefore, aiming at the problem of peak current and return power when single-phase-shift (SPS) control IBDC converter, this paper adopts dual-phase-shift control (DPS) method to establish an optimal efficiency calculation model.

3.1. Principle of Dual-Phase-Shift Control

As shown in Figure 5, a typical IBDC converter circuit consists of two symmetrical H-bridges and high-frequency transformers. Compared with the traditional SPS control, DPS control is to introduce a new phase shift duty cycle between the two diagonal switch tubes of the full bridge on the primary side or on the secondary side. In this paper, the shift ratio D_1 of the primary side in a half period is defined as the internal shift ratio, and the shift ratio D_2 between the two sides in the half period is defined as the external shift ratio. When the internal shift ratio $D_1 = 0$, DPS control becomes traditional SPS control.

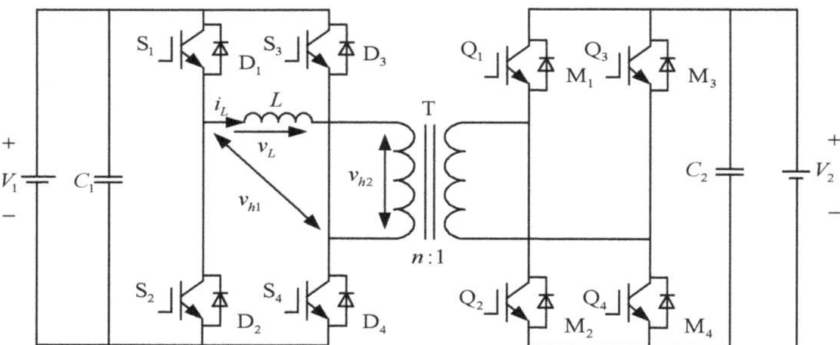

Figure 5. IBDC converter topology.

Figure 6 describes the working state of the converter in a switching cycle, such as the conduction time of the switch tube and diode, and the voltage change. Due to the symmetry of control, the system waveform of t_0–t_4 period is taken as the research object, and the working mode of converter can be divided into five states. According to the analysis of the working status of these five stages in turn, it can be found that the inductance current formula in the t_0–t_4 period is a piecewise linear formula.

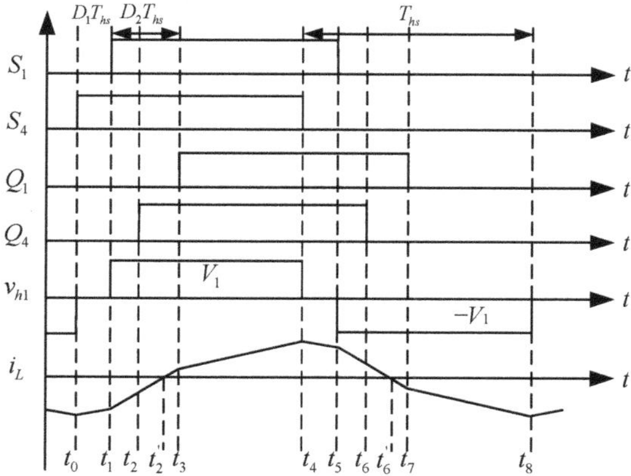

Figure 6. Waveform of dual phase shift control system.

When $0 \leq D_1 \leq D_2 \leq 1$, set $t_0 = 0$, and then other time point can be expressed as $t_1 = D_1 T_{hs}$, $t_2 = D_2 T_{hs}$, $t_3 = (D_1 + D_2)T_{hs}$, $t_4 = T_{hs}$, $t_5 = (1 + D_1)T_{hs}$, $t_6 = (1 + D_2)T_{hs}$, and $t_7 = (1 + D_1 + D_2)T_{hs}$. Voltage regulation ratio was set to $k = V_1/(nV_2)$, and switching frequency was set to $f_s = 1/(2T_{hs})$. According to the symmetry, $i_L(t_0) = -i_L(t_4)$ can be obtained, so the inductance current $i_L(t)$ in each time period can be obtained, and then the following transmission power formula is derived by the inductance current formula:

$$P = \frac{1}{T_{hs}} \int_0^{T_{hs}} v_{h_1} i_L(t)dt = \frac{nV_1V_2}{2f_sL}\left[D_2(1 - D_2) - \frac{1}{2}D_1^2\right] \tag{13}$$

3.2. Transmission Power Loss Model of IBDC Converter

The loss of an IBDC converter under DPS control mainly includes switching loss, conduction loss caused by switching devices, copper loss, and iron loss caused by magnetic components. The loss generated by the switching device is related to its on-state voltage drop and switching frequency. The loss generated by the magnetic element is related to the effective value of the inductance current and the winding resistance.

3.2.1. Switching Loss Model

When the switch tube works in soft switching mode, switching losses can be ignored. When the switch tube is in a hard switching state, due to the overlap of voltage waveform and current waveform in the transient process of opening and closing, the loss is generated [20]. The turn-off loss and turn-on loss are:

$$P_{off} = \frac{f_s t_{off}}{2}\left[(V_1 + V_F)\sum_{t_s \in \Omega_s}|i_L(t_s)| + n(V_2 + V_F)\sum_{t_Q \in \Omega_Q}|i_L(t_Q)|\right] \tag{14}$$

$$P_{on} = \frac{f_s t_{on}}{2}n(V_2 + V_F)(|i_L(t_2)| + |i_L(t_6)|) \tag{15}$$

where t_{off} and t_{on} are switch turn-off time and turn-on time, respectively, and V_F is the forward voltage drop of diode. Ω_s and Ω_Q indicate the set of turn-off moments.

3.2.2. Conduction Loss Model

Since both the switch tube and the diode have a forward voltage drop when they are turned on, the on-state loss will be generated when the current flows, which is manifested in the form of heat.

Based on the odd symmetry of the converter working waveform and the conduction state of the switches and diodes in each stage, the conduction loss of IGBT and diode can be obtained [21]:

$$P_I = \frac{V_{sat}}{T_{hs}}\left(\int_{t_o}^{t}|i_L(t)|dt + 2\int_{t'_2}^{t}|i_L(t)|dt + \int_{t2}^{t3}n|i_L(t)|dt\right) \tag{16}$$

$$P_d = \frac{V_F}{T_{hs}}\left[\int_{t_o}^{t_1}|i_L(t)|dt + 2\int_{t_1}^{t'_2}|i_L(t)|dt + 2\int_{t_0}^{t2}n|i_L(t)|dt + \int_{t2}^{t3}n|i_L(t)|dt + 2\int_{t3}^{t4}n|i_L(t)|dt\right] \tag{17}$$

where V_{sat} represents the on-state voltage drop of IGBT, which is a constant.

3.2.3. Magnetic Components Loss Model

Magnetic components mainly include transformers and auxiliary inductance, and the loss that produced during their work are mainly composed of copper loss and iron loss [22].

During the whole switching period, the current i_L flows through the transformer and the auxiliary inductance, and the copper loss is related to the root mean square of the current i_L:

$$P_{cop} = (R_{tr} + R_{au})I_{rms}^2 \tag{18}$$

where R_{tr} and R_{au} are winding resistance of the transformer and auxiliary inductor, respectively, which are constants. I_{rms} represents the root mean square of the current i_L.

Iron loss of magnetic components is mainly composed of hysteresis loss, eddy current loss, and residual loss. The calculation formula is as follows:

$$P_{iron} = \frac{2m f_s \mu_0^2 N^2 V_e}{g^2}I_{rms}^2 \tag{19}$$

where m is the iron loss coefficient, μ_0 is the vacuum permeability, g is the air gap, N is the turns of the coil, and V_e is the effective volume. These parameters can be found from the parameter table [23].

3.3. Optimal Efficiency Calculation Model

According to Formulas (14)–(19), the switching loss P_{SW}, conduction loss P_{CON}, and transformer and auxiliary inductance loss P_{TA} of the IBDC converter can be obtained, so the total loss P_{loss} can be obtained. The detailed formula derivation is reflected in the Appendix A.

$$P_{loss} = P_{SW} + P_{CON} + P_{TA} \tag{20}$$

Therefore, the total loss P_{loss} under the control of DPS is related to the internal shift ratio D_1 and the external shift ratio D_2. This paper establishes an optimal efficiency model to select the optimal parameters D_1 and D_2 for a specific transmission power.

The efficiency of IBDC converter is defined as the percentage of the ratio of output power to input power:

$$\eta = \frac{P}{P + P_{loss}} \tag{21}$$

It can be seen from Formula (13) that the given transmission power P_0 can be obtained by various combinations of D_1 and D_2, but the total loss of the converter is different under each combination, so its efficiency is also different. In order to obtain the combination of D_1, D_2 corresponding to the minimum total loss of the converter, the Lagrange function can be established:

$$L(D_1, D_2, \lambda) = P_{loss}(D_1, D_2) + \lambda[P(D_1, D_2) - P_0] \tag{22}$$

Formula (22) is solved to find the optimal total loss value. The calculation formula is as follows:

$$\begin{cases} \frac{\partial L}{\partial D_1} = \frac{\partial P_{SW}}{\partial D_1} + \frac{\partial p_{CON}}{\partial D_1} + \frac{\partial p_{TA}}{\partial D_1} + \lambda \frac{\partial p}{\partial D_1} = 0 \\ \frac{\partial L}{\partial D_2} = \frac{\partial P_{SW}}{\partial D_2} + \frac{\partial P_{CON}}{\partial D_2} + \frac{\partial P_{TA}}{\partial D_2} + \lambda \frac{\partial P}{\partial D_2} = 0 \\ \frac{\partial L}{\partial \lambda} = P(D_1, D_2) - P_0 = 0 \end{cases} \tag{23}$$

By substituting Formulas (13) and (19) into Formula (23) and solving the root of the above equations, the optimal combination (D_1, D_2), and the efficiency of the IBDC converter reaches the maximum.

4. Orderly Charging Model of EV Considering IBDC Converter Optimal Efficiency

At the load layer, this paper takes the electric vehicles under the switch station as the research object. In this paper, a double-layer optimal scheduling model for EVs is established based on the distributed scheduling architecture.

4.1. Multi-Level Information Interaction Mechanism

The principle of distributed scheduling is as follows: each switch station can further decompose all transformer area into several areas according to their geographic location, each area is dispatched by aggregators, and all aggregators accept the instructions of the dispatch center.

As shown in Figure 7, under the switch station, electric vehicle aggregators spanning multiple transformer areas are set up as an information bridge between a single electric vehicle and the dispatch center. It not only implements the instructions of the upper-level dispatch center, but also guides the lower-level specific electric vehicle charging time. The aggregators autonomy model facilitates centralized management of electric vehicles in the same area, avoiding the problems of low efficiency and huge data volume caused by the direct dispatch of each electric vehicle by the dispatch center.

The steps of the multi-level interaction mechanism are as follows:

Step 1: The user's charging pile uploads the daily time when the electric vehicle is connected to the system, the state of charge at the time of connection, and the time when the user is expected to leave the system to the aggregator.

Step 2: Each aggregator obtains information about all electric vehicles in the area under its jurisdiction, and the dispatch center formulates demand targets according to the data integration of each aggregator so as to issue a dispatch plan for each time period to each aggregator.

Step 3: Based on the data uploaded by each electric vehicle, each aggregator's scheduling goal not only requires the minimum variance of the load level under the switch station, but also requires the minimum sum of deviations between the actual scheduling results of each aggregator and the scheduling plan determined by the scheduling center.

Step 4: Under each aggregator, by controlling the charging time of all electric vehicles in the area, the deviation between the upper-level scheduling plan and the actual scheduling results of the lower-level electric vehicles is minimized, and step 4 is returned until the given convergence condition is reached.

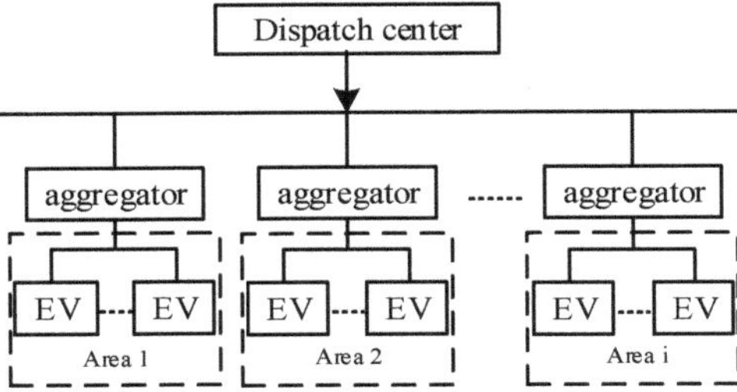

Figure 7. Distributed scheduling structure diagram.

4.2. Upper-Level Dispatch Model Considering Load Balance under the Switch Station

The upper-level dispatch model fully considers the overall load fluctuation level under the switch station, and realizes peak-shaving and valley-filling by formulating a reasonable dispatch plan. Therefore, for the topological structure of the branch layer, the upper-level objective function is established. The specific expression is as follows:

$$
F_{\text{up}} = \frac{1}{T}\sum_{t=1}^{T}\left(P_{\text{L},t} + \sum_{j=1}^{N}x_{j,t} - \frac{1}{T}\sum_{t=1}^{T}\left(P_{\text{L},t} + \sum_{j=1}^{N_0}x_{j,t}\right)\right)^2 + \lambda\sum_{j=1}^{N}\left(\sum_{t=1}^{T}(P_{j,t}-x_{j,t})^2\right) \tag{24}
$$

where $P_{\text{L},t}$ represents the original load level at time t in the network; $x_{j,t}$ represents the scheduling plan of the j-th aggregator at the branch layer at time t; $P_{j,t}$ represents the actual scheduling result of j-th aggregator; N represents the total number of aggregators in a certain switch station; and λ represents the penalty coefficient, which is used to restrict the deviation between the actual scheduling result and the scheduling plan.

In addition to satisfying power flow equation constraints and voltage constraints, the other upper-level model constraints are as follows:

$$
0 \leq x_{j,t} \leq \sum_{i=1}^{n}P_{j,i,\text{ch}}y_{j,i,t} \quad \forall t \in [1,\text{T}] \tag{25}
$$

where $P_{j,i,\text{ch}}$ represents the charging power of the i-th electric vehicle under the j-th aggregator; $y_{j,i,t}$ represents the state of the i-th electric vehicle under the j-th aggregator connecting to the network at time t; the value is 1 when it is connected (may participate in the operation of the distribution network or not), and 0 when it is not connected.

4.3. Lower-Level Scheduling Model

In the lower-level dispatch model, this article considers the transmission power loss of the IBDC converter, and selects the optimal parameters D_1 and D_2 for a specific transmission power. In addition, the target of the lower model is taken as a part of the objective function of the upper-level. Each aggregator receives the dispatching instructions from the dispatch center, and the objective is to minimize the deviation between the actual dispatching results of all electric vehicles, which under the jurisdiction of each aggregator and the dispatching plan. For the j-th aggregator, the objective function is as follows:

$$
F_{\text{low}} = \sum_{t=1}^{T}\left(\sum_{i=1}^{n}P_{j,i,\text{ch}}y_{j,i,t} - x_{j,t}\right)^2 \tag{26}
$$

where $P_{j,i,\mathrm{ch}}$ represents the rated charging power of the *i*-th electric vehicle under the *j*-th aggregator; $y_{j,i,t}$ represents whether the *i*-th electric vehicle under the *j*-th aggregator participates in the dispatching of the distribution network at time *t*.

The constraints are as follows:

$$y_{j,i,t} = 0 \quad t > t_{j,i,\mathrm{dep}} \parallel t < t_{j,i,\mathrm{arr}} \tag{27}$$

where $t_{j,i,\mathrm{dep}}$ and $t_{j,i,\mathrm{arr}}$, respectively, represent the time when the *i*-th electric vehicle under the *j*-th aggregator is connected to the system on the same day (may participate in the distribution network operation or not), and the time when it is off the grid the next day is used to restrict off-grid electric vehicles from participating in distribution network dispatching.

$$S_{j,i,t_{\mathrm{dep}}} \geq 0.9 S_{\mathrm{ev,max}} \tag{28}$$

$$0 \leqslant S_{j,i,t} \leqslant S_{\mathrm{ev,max}} \tag{29}$$

where $S_{\mathrm{ev,max}}$ represents the maximum capacity of electric vehicles. Formula (28) is to meet the next day's driving capacity demand, and Formula (29) is to prevent electric vehicles from overcharging.

$$S_{j,i,t+1} = S_{j,i,t} + \eta_{\mathrm{ch}} P_{j,i,\mathrm{ch}} \Delta t \tag{30}$$

where $S_{j,i,t}$ represents the SOC of the *i*-th electric vehicle at time *t*; Δt represents a period of time, the value is 1; η_{ch} represents the charging efficiency, and this parameter is determined by Formula (23) in Section 3. With a given transmission power, find the optimal D_1 and D_2 to minimize the power transmission loss of the IBDC converter and obtain the optimal efficiency η_{ch}.

4.4. Overall Flow Chart of Hierarchical and Partitioned Optimization

This paper proposes an EV double-layer scheduling model based on the IBDC converter optimal efficiency model, and establishes the hierarchical and partitioned optimization model of the feeder–branch–load layer. The specific flow chart is shown in Figure 8:

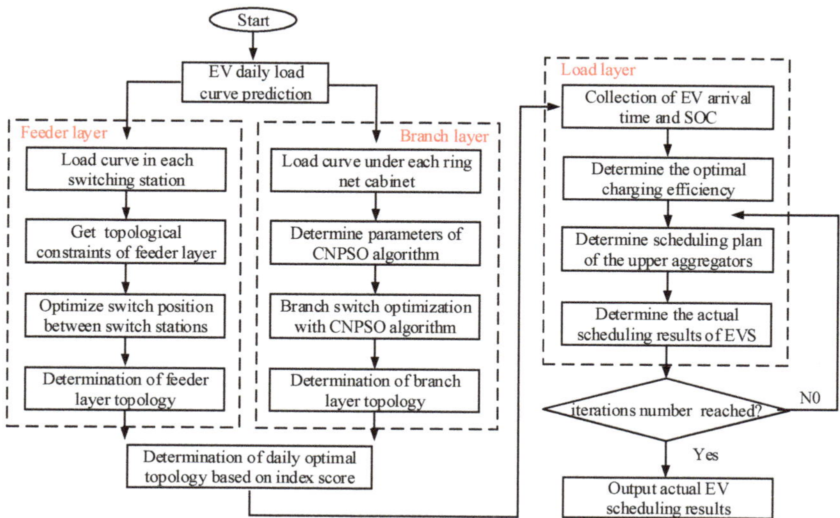

Figure 8. Electric vehicle dispatching strategy diagram considering dynamic topology.

5. Case Study

In order to verify the effectiveness of the proposed dynamic reconfiguration and electric vehicle scheduling model, this paper analyzes the cases in Figures 9 and 10 on MATLAB R2014b. The system structure diagram is 10 kV medium voltage distribution network diagram. In Figure 9, the power point is 35 kV substation, and the shadow part is different switching stations. The network contains five switches on the main feeder. In the initial state, the black switch represents the closed state, and the white switch represents the breaking state. In Figure 10, the switch station contains 13 segmented switches and three tie switches, and three dispatching aggregators are configured. The dispatching aggregators manage 100 electric vehicles, respectively, and the tie switches 14, 15, and 16 are all disconnected before optimization.

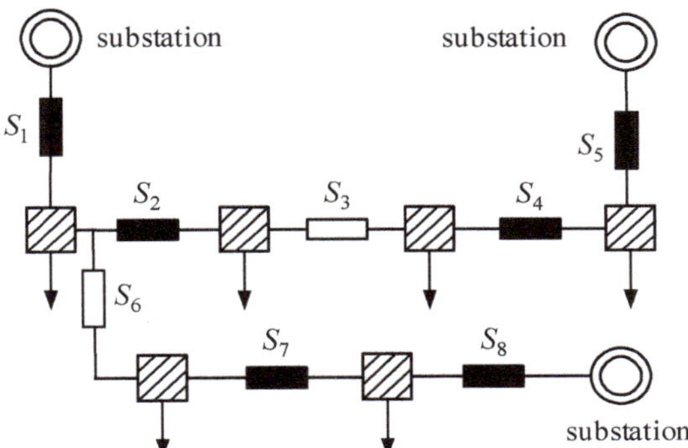

Figure 9. Topology structure of 10 kV feeder layer.

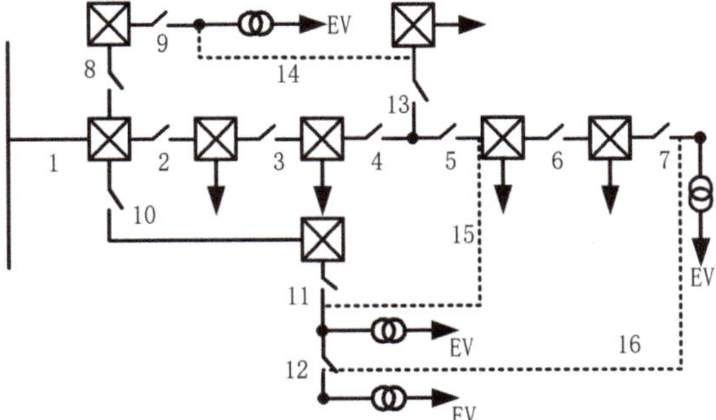

Figure 10. Structure diagram under a switch station.

The parameters of electric vehicle are as follows: the average charging power is 12 kW, the charging efficiency is 90%, the average battery capacity is 60 kWh, the upper limit of battery SOC is 95%, and the lower limit is 5%. For the calculation process and results of the orderly charging model of electric vehicles, this paper adopts the per unit (pu) value to facilitate calculation and explanation.

5.1. Optimization Results of Feeder Layer Reconstruction

The structure is powered by T-type series connection, and the Gurobi solver is called for optimization by analyzing the total load of each switch station in each period. In the actual simulation, the switch station near the substation is generally not used as the transfer object. The final switching operation number is 8, which can achieve load balancing of three substations. The results are shown in Table 1. Therefore, the load balancing level is improved by adjusting the switching state between switching stations, and the load difference of the three substations is minimized in one day.

Table 1. Comparison of feeder layer load balance.

Scenes	Time	Load Balance Degree	Switch State
Before optimization	10:00	26.285 MW	—
	20:00	28.466 MW	—
After optimization	10:00	24.922 MW	S_2, S_7 closed S_3, S_6 open
	20:00	16.235 MW	S_3, S_6 closed S_2, S_7 open

5.2. Topology Reconstruction Results under Switch Station

Population sizepop = 100; iteration number N = 100; inertia weight is dynamic parameters. The threshold value is 0.1. It can be seen from Table 2 that after optimization, the switches 13, 14, and 15 are open during the period of 5:00 to 12:00, the switches 5, 10, and 15 are open during the period of 12:00 to 21:00, and the switches 11, 15, and 16 are open during the period of 21:00~24:00.

Table 2. Switch action before and after reconstruction.

Scenes	Time	Open Switch Number
Before reconfiguration	whole day	14, 15, 16
	5:00	14, 15, 12
After reconfiguration	12:00	5, 10, 15
	21:00	11, 15, 16

Assuming that the load connected to the lines numbered 4 and 5 in the original diagram are office buildings and other loads, the load is large during the day, so the network loss on lines 4 and 5 during the period after 12:00 is relatively large. At this time, it is necessary to open the line switch and the part of the load is transferred to the rest of the line for power supply.

Table 3 shows the changes in the network loss before and after optimization. The network loss is reduced by 17.42%, which shows that the network loss of the branch layer can also be reduced by adjusting the switch state.

Table 3. Switch action before and after reconstruction.

Scenes	Value of Network Loss	Improvement Rate
Before reconfiguration	202.5193 kWh	—
After reconfiguration	139.5191 kWh	31.11%

It can be seen from Figure 11 that the traditional particle swarm optimization algorithm may fall into local optimum and the iteration speed is slow. However, in the case of increasing the population size, the algorithm proposed in this paper makes iterative speed faster by deleting individuals with a higher degree of sharing. The average number of iterations can reach convergence at about 10 times. Moreover, while ensuring the speed, the global optimal solution is ensured by adding a mixed disturbance near the group extreme value. Therefore, the improved algorithm in this paper is more accurate and efficient.

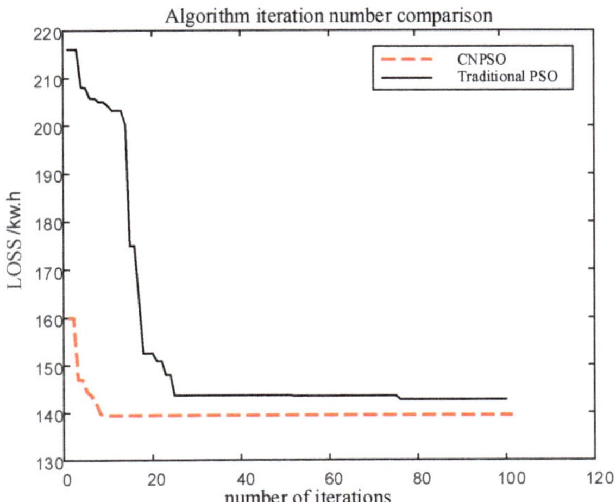

Figure 11. Algorithm iteration number comparison chart.

5.3. Orderly Charging Results of Electric Vehicles

The simulation parameters of the converter in this article are as follows: load R = 100 Ω, DC capacitance $C_1 = C_2 = 2200$ μF, switching frequency $f_s = 20$ kHz, and transformer leakage inductance L = 7.7 μH. It can be seen from Figure 12 that when the power of 60 kW is given, D1 = 0.14, and the efficiency reaches the highest. The output voltage of electric vehicle is 300 V in this paper, so the optimal efficiency in the ordered charging model in this paper is 96.2%.

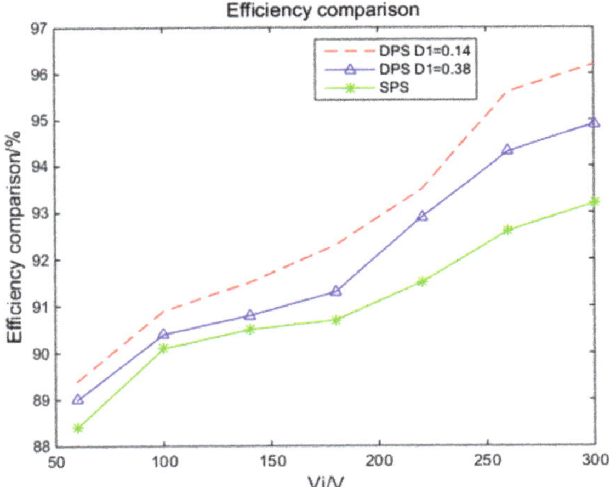

Figure 12. Comparison of converter efficiency under different controls.

The switch station is mostly residential areas. It can be seen from the Figure 13 that electric vehicles were originally charged from 16:00 to 22:00, that is, the electric vehicles are charged immediately when residents arrive home. At this time, it is superimposed with other original loads, resulting in a large load under the switch station during this period of time. In addition, residents use less electricity at night, and electric vehicles are already fully charged, so the load is very small from 0:00 to 5:00, resulting in a large peak-to-valley difference in a day. After optimization, it can be seen from Figure 13 that the peak–valley difference is reduced from the original 16.5026 to 13.3174, which greatly improves the load stability of the branch layer and alleviates the phenomenon of adding peaks on the peak.

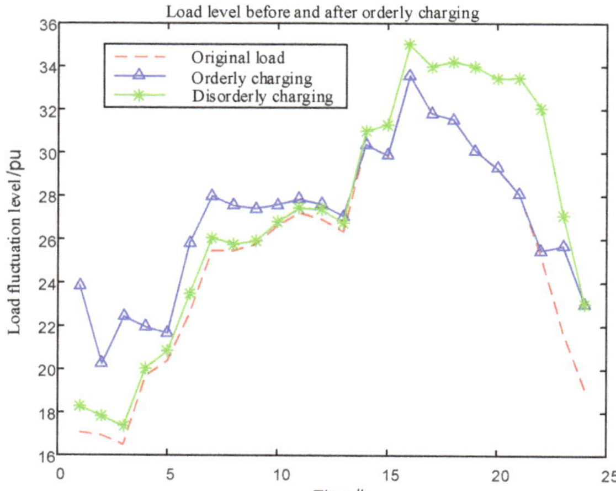

Figure 13. Load level changes before and after EV orderly charging.

Figure 14 shows the initial capacity and the capacity after the final optimization of all electric vehicles in aggregator 1. In this paper, the travel scenarios of electric vehicles are divided into three types by analyzing the arrival time and expected travel time of electric vehicles. Figure 14 illustrates that no matter when the electric vehicle is connected, the optimization model in this article can guarantee the travel demand of the electric vehicle the next day.

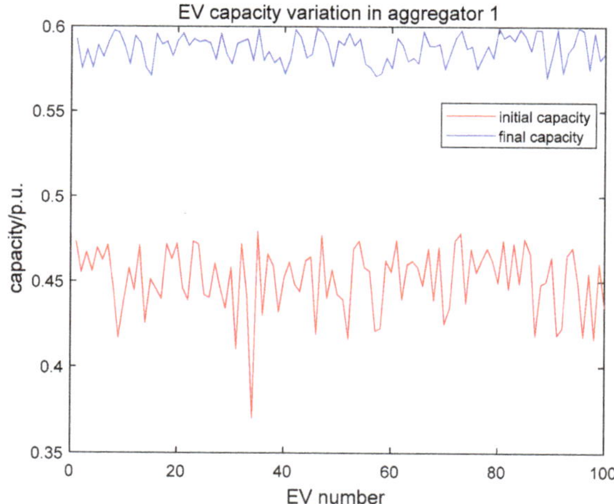

Figure 14. EV capacity variation in aggregator 1.

After the aggregator receives the dispatching instruction from the dispatch center, through double-layer optimization iteration, the actual dispatch results of the three aggregators is shown in Figure 15. The low cost of electricity at night and low load levels cause the charging behavior to be more frequent from 0:00 to 10:00 and after 22:00. By formulating a reasonable orderly charging strategy, the original disordered state of "plug and play" is changed. On the premise of meeting the demand of all EVs on the next day, the EVs connected during the peak power consumption period are arranged to be charged at night, which not only improves the load level at night, but also eases the power shortage during the peak power consumption period.

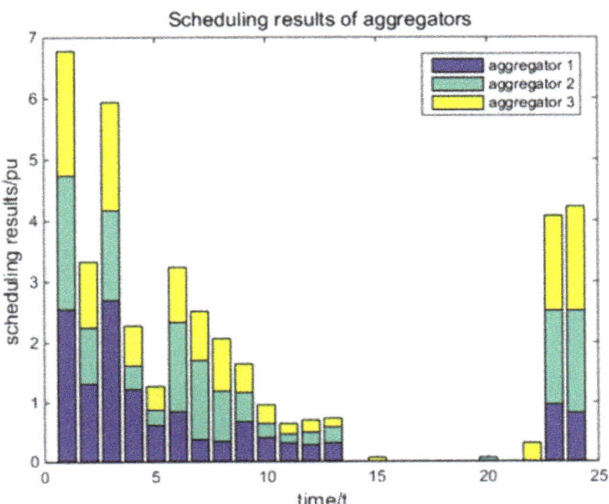

Figure 15. The actual scheduling results of three aggregators.

6. Conclusions

With the access of large-scale EVs in the future, it will inevitably increase the network loss of the medium-voltage distribution network and aggravate the imbalance of the distribution network load. In order to improve this situation, this paper proposes a three-layer optimization model from the perspectives of network structure and load itself. Each layer has its own findings with different priorities. The specific conclusions are as follows.

Firstly, at the feeder layer, this paper takes the connection relationship between switching stations as the research object, and establishes the dynamic reconstruction model between stations. By solving this model, it realizes the equalization of the load carried by the substation and avoids the occurrence of continuous heavy load in a certain substation. Secondly, at the branch layer, this paper takes the topology structure under a certain switch station as the research object, and a dynamic reconstruction plan is formulated for the topology structure. By solving this model, the branch layer network loss is effectively reduced, and the loss reduction rate reaches 31.11%. At the same time, this paper adopts the CNPSO algorithm to adapt to the larger branch topology scale in the future. This algorithm avoids the algorithm trapped in local optimum and improves the iterative speed by nearly 20%. Finally, at the load layer, this paper takes the electric vehicles under the switch station as the research object. The first step is to establish an IBDC converter optimal efficiency model, accurately calculating the charging power of electric vehicles in each period. The second step is to establish a double-layer distributed optimization scheduling model, which not only effectively reduces the difficulty of directly dispatching large-scale electric vehicles by dispatch center, but also formulates a reasonable charging strategy, achieving a 19.3% reduction in peak-to-valley difference. The findings of this paper are based on the consideration of all the major links in the medium-voltage distribution network, including branch layer network loss and optimal charging efficiency, which are rarely studied before. All of these should be present when solving the problems of medium-voltage distribution network loss and load fluctuation.

Author Contributions: The individual contributions of the authors are specified as follows: conceptualization: Q.Z. and J.Z. (Jian Zhao); methodology: Q.Z.; software: H.J.; validation: L.T., J.Z. (Jian Zhao), and H.J.; formal analysis: H.J.; writing—original draft preparation: Q.Z. and H.J.; writing—review and editing: X.W.; supervision: J.Z. (Jinhui Zhou) All authors have read and agreed to the published version of the manuscript.

Funding: This research was funded by National Key R&D Program of China (2018YFB0905105); National Natural Science Foundation of China (Grant NO. 51907114) and Science and Technology Project of State Grid Zhejiang Electric Power Co., LTD. (5211DS18002S).

Acknowledgments: The authors would like to acknowledge the support of the National Natural Science Foundation of China (Grant NO. 51907114) and Science and Technology Project of State Grid Zhejiang Electric Power Co., LTD. (5211DS18002S).

Conflicts of Interest: The authors declare no conflict of interest.

Appendix A

The detailed formulas the switching loss P_{SW}, conduction loss P_{CON}, and transformer and auxiliary inductance loss P_{TA} are as follows:

$$\begin{cases} P_{SW} = m_1\{m_2[k(1-D_1)+2D_2-1]+m_3[k(D_1-2D_2+1)-m_4D_1+m_5(D_1-1)]\} \\ P_{CON} = h_1\left[h_2D_1^2+h_3D_2^2+h_4D_2(1-D_1-D_2)+h_5(1-D_2)(D_2-D_1)+2D_2\right] \\ P_{TA} = w_1\left[(k-1)^2(1+2D_1)(1-D_1)^2-4kD_2^2(3D_1+D_2-3)+4k(D_1-D_2)^3\right] \end{cases} \tag{A1}$$

where m_1–m_5, h_1–h_5 and w_1 are constants that obtained by formulas (14)–(19).

$$\begin{cases} m_1 = \frac{nV_2}{4L} \\ m_2 = 2(V_1+V_F)t_{off} \\ m_3 = n(V_2+V_F) \\ m_4 = 2kt_{off} \\ m_5 = t_{off}+t_{on} \end{cases} \tag{A2}$$

$$\begin{cases} h_1 = \frac{nV_2}{2f_sL} \\ h_2 = \frac{(n+1)k}{4} \\ h_3 = (1-k)V_F \\ h_4 = (k+1)(V_{sat}+nV_F) \\ h_5 = 2n(k-1) \end{cases} \tag{A3}$$

$$w_1 = \frac{n^2V_2^2}{48f_sL^2}\left(R_{tr}+R_{au}+\frac{2mf_s\mu_0^2N^2V_e}{g^2}\right) \tag{A4}$$

where t_{off} and t_{on} are switch turn-off time and turn-on time, respectively, and V_F is the forward voltage drop of diode. R_{tr} and R_{au} are winding resistance of the transformer and auxiliary inductor, respectively. V_e is the effective volume. V_{sat} represents the on-state voltage drop of IGBT. The remaining parameters are the device's own parameters, which have also been explained in the main text.

Appendix B

The value of the inertia weight value will affect the particle motion state. In order to prevent falling into the local optimum, it is necessary to continuously adjust the value according to the particle fitness value to increase the diversity of the particles. The specific steps are as follows:

Step 1: Calculate the fitness value of all particles, from which the entropy value between the particles is calculated, which is used to evaluate the distribution state between the particles. The formula for solving the entropy is as follows:

$$\Delta = \sum_{i=1}^{n} \left(\frac{fit(x_i) - E(fit)}{pu} \right) \tag{A5}$$

where *fit*(*x*) represents the fitness function; *E*(*fit*) represents the average fitness value of all particles; and *pu* is the normalization factor.

Step 2: When the entropy value is less than a given value, the concentration of particles is large, and the inertia weight needs to be adjusted to reduce the particle concentration. The update formula is as follows:

$$\omega(t) = \begin{cases} S(t) = 1 - \dfrac{1}{1+e^{\frac{t}{T}}} \\ \omega_{\max} - (\omega_{\max} - \omega_{\min}) \times S(t) & \Delta > 0.5 \\ \omega_{\max} - (\omega_{\max} - \omega_{\min}) \times S(t) + \overline{\omega} \times r & \Delta \leq 0.5 \end{cases} \tag{A6}$$

where *T* is the maximum number of iterations; *t* is the current iteration; $\overline{\omega}$ is the average of the maximum and minimum inertia weights.

References

1. Shafiee, S.; Fotuhi-Firuzabad, M.; Rastegar, M. Investigating the Impacts of Plug-in Hybrid Electric Vehicles on Power Distribution Systems. *IEEE Trans. Smart Grid* **2013**, *4*, 1351–1360. [CrossRef]
2. Veldman, E.; Verzijlbergh, R.A. Distribution Grid Impacts of Smart Electric Vehicle Charging from Different Perspectives. *IEEE Trans. Smart Grid* **2015**, *6*, 333–342. [CrossRef]
3. Mahmud, K.; Hossain, M.J.; Ravishankar, J. Peak-Load Management in Commercial Systems with Electric Vehicles. *IEEE Syst. J.* **2019**, *13*, 1872–1882. [CrossRef]
4. Beza, T.M.; Huang, Y.-C.; Kuo, C.-C. A Hybrid Optimization Approach for Power Loss Reduction and DG Penetration Level Increment in Electrical Distribution Network. *Energies* **2020**, *13*, 6008. [CrossRef]
5. Chen, Q.; Wang, W.; Wang, H.; Wu, J.; Wang, J. An Improved Beetle Swarm Algorithm Based on Social Learning for a Game Model of Multi objective Distribution Network Reconfiguration. *IEEE Access* **2020**, *8*, 200932–200952. [CrossRef]
6. Siti, M.W.; Nicolae, D.V.; Jimoh, A.A.; Ukil, A. Reconfiguration and Load Balancing in the LV and MV Distribution Networks for Optimal Performance. *IEEE Trans. Power Deliv.* **2007**, *22*, 2534–2540. [CrossRef]
7. Prasad, K.; Ranjan, R.; Sahoo, N.C.; Chaturvedi, A. Optimal reconfiguration of radial distribution systems using a fuzzy mutated genetic algorithm. *IEEE Trans. Power Deliv.* **2005**, *20*, 1211–1213. [CrossRef]
8. Yang, H.; Xie, X.; Vasilakos, A.V. Noncooperative and Cooperative Optimization of Electric Vehicle Charging Under Demand Uncertainty: A Robust Stackelberg Game. *IEEE Trans. Veh. Technol.* **2016**, *65*, 1043–1058. [CrossRef]
9. Lee, C.; Liu, C.; Mehrotra, S.; Bie, Z. Robust Distribution Network Reconfiguration. *IEEE Trans. Smart Grid* **2015**, *6*, 836–842. [CrossRef]
10. Azizivahed, A. Energy Management Strategy in Dynamic Distribution Network Reconfiguration Considering Renewable Energy Resources and Storage. *IEEE Trans. Sustain. Energy* **2020**, *11*, 662–673. [CrossRef]
11. Wang, J.; Wang, W.; Wang, H.; Zuo, H. Dynamic Reconfiguration of Multi objective Distribution Networks Considering DG and EVs Based on a Novel LDBAS Algorithm. *IEEE Access* **2020**, *8*, 216873–216893. [CrossRef]
12. Wagenaars, P.; Wouters, P.A.A.F.; van der Wielen, P.C.J.M.; Steennis, E.F. Influence of Ring Main Units and Substations on Online Partial-Discharge Detection and Location in Medium-Voltage Cable Networks. *IEEE Trans. Power Deliv.* **2011**, *26*, 1064–1071. [CrossRef]
13. Hua, L.; Wang, J.; Zhou, C. Adaptive Electric Vehicle Charging Coordination on Distribution Network. *IEEE Trans. Smart Grid* **2014**, *5*, 2666–2675.
14. Song, Y.; Zheng, Y.; Hill, D.J. Optimal Scheduling for EV Charging Stations in Distribution Networks: A Convexified Model. *IEEE Trans. Power Syst.* **2017**, *32*, 1574–1575. [CrossRef]
15. Zhao, J.; Wan, C.; Xu, Z. Spinning Reserve Requirement Optimization Considering Integration of Plug-In Electric Vehicles. *IEEE Trans. Smart Grid* **2016**, *8*, 2009–2021. [CrossRef]
16. Khan, S.U.; Mehmood, K.K.; Haider, Z.M.; Rafique, M.K.; Khan, M.O.; Kim, C.-H. Coordination of Multiple Electric Vehicle Aggregators for Peak Shaving and Valley Filling in Distribution Feeders. *Energies* **2021**, *14*, 352. [CrossRef]
17. Wu, J. Multi-agent modeling and analysis of EV users' travel willingness based on an integrated causal/statistical/behavioral model. *J. Mod. Power Syst. Clean Energy* **2018**, *6*, 1255–1263. [CrossRef]
18. Ayachit, A.; Kazimierczuk, M.K. Averaged Small-Signal Model of PWM DC-DC Converters in CCM Including Switching Power Loss. *IEEE Trans. Circuits Syst. II-Express Briefs* **2019**, *66*, 262–266. [CrossRef]
19. Soltau, N.; Eggers, D.; Hameyer, K.; De Doncker, R.W. Iron Losses in a Medium-Frequency Transformer Operated in a High-Power DC–DC Converter. *IEEE Trans. Magn.* **2014**, *50*, 953–956. [CrossRef]

20. Cheng, H.; Gao, Q.M.; Zhu, J.B.; Yang, X.K.; Wang, C. Dynamic Modeling and Minimum Backflow Power Controlling of the Bi-Directional Full-Bridge DC-DC Converters Based on Dual-Phase-Shifting Control. *Trans. China Electrotech. Soc.* **2014**, *29*, 245–253.
21. Roshen, W.A. A Practical, Accurate and Very General Core Loss Model for Nonsinusoidal Waveforms. *IEEE Trans. Power Electron.* **2007**, *22*, 30–40. [CrossRef]
22. Oggier, G.G.; GarcÍa, G.O.; Oliva, A.R. Switching Control Strategy to Minimize Dual Active Bridge Converter Losses. *IEEE Trans. Power Electron.* **2009**, *24*, 1826–1838. [CrossRef]
23. Inoue, S.; Akagi, H. A Bidirectional DC-DC Converter for an Energy Storage System with Galvanic Isolation. *IEEE Trans. Power Electron.* **2007**, *22*, 2302–2303. [CrossRef]

Article

A Novel Hybrid GWO-LS Estimator for Harmonic Estimation Problem in Time Varying Noisy Environment

Muhammad Abdullah [1], Tahir N. Malik [2], Ali Ahmed [1], Muhammad F. Nadeem [1,3], Irfan A. Khan [3,*] and Rui Bo [4]

[1] Department of Electrical Engineering, University of Engineering and Technology Taxila, Taxila 47080, Pakistan; m.abdullahuet@gmail.com (M.A.); engr.ranaali332@gmail.com (A.A.); faisal.nadeem@uettaxila.edu.pk (M.F.N.)

[2] Department of Electrical Engineering, HITEC University Taxila, Taxila 47080, Pakistan; tahir.nadeem@hitecuni.edu.pk

[3] Clean and Resilient Energy Systems (CARES) Research Lab., Texas A&M University, College Station, TX 77553, USA

[4] Department of Electrical and Computer Engineering, Missouri University of Science and Technology, Rolla, MO 65409, USA; rbo@mst.edu

[*] Correspondence: irfankhan@tamu.edu

Abstract: The power quality of the Electrical Power System (EPS) is greatly affected by electrical harmonics. Hence, accurate and proper estimation of electrical harmonics is essential to design appropriate filters for mitigation of harmonics and their associated effects on the power quality of EPS. This paper presents a novel statistical (Least Square) and meta-heuristic (Grey wolf optimizer) based hybrid technique for accurate detection and estimation of electrical harmonics with minimum computational time. The non-linear part (phase and frequency) of harmonics is estimated using GWO, while the linear part (amplitude) is estimated using the LS method. Furthermore, harmonics having transients are also estimated using proposed harmonic estimators. The effectiveness of the proposed harmonic estimator is evaluated using various case studies. Comparing the proposed approach with other harmonic estimation techniques demonstrates that it has a minimum mean square error with less complexity and better computational efficiency.

Keywords: grey wolf optimizer; electrical harmonics; harmonic estimation; total harmonic distortion

Citation: Abdullah, M.; Malik, T.N.; Ahmed, A.; Nadeem, M.F.; Khan, I.A.; Bo, R. A Novel Hybrid GWO-LS Estimator for Harmonic Estimation Problem in Time Varying Noisy Environment. *Energies* **2021**, *14*, 2587. https://doi.org/10.3390/en14092587

Academic Editor: Miguel Castilla

Received: 12 March 2021
Accepted: 23 April 2021
Published: 1 May 2021

Publisher's Note: MDPI stays neutral with regard to jurisdictional claims in published maps and institutional affiliations.

1. Introduction

Recently, estimation and mitigation of electrical harmonics have attained much attention due to excessive integration of non-linear, power electronic components and renewable energy sources in an Electrical Power System (EPS) [1,2]. Harmonics can be described as distortion in the fundamental voltage/current waveform of the electrical signal. In an EPS, a higher value of Total Harmonic Distortion (THD) can cause various drawbacks like deterioration of electrical components, increased power loss, and interferences for communication systems. The power quality of EPS is directly influenced by electrical harmonics. Hence, it is necessary to properly estimate and mitigate the electrical harmonics for smooth and stable operation of EPS. Moreover, various regulations and standards have been developed for harmonic levels by authorized organizations like the International Electrotechnical Commission (IEC) and Institute of Electrical and Electronics Engineers (IEEE) [3,4].

So far, various studies have been performed to model, eliminate, measure, and estimate electrical harmonics to mitigate their adverse effects. The essential problem among mentioned studies is the estimation of electrical harmonics, and different methodologies were proposed in literature to solve this issue. The accurate approximation of electrical harmonics in EPS is an extremely complex and multimodal problem [5]. Moreover, electrical harmonics are getting more prominent due to the continuous integration of non-linear power electronics equipment. Time varying noisy environment makes this problem even more complex and dynamic. So, the solution to the harmonic estimation (HE) problem needs to be upgraded [6,7]. The harmonic estimation problem deals with two types of approximations. The first one is the detection of harmonics amplitude and the second one is the detection of the harmonics phase. Amplitude estimation of particular harmonics is a linear problem, while phase and frequency estimation of particular harmonics is a nonlinear problem [8,9]. Frequency estimation of power system harmonics and their transient analysis are also sometimes resolved in harmonic estimation problem [4,10,11].

In literature, researchers solved HE problem via various mathematical, statistical, and heuristic methodologies stated in Figure 1. Initially, mathematical techniques like Fourier transform, fast Fourier transform, and discrete Fourier transform are applied, but their disadvantages, like spectral leakage, picket fence, and those which were only applicable for static signals, lead the researchers to explore more efficient tools for HE problem [12–14]. Partial accuracy was achieved by applying Hilbert and wavelet transformation [11,15]. To achieve greater accuracy, artificial intelligence techniques like Fuzzy Logic, Neural Network, Expert systems, and ADALINE are also used [16,17]. Although they had accurate results but their computational efficiency is very low.

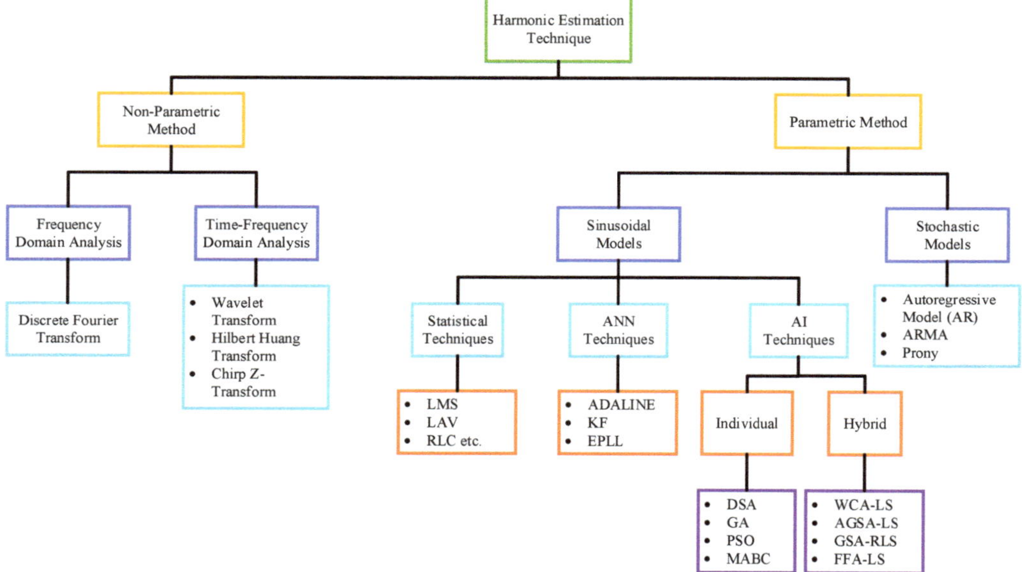

Figure 1. Classification of methodologies used for HE problem.

To enhance computational efficiency, several statistical techniques like Kalman filter (KF), Least square (LS), Least mean square (LMS), Recursive least square (RLS), and Least average value (LAV) were adopted to solve HE problem [18–22]. These techniques had a simple structure, linear behavior, and higher efficiency than mathematical and AI techniques. However, their flaws, like requirement of system information and fine-tuning of the system make them difficult to be utilized.

Recently, hybrid statistical and nature inspired meta-heuristic techniques have acquired huge attention of researchers due to their self-adaptive nature, less computational efficiency, smooth structure, and solving complex engineering problems. Different strategies were reported in literature for addressing HE problem, i.e., Local Ensemble Kalman filter (LETKF) [23,24], Kalman filter and least error square (KFLES) [25], Phase-locked loop (PLL) [24,26], Particle swarm-Passive congregation (PSOPC) [27,28], Genetic Algorithm (GA) [29], Bacterial Foraging-Recursive (BFO-RLS) [30,31], Bacterial Swarming [32],

Firefly algorithm (FFALS) [33], Artificial bee (ABCLS) [34], Biogeography-RLS [35], Differential search [36], Gravity search-RLS [37], A modified ABC (MABC) [3], and Frequency shifting-filtering method (FSF) [38]. GWO equipped with evolutionary operators [39] is also implemented for HE problems but it has large computational time due to the addition of tournament, selection, crossover, and mutation operators. Although the accuracy and computational efficiency of the aforementioned techniques are better but they offer complex structures with large controlling parameters. Moreover, above-mentioned studies have not estimated the frequency component of the signals, and the effectiveness of proposed estimators have been validated only on steady state conditions. However, few studies in the literature estimate the frequency component of the electrical signals [40,41], but they have not included transient conditions to authenticate the performance of HE estimator. While there is a dearth of studies [4,10,11] that considered transient conditions in HE problem, the methodologies utilized in the referred studies are based on statistical techniques that provide better accuracy, but their computational efficiency is very low. The accurate estimation of harmonics in an electrical signal (both in steady and transient state) can be further improved with the latest advancements in solution tools.

From the literature survey, it could be observed that each methodology implemented on the HE problem has its own pros and cons. Some techniques had optimal results but poor computational efficiency and some provide great computational efficiency but compromise accuracy. The authenticity of the literature harmonic estimators in the dynamic and composite power system, tends to be lessening. The investigators are advancing towards an updated estimator that can perform well in both steady and transient conditions with less complexity level, more accuracy, and low computational efficiency. So, it still needs a proper solution that can tackle the HE problem with better accuracy, having less computational time than the state-of-the-art techniques. The major contribution of this paper is as follows:

- A novel Grey Wolf Optimizer (GWO) and LS-based hybrid estimator is proposed for accurate estimation of harmonics in EPS.
- The proposed GWO-LS has the ability to cope with HE problems timely in the modern dynamic and complex smart grid/system.
- GWO is utilized for accurate estimation of non-linear part, and LS is used for accurate estimation of the linear part of HE problems.
- GWO is also utilized for the accurate estimation of frequency components of integral, sub, and inter harmonics.
- The transient analysis of the case studies is done to ascertain the effectiveness of the proposed GWO-LS harmonic estimator.
- The proposed harmonic estimator performed better in dynamic and noisy signals, reduces complexity level, and improves the presented harmonic estimator's computational efficiency and accuracy.
- The effectiveness of the proposed estimator is evaluated on standard test systems used by various researchers.

The rest of the manuscript is organized as: Section 2 discusses the HE problem formulation; Section 3 describes the proposed harmonic estimator. Various case studies are stated in Section 4, the final conclusion is given in Section 5, and future suggestions are conferred in Section 6.

2. Problem Formulation of Harmonic Estimation

The estimation of electrical voltage or current signals having electrical harmonics of dynamic nature is computed in two stages. The first stage deals with the non-linear approximation of harmonic's phase and frequency by GWO, while the second stage deals with the harmonic's amplitude approximation. Generally, an electrical signal containing harmonics is represented by the summation of sine-cosine functions having higher frequencies values that are integral multiples of the fundamental frequency and is given as [3],

$$Y(t) = \sum_{i=1}^{I} K_i \sin(\omega_i t + \varphi_i) + K_{dc} exp(-\gamma_{dc} t) \tag{1}$$

where i represents harmonic order, K_i shows harmonic amplitude, ω_i indicates angular frequency while φ_i states harmonic's phase. DC offset is shown by $K_{dc} exp(-\gamma_{dc} t)$ and ω_i is given as,

$$\omega_i = 2\pi \times f \tag{2}$$

The entire model of a signal having noise is given by [37]:

$$Y(t) = \sum_{i=1}^{I} K_i \sin(\omega_i t + \varphi_i) + K_{dc} exp(-\gamma_{dc} t) + N_t \tag{3}$$

here, N_t stands for total noise in a specific signal. The digital representation of the above signal is represented as:

$$Y(sT_s) = \sum_{i=1}^{I} K_i \sin(\omega_i sT_s + \varphi_i) + K_{dc} exp(-\gamma_{dc} sT_s) + N_{sT_s} \tag{4}$$

where s shows sample number and T_s represents sampling time. Applying trigonometric identity on the above equation yields:

$$Y(s) = \sum_{i=1}^{I} [K_i \sin(\omega_i sT_s)\cos\varphi_i + K_i \cos(\omega_i sT_s)\sin\varphi_i] + K_{dc} exp(-\gamma_{dc} sT_s) + N_{sT_s} \tag{5}$$

Expanding decaying offset and neglecting higher frequency terms update the above equation as,

$$Y(s) = \sum_{i=1}^{I} [K_i \sin(\omega_i sT_s)\cos\varphi_i + K_i \cos(\omega_i sT_s)\sin\varphi_i] + K_{dc} - K_{dc}\gamma_{dc} sT_s + N_{sT_s} \tag{6}$$

The equation to be estimated for the estimated signal becomes:

$$\hat{Y}(s) = X.S(s)^T \tag{7}$$

$$X = (K_1\cos\varphi_1, K_1\sin\varphi_1 \cdots\cdots K_i\cos\varphi_i, K_i\sin\varphi_i, K_{dc}, K_{dc}\gamma_{dc}, 1) \tag{8}$$

$$S(s)^T = \sum_{i=1}^{I} [\sin(\omega_1 sT_s), \cos(\omega_1 sT_s) \cdots\cdots \sin(\omega_i sT_s), \cos(\omega_i sT_s)], 1, -sT_s, N_{sT_s} \tag{9}$$

where $S(s)^T$ and X is the matrix of known and unknown parameters, respectively, while X is updated successfully to approximate signal $\hat{Y}(s)$. The amplitude and phase of an unknown ith harmonic with the decaying dc component after the X matrix is updated can be as:

$$K_i = \sqrt{\varphi_{2i}^2 + \varphi_{2i-1}^2} \tag{10}$$

$$\varphi_i = \tan^{-1}\left(\frac{\varphi_{2i}}{\varphi_{2i-1}}\right) \tag{11}$$

$$K_{dc} = \varphi_{2i+1} \tag{12}$$

$$\varphi_{dc} = \left(\frac{\varphi_{2i+2}}{\varphi_{2i+1}}\right) \tag{13}$$

Finally, the objective function for the harmonic estimation problem is formulated as [6]:

$$S = \min \sum_{i=1}^{I} MSE_i^2(i) = \min \sum_{i=1}^{I} \left[Y(s) - \hat{Y}(s) \right]^2 \tag{14}$$

here, $Y(s)$ is our actual power signal while $\hat{Y}(s)$ is the final approximated signal by GWO-LS.

3. Proposed Methodology

3.1. Grey Wolf Optimizer

GWO is a nature inspired-heuristic algorithm proposed by Seyed Ali Mirjalili [42] and is widely employed to obtain the optimal solution for various optimization problems. GWO is inspired by the social behavior of grey wolves; naturally, they live in a group form called packs, and the members of each group vary from 5 to 12. Strong social hierarchy is a unique feature of grey wolves. This social hierarchy is based on four ranks of wolves that are termed as alpha (α), beta (β), delta (δ), and omega (ω) wolf. The 'α' is the superior of all because of its leadership quality, managerial skills, and decision-making power, while 'β' wolf is inferior to α and acts as its adviser but discipliner for the rest of the pack. The third level belongs to delta wolves, and they submit themselves to both alpha and beta wolves. They are also helpful in hunting, care-taking of the whole pack. The lowest level of the pack is an omega wolf having lethargic nature.

Another fascinating phenomenon of grey wolves is their unique hunting mechanism, which undergoes five steps: (1) tracking, (2) chasing, (3) encircling, (4) harassing, and (5) attacking. Figure 2 displays the complete hunting process of grey wolves. Figure 2A shows the wolf tracking and chasing, while Figure 2B–D reveals how wolves encircle and harass their prey. Similarly, the attacking behavior of wolves is depicted in Figure 2E. The encircling behavior of the grey wolf is modeled using the following equations.

$$D = \left| \vec{C}.\vec{P}_p(y) - \vec{P}(y) \right| \tag{15}$$

$$\vec{P}(y+1) = \vec{P}_p(y) - \vec{A} \times \vec{D} \tag{16}$$

where y presents the current iteration, D shows the distance between specific wolf and prey while $\vec{A}$ and $\vec{C}$ are the control variables for updating iteration. $\vec{P}$ indicates the best location of grey wolves (Local best) while $\vec{P}_p$ depicts the position of prey with respect to the grey wolf (Global best). The coefficient vectors $\vec{A}$ and $\vec{C}$ are computed as follows:

$$\vec{A} = 2\vec{a}.c_1 - \vec{a} \tag{17}$$

$$\vec{C} = 2.c_2 \tag{18}$$

where $\vec{a}$ is computed using $\vec{a} = 2 - 1 * (2/max\_iter)$, and its value decreases from 2 to 0. $\vec{C}$ is a control variable having a value in the range of 0–2 while c_1 and c_2 are two random numbers between $[0, 1]$. The hunting mechanism of the grey wolf is modeled by:

$$\vec{D}_\alpha = \left| \vec{C}_1.\vec{P}_\alpha - \vec{P} \right| \tag{19}$$

$$\vec{D}_\beta = \left| \vec{C}_1.\vec{P}_\beta - \vec{P} \right| \tag{20}$$

$$\vec{D}_\delta = \left| \vec{C}_1.\vec{P}_\delta - \vec{P} \right| \tag{21}$$

$$\vec{P_1} = \left| \vec{P_\alpha} - \vec{A_1}.\vec{D_\alpha} \right| \tag{22}$$

$$\vec{P_2} = \left| \vec{P_\beta} - \vec{A_1}.\vec{D_\beta} \right| \tag{23}$$

$$\vec{P_3} = \left| \vec{P_\delta} - \vec{A_1}.\vec{D_\delta} \right| \tag{24}$$

$$\vec{P}(y+1) = \frac{\vec{P_1} + \vec{P_2} + \vec{P_3}}{3} \tag{25}$$

here, P_α and D_α represent the position of α from the prey and distance between α and prey, respectively. Similarly, P_β, P_δ, D_β, and D_δ describe the position and distance of respective wolves. The grey wolves terminate the hunting step by attacking the prey, and the attacking phenomenon relates to the exploitation phase of the GWO algorithm. The value of a decreases from 2 to 0 to model the attacking behavior. The position vector A is a random value in the range of $[-2a, 2a]$. When the value of variable A lies in $[-1, 1]$, the new position of the searching agent can be generated at any point between its current location and the prey location. If $|A| < 1$ (convergence), this value compels the wolf for attacking the prey, while if $|A| > 1$ (divergence) wolves search more for the better prey (exploration process) as shown in Figure 3. Another coefficient vector C supports the searching mechanism of GWO, and it describes the obstacles which can occur in nature during the hunting step. The value of C allows the wolf to prevent obstacles and approach the prey. Therefore, we can say that A and C are the control variables here.

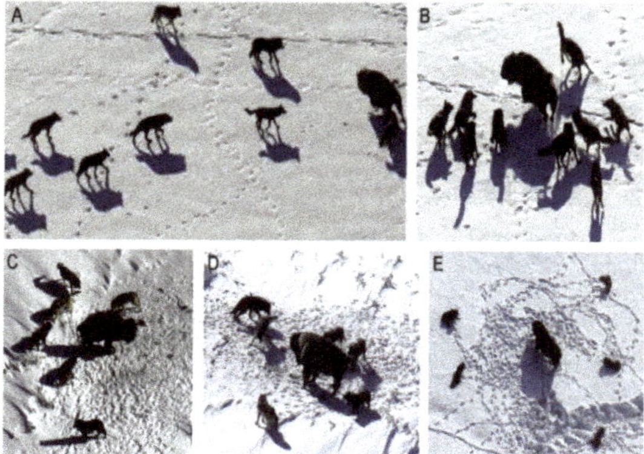

Figure 2. Prey Hunting mechanism of Grey Wolves [42]. (**A**) tracking and chasing (**B–D**) encircling and harassing and (**E**) attacking.

Figure 3. Exploration and Exploitation in GWO.

3.2. Hybrid GWO-LS Harmonic Estimator

In this paper, a hybrid of statistical technique Least square method (LSM) [43] and a meta-heuristic technique GWO is proposed to solve the HE problem. The purpose of hybridizing two different nature techniques is to minimize the computational burden, increasing computational efficiency and accurate estimation of harmonics. The prominent feature of GWO is its social hierarchy and is well adapted for solving complex problems, and its working methodology is explained in the previous section, while the selection of LSM to develop the proposed harmonic estimator is based on various reasons. Firstly, it reduces the computational time of the GWO. Secondly, it moves to an exact solution as a statistical technique rather than trapping in local optima. So, it provides accurate results in linear estimation (Estimation of harmonic amplitude) of the HE problem. Thirdly it reduces the computational burden of the GWO and improves the convergence characteristics of GWO-LS as a whole. In the proposed hybrid harmonic estimator, LSM is utilized for accurate estimation of the linear part (harmonic amplitude) of the HE problem, while GWO is utilized for a proper approximation of the non-linear part (harmonic phase and frequency) of the HE problem.

For proper estimation of harmonics in the desired signal, H_t which is a matrix of grey wolves having t number of searching agents are initialized according to a number of power system harmonics to be estimated. The objective function of the HE problem is to minimize Mean Square Error (MSE). First of all, a matrix of t number of searching agents (Grey Wolves) is defined where the location of a single searching agent consists of t number of harmonics.

$$H_i = \left(H_1^i, H_2^i, H_3^i \cdots \cdots , H_n^i \right) \tag{26}$$

Numerous searching agents combine to form an agent matrix H_t.

$$H_t = \begin{bmatrix} H_1^1 & H_2^1 & H_3^1 & \cdots & H_i^1 \\ H_1^2 & H_2^2 & H_3^2 & \cdots & H_i^2 \\ H_1^3 & H_2^3 & H_3^3 & \cdots & H_i^3 \\ \vdots & \vdots & \vdots & \vdots & \vdots \\ H_1^t & H_2^t & H_3^t & \cdots & H_i^t \end{bmatrix} \tag{27}$$

here, t indicates the population size of grey wolves, and i is the total harmonics number in a single search agent. A single searching agent consists of various locations, and these locations are equal to the total number of harmonics in a signal. The location of each search agent is initialized randomly as follows:

$$H_i^t = B_L + rand(B_U - B_L) \tag{28}$$

where B_L and B_U are the lower and upper bounds specified by the HE problem, respectively. The GWO-LS estimator aims to properly approximate the harmonics' amplitude and phase of the harmonics in a time-varying noisy environment.

The error signal, which is obtained by the comparison of the actual signal and approximated signal, is fed to the optimization framework. This framework operates to derogate the estimation error until best result is obtained. The proposed harmonic estimator GWO-LS is said to be the least complex estimator due to the smaller number of control variables. It has only two control variables irrespective of GA-LS having four, PSO-LS having five, BFO having four, F-BFO-LS having six, BFO-RLS having four, BBO-RLS having four, GSA-RLS having five, and MABC having four controlling variables that improve the computational time ultimately. The detail description of GWO-LS tuning parameters is given in Table 1 while the detailed flowchart of GWO-LS is depicted in Figure 4.

Table 1. Tuning Parameters for GWO-LS framework.

Control Variables	Range	Description
a	(2–0)	Linearly decreased 2 to 0 over the course of iterations A is the random value in the interval $[-2a, 2a]$.
A	$(-2a, 2a)$	If A < 1, then the Exploitation process occurs; otherwise, exploration process.
C	(0–2)	It supports the exploration process and models the obstacles which can occur in nature during the hunting step.

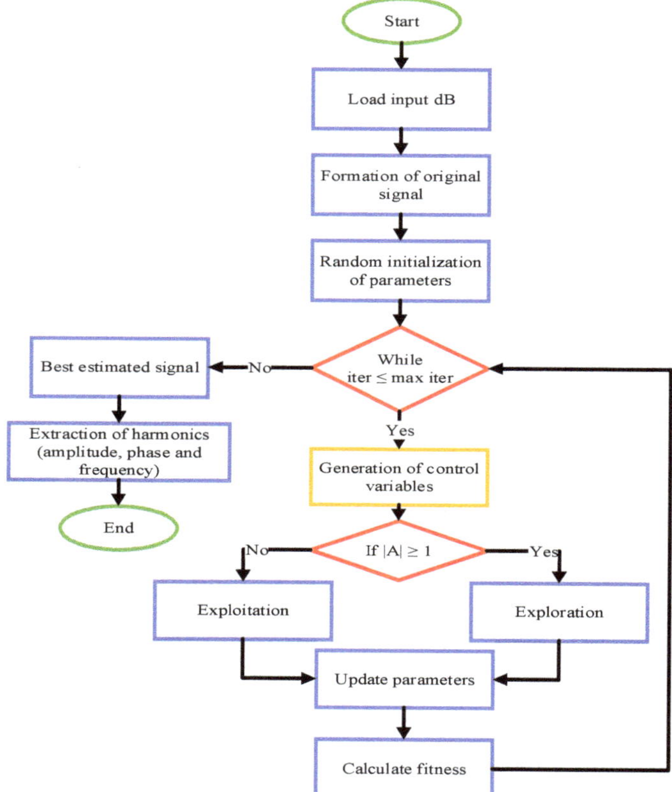

Figure 4. Flowchart of GWO-LS Harmonic Estimator.

3.3. Computational Procedure

The proposed methodology to solve the HE problem comprises of following steps:

- Load input database.
- A signal named "original signal" is formulated by utilizing an input database.
- Initialization of HE and GWO-LS parameters.
- GWO-LS is applied for updating unknown parameters.
- Formation of estimated signal by updated parameters.
- Comparison of the original signal and estimated signal to evaluate objective function (MSE).

4. Simulation Results and Discussion

In this paper, two case studies are utilized to validate the effectiveness of the proposed harmonic estimator. These case studies are widely used in literature for the comparative analysis of the HE problem and are given as:

1. *Test System I: Estimating integral harmonics without including Sub and Inter harmonics in time-varying noisy environments.*
2. *Test System II: Estimation of integral harmonics including Sub and Inter harmonics in a time-varying noisy environment.*

The signal-to-noise ratio levels are selected as 10 dB, 20 dB, and 40 dB for a fair comparison of the proposed estimator with literature available estimators. In addition to creating time varying noisy environment, the constructed signals are made more complex for estimation by including transients at different time intervals. Moreover, three different performance indicators are utilized to examine the performance of the proposed GWO-LS estimator in comparison with the state-of-the-art estimator available in the literature. The selected performance indicators are given as:

- Mean Square Error (MSE) which can be computed using (13).
- Residual Sum of Squares (RSS) and is calculated as:

$$ RSS = \sum \left[Y(s) - \hat{Y}(s) \right]^2 \tag{29} $$

- Performance Index (PER) is another evaluation indicator used in this study for the comparison of GWO-LS with the state of art techniques and is determined as:

$$ PER = \frac{\sum \left[Y(s) - \hat{Y}(s) \right]^2}{Y^2} \times 100\% \tag{30} $$

The simulation work is carried out in MATLAB 2019a programming environment on a personal computer having Windows 8 operating system with a 2.30 GHz processor and 6GB RAM.

4.1. Test System I: Estimation of Integral Harmonics without Including Sub and Inter Harmonics in Time-Varying Noisy Environment

The test signals generated by variable frequency drives (VFDs), electric arc furnaces, and power electronics devices are used for the estimation of integral harmonics in the current study. These signals are extensively used in literature for the comparative analysis of various harmonic estimators. The data of frequency, phase, and amplitude required for these signals is given in Table 2. The original test signal, which is a continuous signal, is generated using data provided in Table 1 [3,39]. The test signal is modeled and discretized by a renowned Nyquist criterion having 64 samples in one cycle, and the sampling frequency is set to be 3.2 kHz. The GWO-LS framework has been applied to estimate harmonics in this test signal having multiple levels of adaptive noise with the inclusion of decaying DC offset. The time-varying noisy environment is created by adding the different random noises and DC offset values in the original signal. To validate the performance of presented estimator in approximating dynamic parameters, a short transient is produced in 5th harmonic from 0.12 s to 0.26 s.

Table 2. Harmonic Contents of Actual Signal.

Harmonics Number	Frequency (Hz)	Amplitude (p.u)	Phase (Degree)
1	50	1.5	80
3	150	0.5	60
5	250	0.2	45
7	350	0.15	36
11	550	0.1	30

The simulation parameters for the HE problem and GWO-LS are stated in Table 3. These values are taken from the literature harmonic estimators [3,39,43]. Moreover, the parametric values of Table 3 are selected in such a way so that a fair comparison between the proposed and literature harmonic estimator can be performed. The searching agent is an important simulation parameter. Its value is selected as 50 because the MSE comes out to be least for this value. The variation of Evaluation parameters with respect to searching agents is shown in Figure 5.

Table 3. Case Study I Parameters for Simulation.

Model Parameters	Parametric Value
Number of Iterations	1000
Grey wolves (Searching Agents)	50
Number of trials	25
Nyquist Criterion Samples per cycle	64
Sampling frequency	3.2 kHz
Noise levels in dB	40, 20, 10
DC Offset	0.5 exp $(-5t)$
Number of Iterations	1000

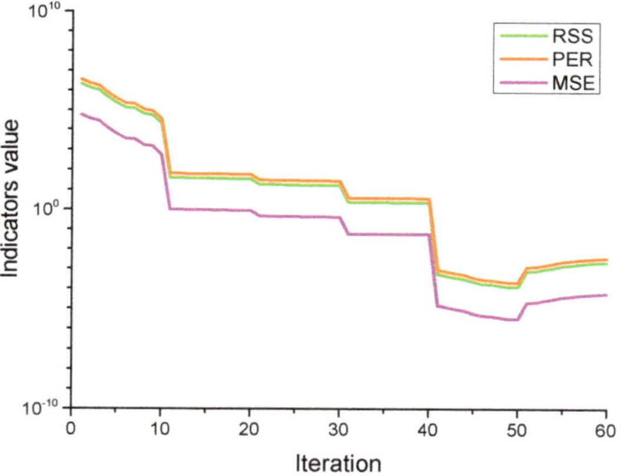

Figure 5. Variation of evaluation parameters of RSS, PER and MSE with respect to searching agents.

Two signals are generated using the proposed harmonic estimator; the first one is the actual original signal made from input data with five integral harmonics. The second signal is an approximated one that is obtained by GWO-LS containing the approximated harmonics amplitudes and phases. These two signals are compared via MSE. The actual and approximated harmonics are then compared with respect to the percentage error. The simulation is carried out on four signals which are actual signals with no noise, a signal having 40 dB, 20 dB, and 10 dB noises (SNRs). The output values of best and worst harmonics amplitudes and phases with their percentage errors from actual harmonic values are tabulated in Table 4.

Table 4. Case Study I Parameters after Simulation.

Model Parameters	Parametric Value
Best MSE	2.01×10^{-5}
Worst MSE	2.31×10^{0}
Average MSE	3.16×10^{-1}
Standard deviation	6.47×10^{-1}
Total Harmonic Distortion	1.43×10^{-1}

The comparison of actual and estimated signals is shown in Figure 6 at different SNRs. It is evident from Figure 6a that the estimated signal using GWO-LS exactly matches the original signal. However, it can be observed from Figure 6c,d that the signals become corrupted with the addition of different noise levels and decaying DC offset. In the presence of a 40 dB noise level, the proposed GWO-LS estimator accurately approximates the original signal while small deviations are observed in the original and the approximated signal when added noise is 20 dB. The estimation of the signal is more difficult in the presence of a 10 dB noise level, but a comparison of results from Table 5 indicates that the GWO-LS estimator outclasses the state of art techniques in such complex signal estimation.

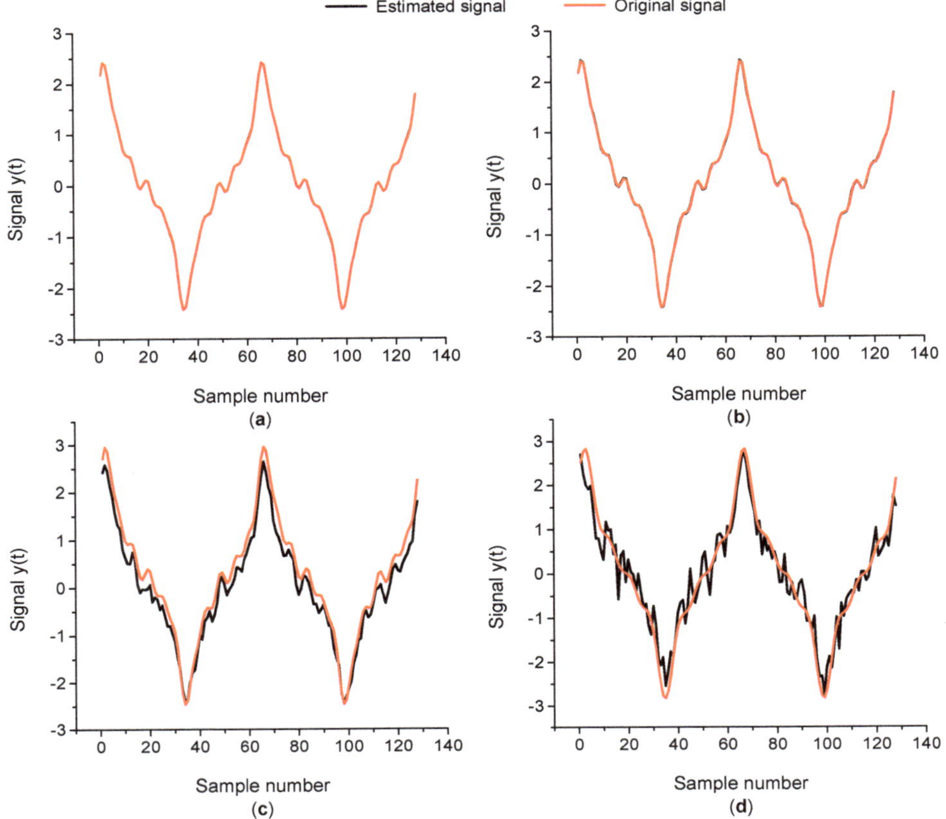

Figure 6. Actual and Estimated signal comparison (**a**) non-noisy, (**b**) 40 dB noise level, (**c**) 20 dB noise level, and (**d**) 10 dB noise level.

Table 5. GWO-LS Numerical Comparison for Case Study I.

Techniques	Parameter	1st Harmonic	3rd Harmonic	5th Harmonic	7th Harmonic	11th Harmonic	Computational Time (s)
Test Signal	Frequency (Hz)	50	150	250	350	550	-
	Amplitude (per unit)	1.5	0.5	0.20	0.15	0.1	
	Phase (degree)	80	60	45	36	30	
F-BFO-LS	Amplitude (per unit)	1.49×10^0	4.89×10^{-1}	2.08×10^{-1}	1.47×10^{-1}	1.02×10^{-1}	10.532
	Percentage Error (%)	7.33×10^{-1}	4.89×10^{-1}	4.00×10^0	2.13×10^0	1.90×10^0	
	Phase (degree)	7.99×10^1	6.12×10^1	4.72×10^1	3.67×10^1	3.05×10^1	
	Percentage Error (%)	1.75×10^{-1}	1.93×10^0	4.93×10^0	1.83×10^0	1.73×10^0	
BFO-RLS	Amplitude (per unit)	1.50×10^0	4.92×10^{-1}	2.01×10^{-1}	1.48×10^{-1}	1.02×10^{-1}	9.345
	Percentage Error (%)	1.95×10^{-1}	1.59×10^0	4.54×10^{-1}	1.41×10^0	1.48×10^0	
	Phase (degree)	7.99×10^1	5.91×10^1	4.63×10^1	3.64×10^1	3.01×10^1	
	Percentage Error (%)	1.06×10^{-1}	1.54×10^0	2.84×10^0	1.24×10^0	2.14×10^{-1}	
BBO-RLS	Amplitude (per unit)	1.50×10^0	5.00×10^{-1}	2.01×10^{-1}	1.49×10^{-1}	9.99×10^{-2}	5.852
	Percentage Error (%)	1.05×10^{-1}	7.85×10^{-2}	4.45×10^{-1}	9.56×10^{-1}	1.00×10^{-1}	
	Phase (degree)	8.00×10^1	5.95×10^1	4.55×10^1	3.61×10^1	3.00×10^1	
	Percentage Error (%)	6.25×10^{-2}	7.45×10^{-1}	1.16×10^0	3.28×10^{-1}	4.10×10^{-2}	
GSA-RLS	Amplitude (per unit)	1.50×10^0	5.00×10^{-1}	2.01×10^{-1}	1.50×10^{-1}	9.99×10^{-2}	5.6545
	Percentage Error (%)	9.45×10^{-2}	5.52×10^{-2}	3.55×10^{-1}	7.56×10^{-1}	9.00×10^{-2}	
	Phase (degree)	8.00×10^0	5.96×10^1	4.55×10^1	3.61×10^1	3.00×10^1	
	Percentage Error (%)	5.15×10^{-2}	6.55×10^{-1}	1.11×10^0	3.08×10^{-1}	3.25×10^{-2}	
MABC	Amplitude (per unit)	1.50×10^0	5.00×10^{-1}	2.00×10^{-1}	1.50×10^{-1}	1.00×10^{-1}	1.0110
	Percentage Error (%)	5.30×10^{-2}	4.41×10^{-2}	1.39×10^{-1}	3.19×10^{-2}	1.84×10^{-2}	
	Phase (degree)	8.00×10^1	6.01×10^1	4.52×10^1	3.60×10^1	3.00×10^1	
	Percentage Error (%)	2.50×10^{-2}	2.09×10^{-1}	3.64×10^{-1}	3.60×10^{-3}	1.40×10^{-1}	
GWO-LS	Amplitude (per unit)	1.50×10^0	4.99×10^{-1}	2.00×10^{-1}	1.50×10^{-1}	9.99×10^{-2}	0.571
	Percentage Error (%)	2.17×10^{-2}	1.83×10^{-2}	7.65×10^{-2}	1.33×10^{-2}	3.50×10^{-3}	
	Phase (degree)	8.00×10^1	6.00×10^1	4.50×10^1	3.60×10^1	3.00×10^1	
	Percentage Error (%)	7.70×10^{-3}	1.48×10^{-2}	6.60×10^{-2}	2.70×10^{-3}	8.33×10^{-2}	

The convergence characteristics of the proposed GWO-LS estimator using selected indicators at different noisy signals are shown in Figure 7. The proposed estimator converges in less than 120 iterations for a non-noisy signal estimation. Similarly, GWO-LS takes 145, 162, and 184 iterations to converge for the 40 dB, 20 dB and 10 dB signal, respectively. Figure 7 indicates that the number of iterations to converge increases as the signal's noise level increases. It is evident from this Figure that the proposed harmonic estimator takes a higher time for the convergence of signal having a higher noise level. The computational time comparison of GWO-LS with the literature techniques is given in Table 6. It can be observed that the computational efficiency of the proposed harmonic estimator is much higher.

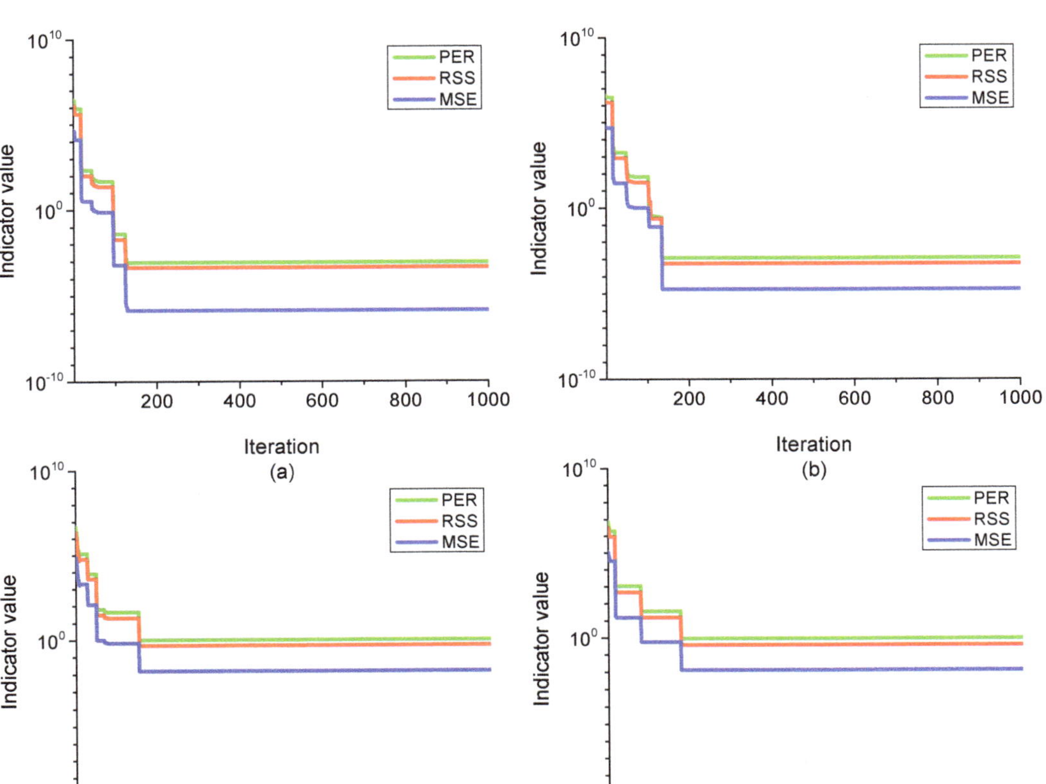

Figure 7. The convergence characteristics of the GWO-LS framework for the case study I (**a**) non-noisy signal (**b**) 40 dB (**c**) 20 dB and (**d**) 10 dB.

The convergence behavior of GWO-LS for estimation of integral harmonic's amplitude and phases is shown in Figure 8. It can be seen from Figure 8a that the proposed harmonic estimator speedily converges for the first harmonic, whereas a large number of iterations are required to converge for the seventh harmonic. Figure 8b depicts that the proposed GWO-LS framework requires almost the same number of iterations for all harmonics. The estimation of amplitude and phase of the seventh harmonic is much more complex than the rest of the harmonic contents under a non-noisy and noisy environment within the framework of time-varying nature.

Table 6. GWO-LS Numerical Comparison for Case Study II.

Techniques	Parameter	Sub Harmonic	1st Inter Harmonic	2nd Inter Harmonic	Computational Time (s)
Test Signal	Frequency (Hz)	20	180	230	-
	Amplitude (per unit)	0.505	0.25	0.35	
	Phase (degree)	75	65	20	
F-BFO-LS	Amplitude (per unit)	5.21×10^{-1}	2.61×10^{-1}	3.71×10^{-1}	13.253
	Percentage Error (%)	3.25×10^{0}	4.40×10^{0}	6.00×10^{0}	
	Phase (degree)	7.46×10^{1}	6.43×10^{1}	1.97×10^{1}	
	Percentage Error (%)	5.20×10^{-1}	1.03×10^{0}	1.39×10^{0}	
BFO-RLS	Amplitude (per unit)	5.11×10^{-1}	2.58×10^{-1}	3.64×10^{-1}	12.837
	Percentage Error (%)	1.19×10^{0}	3.24×10^{0}	3.97×10^{0}	
	Phase (degree)	7.48×10^{1}	6.53×10^{1}	1.99×10^{1}	
	Percentage Error (%)	2.53×10^{-1}	5.30×10^{-1}	6.61×10^{-1}	
BBO-RLS	Amplitude (per unit)	4.94×10^{-1}	2.46×10^{-1}	3.50×10^{-1}	6.7525
	Percentage Error (%)	1.13×10^{0}	1.65×10^{0}	7.88×10^{-2}	
	Phase (degree)	7.49×10^{1}	6.52×10^{1}	2.00×10^{1}	
	Percentage Error (%)	9.05×10^{-2}	2.63×10^{-1}	1.12×10^{-1}	
GSA-RLS	Amplitude (per unit)	4.94×10^{-1}	2.03×10^{-1}	3.50×10^{-1}	6.1575
	Percentage Error (%)	1.11×10^{0}	1.45×10^{0}	6.58×10^{-2}	
	Phase (degree)	7.49×10^{1}	6.50×10^{1}	2.00×10^{1}	
	Percentage Error (%)	7.55×10^{-2}	2.25×10^{-1}	1.04×10^{-1}	
MABC	Amplitude (per unit)	5.05×10^{-1}	2.50×10^{-1}	3.50×10^{-1}	1.4860
	Percentage Error (%)	4.68×10^{-2}	1.33×10^{-2}	1.10×10^{-1}	
	Phase (degree)	7.50×10^{1}	6.49×10^{1}	2.00×10^{1}	
	Percentage Error (%)	6.15×10^{-2}	9.64×10^{-2}	2.00×10^{-1}	
GWO-LS	Amplitude (per unit)	5.05×10^{-1}	2.50×10^{-1}	3.50×10^{-1}	1.17976
	Percentage Error (%)	2.16×10^{-2}	9.11×10^{-3}	1.11×10^{-3}	
	Phase (degree)	7.50×10^{1}	6.50×10^{1}	2.00×10^{1}	
	Percentage Error (%)	1.90×10^{-2}	4.07×10^{-2}	1.62×10^{-1}	

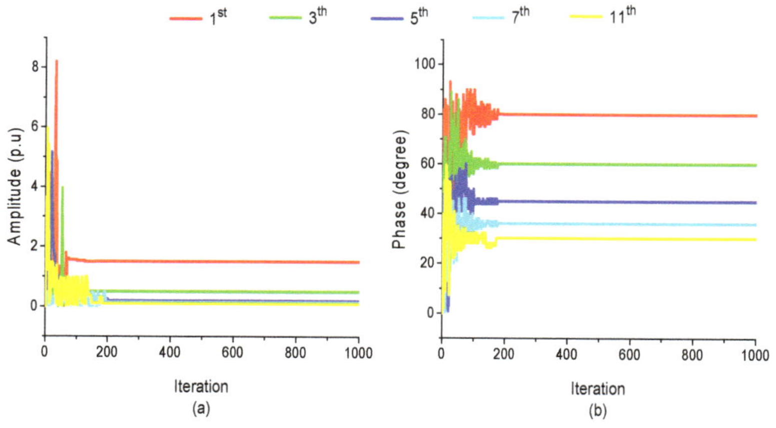

Figure 8. Variation in estimated (**a**) amplitude and (**b**) phase in the course of iterations for the first 5 odd harmonics.

To evaluate the proposed GWO-LS harmonic estimator's effectiveness, the obtained results are compared with available harmonic estimators in literature, including F-BFO-LS, BFO-RLS, and BBO-RLS GSA-RLS and MABC in terms of percentage error and computational time. The comparison of GWO-LS with literature techniques is tabulated in Table 5. The percentage error of GWO-LS for the 1st, 3rd, 5th, 7th, and 11th harmonic phase are 7.70×10^{-3}, 1.48×10^{-2}, 6.60×10^{-2}, 2.7×10^{-3}, and 8.33×10^{-2}, respectively. The minimum error is achieved for 1st, 3rd, 5th, 7th, and 11th using GWO-LS in comparison with other harmonic estimators. The comparison between the proposed and literature harmonic estimator in case of harmonics amplitude and phase can be better seen in Figures 9 and 10. It is clear from both the Figures that the proposed GWO-LS estimator has the least percentage error for all the harmonic contents. The behavior of the proposed estimator for approximating frequency components of integral harmonics is demonstrated by Figure 11. It can be clearly seen from Figure 11 that GWO-LS takes a greater number of iterations for estimating frequency than the harmonics amplitude and phase estimation. The proposed harmonic estimators approximate the fundamental frequency in 284 iterations, frequency of 3rd, 5th, 7th, and 11th harmonic in 266, 257, 261, and 276 number of iterations, respectively. It indicates that estimation of fundamental frequency takes higher iterations in case of integral harmonics approximation. It can be concluded form the pictorial and tabular analysis that GWO-LS accurately estimates the frequency components of integral harmonics along with the amplitude and phase of the harmonics.

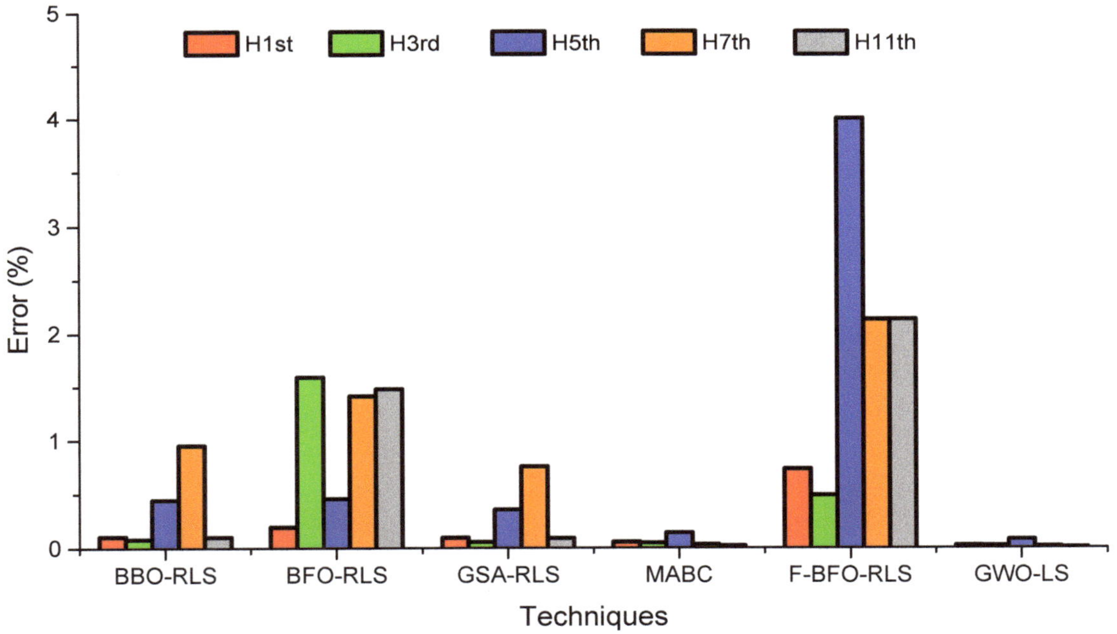

Figure 9. Comparison of amplitudes error for case study I.

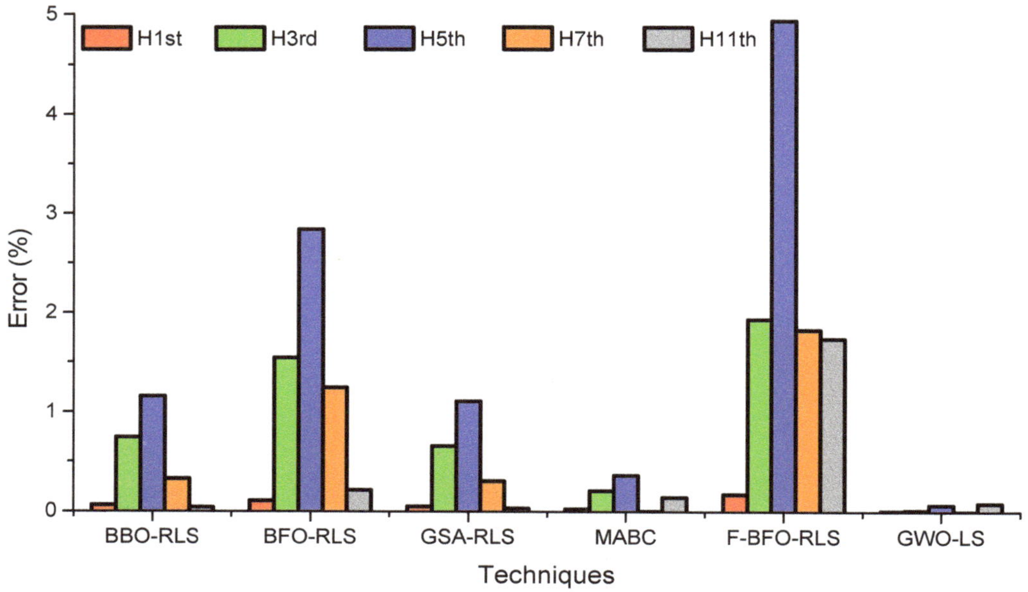

Figure 10. Comparison of phase errors for case study I.

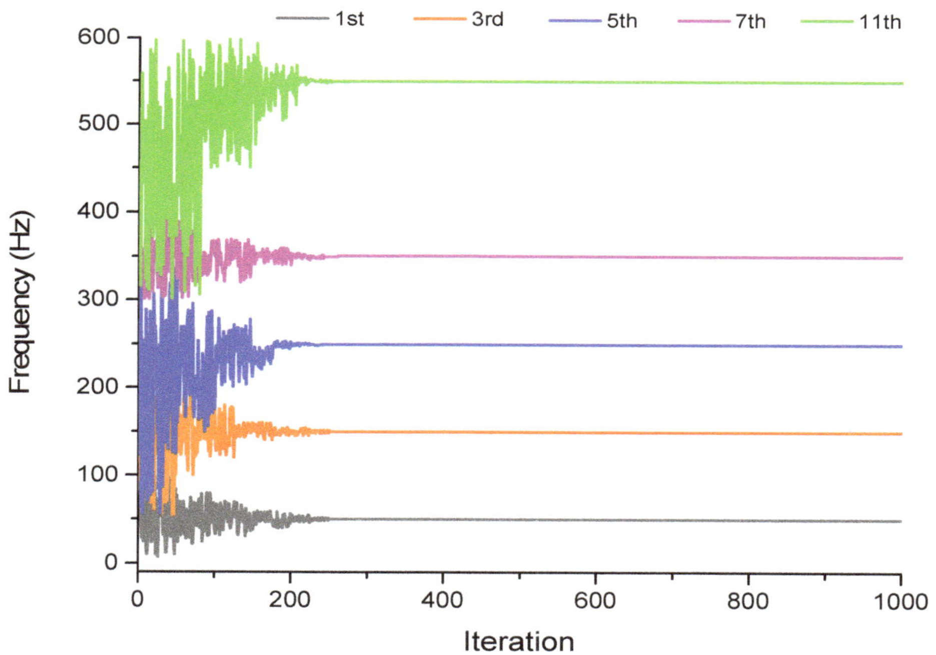

Figure 11. Frequency estimation for case study I.

Figure 12 demonstrates the GWO-LS behavior for estimating the amplitude of the harmonics in both steady and transient conditions. It is evident from Figure 12 that GWO-LS gives accurate estimation results in steady-state cases while a slight variation is observed for the transient case. The proposed estimator shows estimation variation from 0.17 s to 0.20 s clearly visible in Figure 12b. Similarly, the computational time of GWO-LS to accurately estimate a signal is 0.572 s. It can be observed from Table 5 that the time required to estimate a signal in a noisy environment is minimum for GWO-LS in comparison with all other techniques and it estimates the harmonic amplitudes and phases accurately in the least computational time.

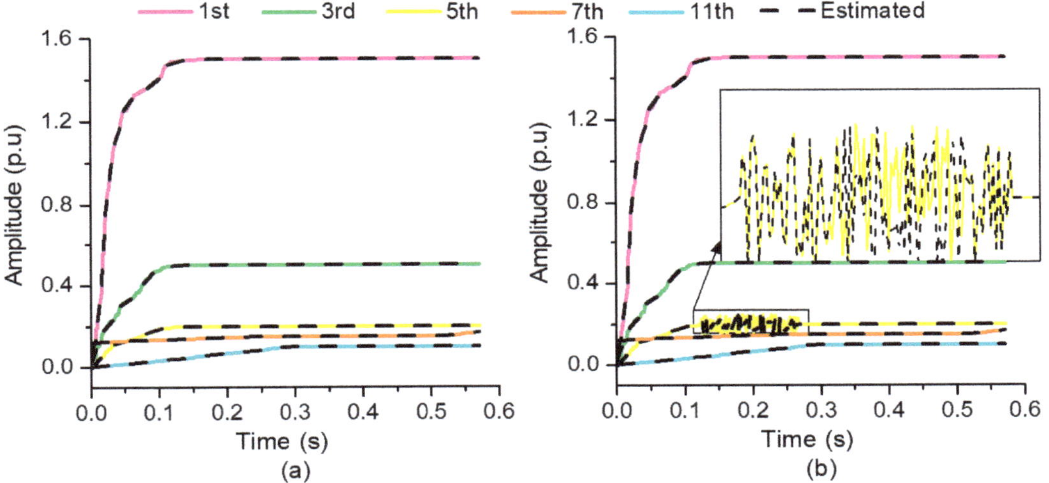

Figure 12. Estimation in harmonics amplitude (**a**) steady state condition (**b**) transient in 5th harmonic.

4.2. Test System II: Estimation of Integral Harmonics Including Sub and Inter Harmonics in a Time-Varying Noisy Environment

In this test system, the power signal is made more complex and distorted by the inclusion of sub and inter harmonics. These sub and inter harmonics have amplitude magnitudes of 0.505, 0.25, and 0.35, respectively, while their phase are 75, 65, and 20, having the frequency of 20, 180, and 230 Hz, respectively [3,38]. The resultant power signal is simulated under both noisy and non-noisy environments. All the other harmonic contents (Amplitudes and phases of integral harmonics) and model evaluation setup remains the same as presented in the previous case study. However, to validate the performance of the presented estimator in approximating dynamic parameters, a short transient is produced in 1st inter harmonic from 0.38 s to 0.5 s.

The numerical values obtained after simulation for this case study are presented in Table 7. If we compare this Table with Table 4 we can see that all values of MSE become higher after considering sub and inter harmonics which indicates that consideration of such harmonics is difficult to estimate. The variation of MSE for the case study for different values of searching agents is presented in Figure 13. It can be seen that the minimum value of MSE is achieved for the 50 number of searching agents. The MSE for this case study is greater than that of case study I due to the inclusion of sub and inter harmonics.

Table 7. Case Study II Parameters after Simulation.

Model Parameters	Parametric Value
Best MSE	4.30235×10^{-5}
Worst MSE	4.05×10^{0}
Average MSE	3.89×10^{-1}
Standard deviation	9.06×10^{-1}

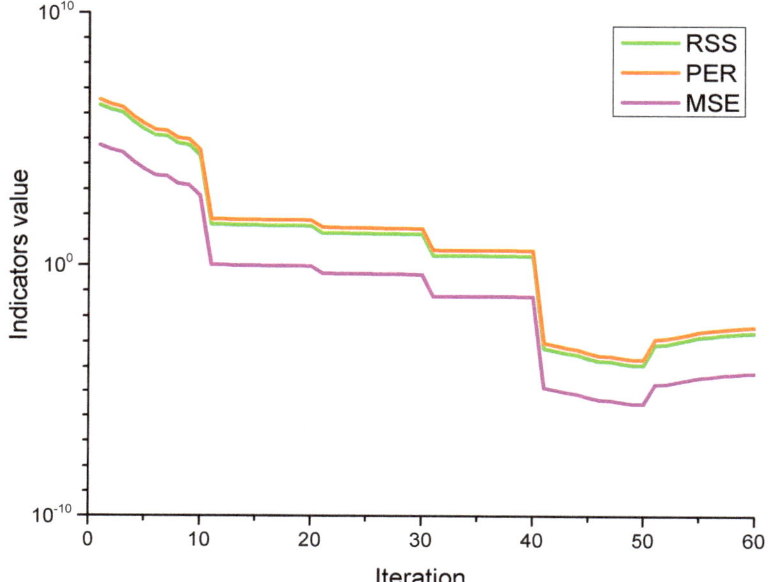

Figure 13. Searching agents for case study-II.

The pictorial form of the original and estimated signal for the GWO-LS estimator can be seen in Figure 14. It can be observed that the power signal is completely changed after the inclusion of sub and inters harmonics. The GWO-LS estimates the non-noisy and 40 dB signal exactly with great accuracy, as shown in Figure 14. The power signal having 10 dB noise is much difficult to estimate than the others due to higher noise. However, still proposed estimator provides better results than estimators presented in the literature.

The convergence characteristics for this case study are shown in Figure 15. Figure 15 indicates that the number of iterations to converge increases as the signal's noise level increases. It is evident from this Figure that the proposed harmonic estimator takes a higher time for the convergence of signal having a higher noise level. Moreover, the presented harmonic estimator takes greater time to converge when compared to case study I. The GWO-LS takes fewer iterations, i.e., 136, to converge for non-noisy signal, but the number of iterations increases with the addition of sub and inter harmonics. GWO-LS takes maximum iterations for 10 dB noisy cases that are 190 iterations.

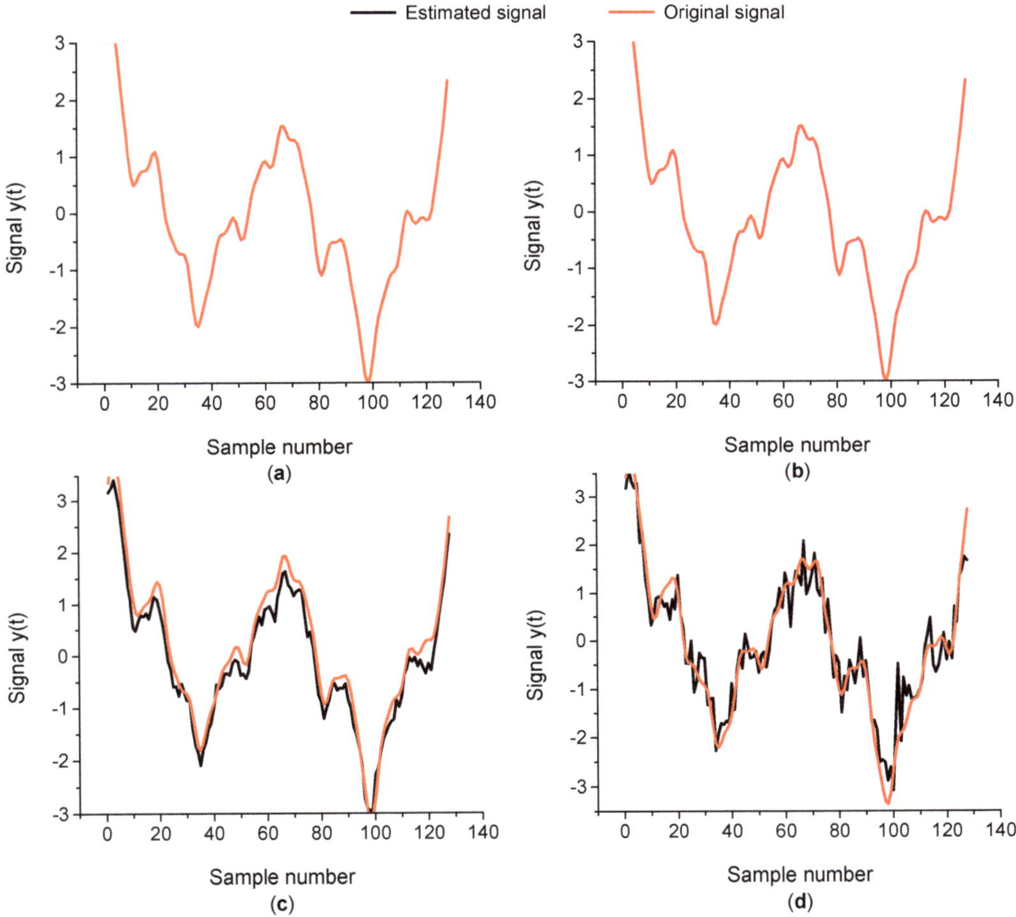

Figure 14. Actual and Estimated signal comparison (**a**) non-noisy, (**b**) 40 dB noise level, (**c**) 20 dB noise level and (**d**) 10 dB noise level.

The convergence behavior of GWO-LS estimator for the approximation of sub and inter harmonics over the course of iterations is described in Figure 16. The amplitude and phase of subharmonic take more iteration to converge and more difficult than the amplitude and phase of inter harmonics. Sub and inter harmonics are hard to approximate due to their fractional frequency nature. The amplitudes of subharmonics and second inter harmonic converge in 314 and 323 iterations, while first inter harmonic takes 289 iterations to converge. The phase of sub and inter harmonics converge in 300, 272, and 288 iterations, respectively. The presented estimator takes more iterations to converge for approximating sub and inter harmonics amplitudes and phases than the integral harmonics under time varying noisy environment but still it gives better results than the state of art techniques.

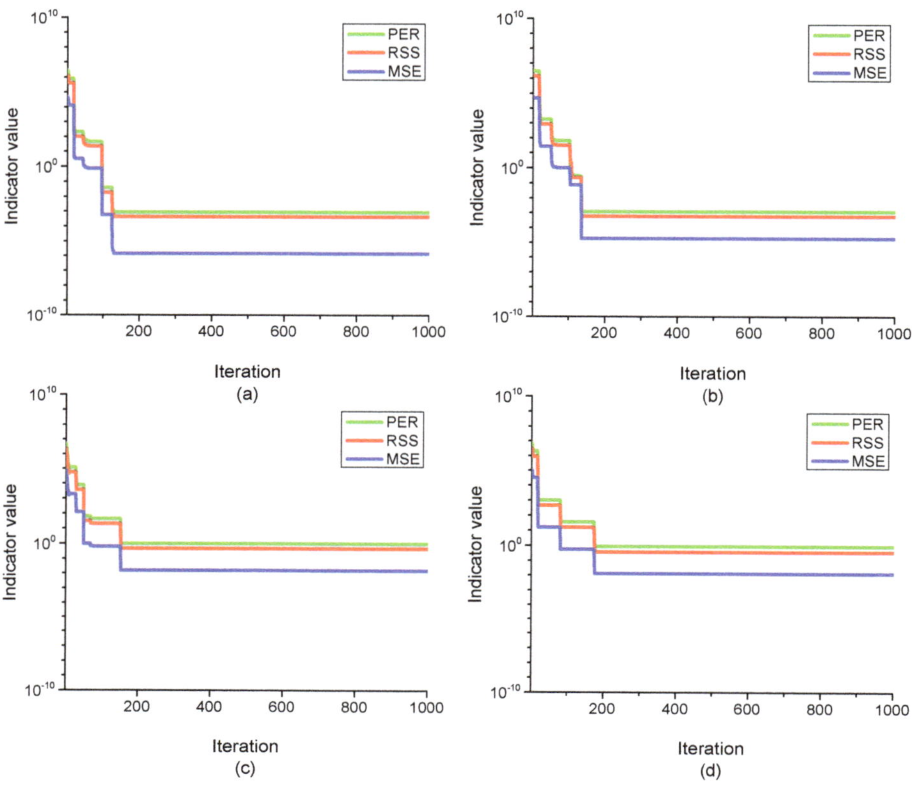

Figure 15. The convergence characteristics of the GWO-LS framework for case study II (**a**) non-noisy signal (**b**) 40 dB (**c**) 20 dB and (**d**) 10 dB.

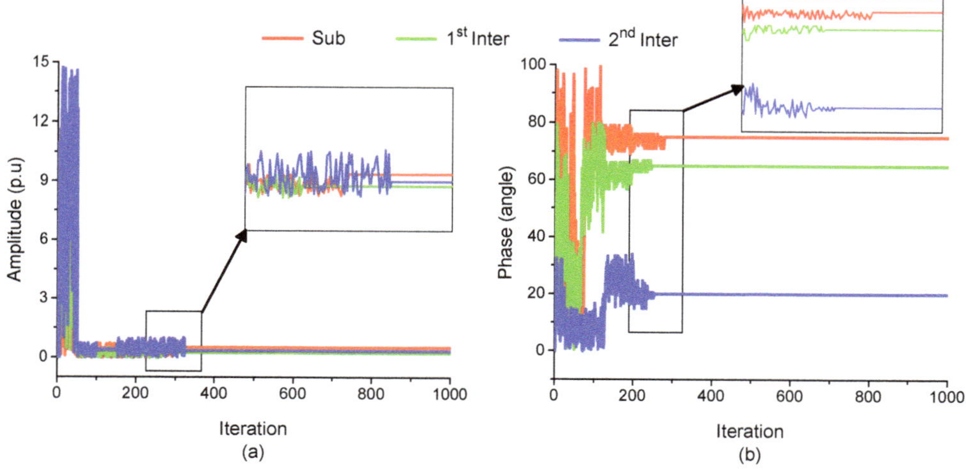

Figure 16. Variation in estimated (**a**) amplitude and (**b**) phase in the course of iterations for the sub and inter harmonics.

The behavior of the proposed estimator for approximating frequency components of sub and inter harmonics is demonstrated by Figure 17. It can be clearly seen from Figure 17 that GWO-LS takes a greater number of iterations for estimating frequency than the harmonics amplitude and phase estimation. The proposed harmonic estimators approximate the sub frequency in 317 iterations, frequency of first and second inter harmonics in 298 and 264 number of iterations, respectively. It reveals that estimation of sub harmonic frequency takes higher iterations while considering all harmonics. Figure 18 demonstrates the GWO-LS behavior for estimating the amplitude of the harmonics in both steady and transient conditions. It is evident from Figure 18 that GWO-LS gives accurate estimation results in steady state cases while a slight variation is observed for the transient case. The proposed estimator shows estimation variation from 0.45 to 0.5 s clearly visible in Figure 18b.

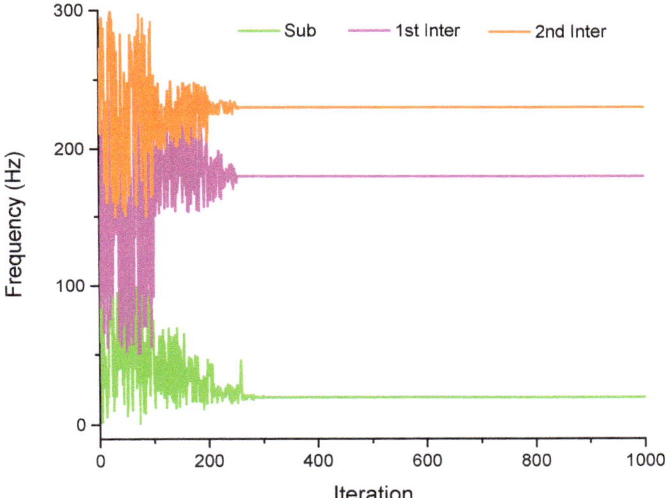

Figure 17. Frequency estimation for case study II.

The comparison of the GWO-LS framework with the literature techniques in terms of percentage error between original and approximated harmonic amplitudes and phases is plotted in Figures 19 and 20. Both the Figures demonstrate that the proposed estimator has the minimum percentage error while estimating integral, sub and inter harmonics. Moreover, the comparison of the results of the GWO-LS estimator is also stated in tabular form as Table 6. It is evident from the Table 6 analysis that GWO-LS harmonic estimator is computationally far better than literature estimators by taking 1.17976 s for the approximation of power signal having a sub and inter harmonics. The percentage error between the original harmonic amplitude and phase and the GWO-LS also proved the proposed methodology's effectiveness. The performance comparison of the presented harmonic estimator with the state of art techniques in terms of performance index also proved the effectiveness of the GWO-LS in the estimation of the sub, inter, and integral harmonics demonstrated in Figure 21 and Table 8. It can be derived from Table 8 and Figure 21 that the proposed estimator has 3.11×10^{-4}, 2.61×10^{-2}, 1.83×10^{-1} performance indexes for the 40 dB, 20 dB, and 10 dB cases; respectively, which are better than the literature results. Hence, we can conclude that the given harmonic estimator gave promising results in harmonic parameters estimation (sub, inter, and integral harmonics) under a highly noisy and time-varying environment.

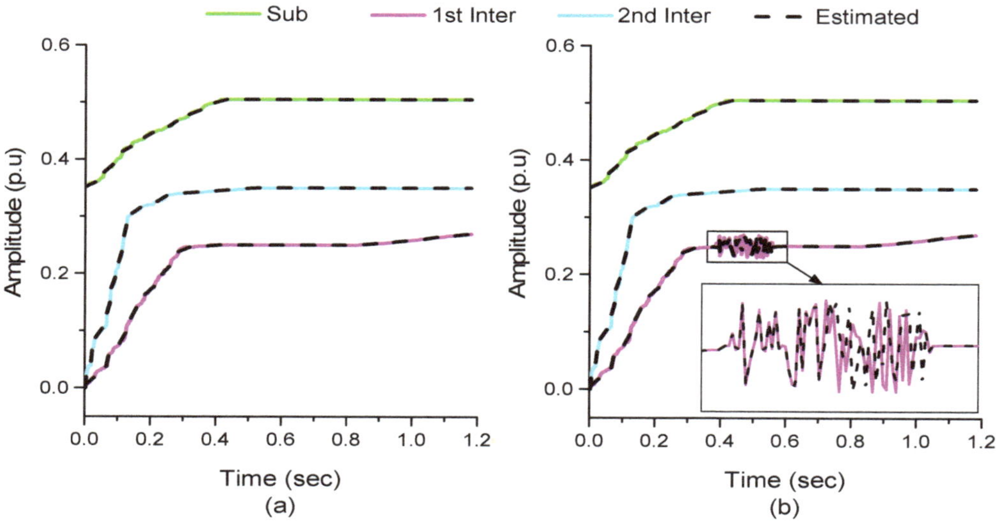

Figure 18. Estimated harmonic amplitude (**a**) steady state condition and (**b**) transient in 1st inter harmonic.

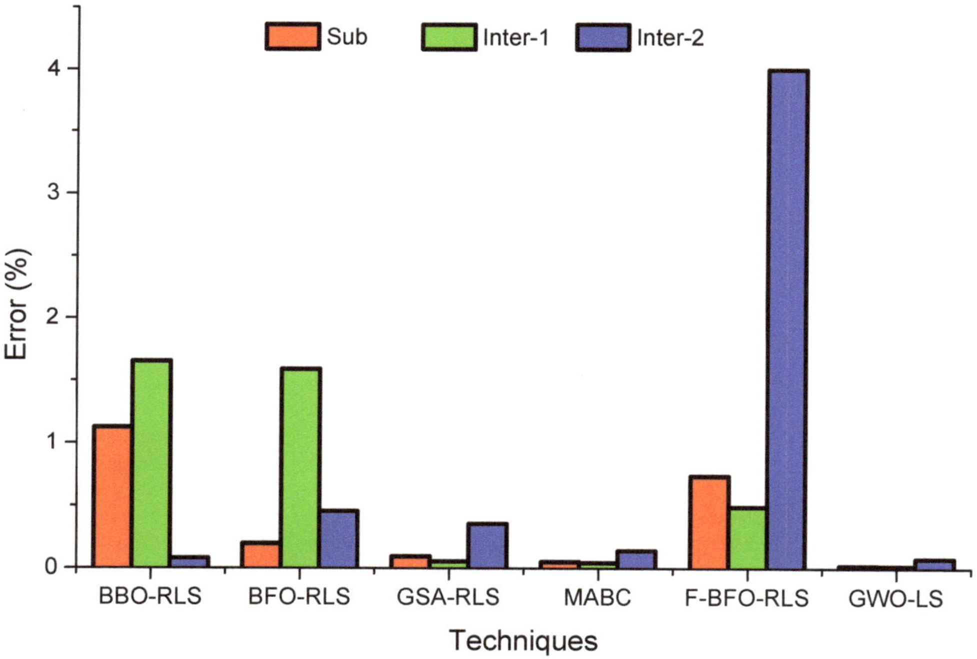

Figure 19. Comparison of amplitudes erros for case study II.

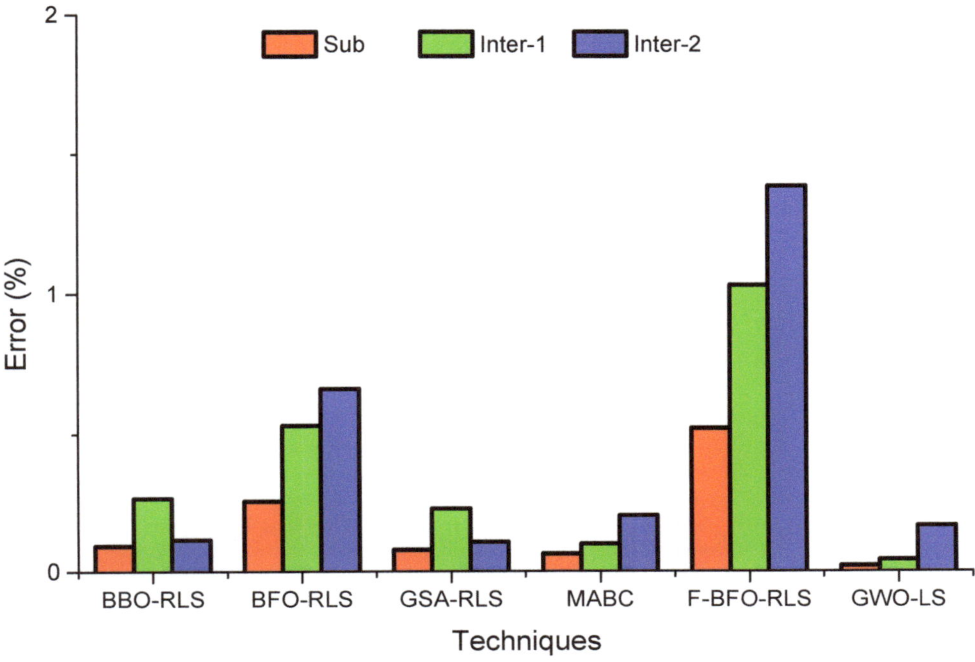

Figure 20. Comparison of phase error for case study-II.

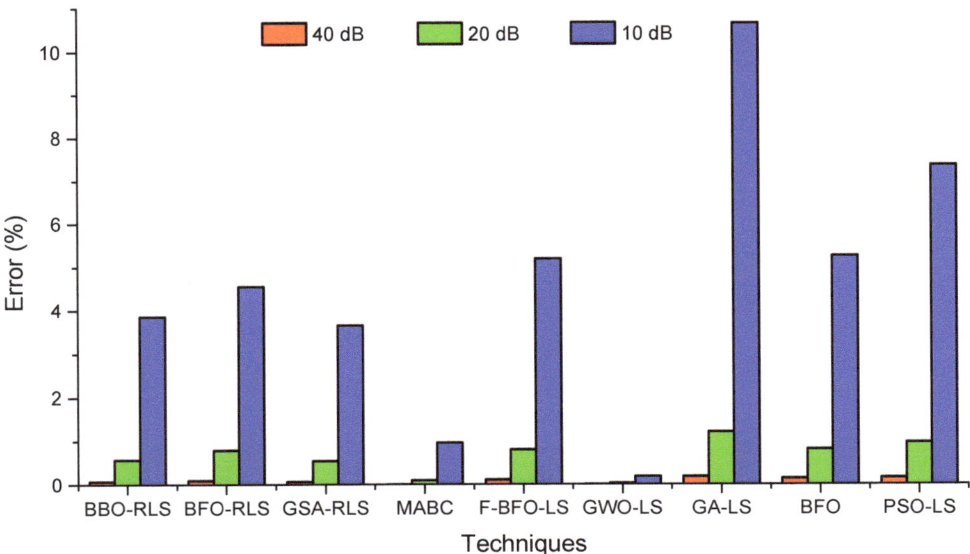

Figure 21. Comparison of performance index for case study II.

Table 8. PER comparison of GWO-LS with Literature harmonics estimator for Noisy signals.

Harmonics Number	Frequency (Hz)	Amplitude (p.u)	Phase (Degree)
	40 dB	20 dB	10 dB
GA-LS	1.83×10^{-1}	1.20×10^{0}	1.07×10^{1}
PSO-LS	1.57×10^{-1}	9.55×10^{-1}	7.36×10^{0}
BFO	1.38×10^{-1}	8.07×10^{-1}	5.25×10^{0}
F-BFO-LS	1.12×10^{-1}	8.02×10^{-1}	5.19×10^{0}
BFO-RLS	9.23×10^{-2}	7.87×10^{-1}	4.55×10^{0}
BBO-RLS	7.50×10^{-2}	5.74×10^{-1}	3.86×10^{0}
GSA-RLS	6.52×10^{-2}	5.48×10^{-1}	3.65×10^{0}
MABC	9.54×10^{-4}	9.54×10^{-2}	9.54×10^{-1}
GWO-LS	3.11×10^{-4}	2.61×10^{-2}	1.83×10^{-1}

5. Conclusions

In this paper, a hybrid harmonic estimator GWO-LS is proposed to estimate harmonics in a power signal. In the proposed harmonic estimator, GWO and LS estimate the non-linear part (phase and frequency) and linear part (amplitude) of a power signal, respectively. For validation of the proposed estimator, two test systems are adopted, among which the first test system contains power harmonics while the second test system considers sub and inter harmonics. Moreover, three different performance indicators are utilized to examine the effectiveness of the proposed approach. The obtained results were compared with recently developed harmonic estimators BFO, F-BFO-LS, ABC-LS, BBO-RLS, GSA-RLS, and MABC. Comparative evaluation of results in terms of percentage error, performance index, and computational time depicts that the proposed GWO-LS estimator estimates the harmonics in a power signal more accurately. The proposed estimator provides less computational complexity, which is essential for the detection of harmonics timely to maintain power quality. The proposed GWO-LS estimator requires 43.5% and 20.6% less time to accurately estimates harmonics in comparison with MABC. Moreover, the obtained percentage error for phase and amplitude indicates that GWO-LS estimates harmonics in a power signal with better accuracy. Furthermore, the proposed estimator accurately estimates harmonics having transient at different time intervals. In addition, the proposed estimator's PER values are much smaller than the state of art techniques, which clearly verifies that the developed methodology performs the HE problem more accurately and rapidly with a simple structure having fewer controlling parameters.

6. Future Recommendations

- Practical and industrial implementation of proposed research for accurate estimation of electrical harmonics amplitude and phase.
- Proposed harmonic estimator can be helpful in designing active filters to nullify the effects of harmonics thus improving power quality.
- Determine the emission level of higher harmonic components and identifying the source of voltage distortion in the power supply system of industrial enterprises.
- Detailed Analysis of the Steady and transient Conditions of the electrical signals having sub, inter, and integral harmonics.

Author Contributions: Conceptualization and draft writing, M.A.; methodology, T.N.M.; software implementation, A.A.; results validation and formal analysis, M.F.N.; investigation and data curation, I.A.K.; methodology, proofread and edit, R.B. All authors have read and agreed to the published version of the manuscript.

Funding: This research received no external funding.

Data Availability Statement: Not Applicable.

Conflicts of Interest: The authors declare no conflict of interest.

References

1. Ahmed, A.; Faisal, M.; Khan, N.; Khan, I.; Member, S.; Alquhayz, H.; Khan, M.A.; Kiani, A.T. A Novel Framework to Determine the Impact of Time Varying Load Models on Wind DG Planning. *IEEE Access* **2021**, *9*, 1–16. [CrossRef]
2. Klyuev, R.V.; Fomenko, O.A.; Gavrina, O.A.; Sokolov, A.A.; Sokolova, O.A.; Dzeranov, B.V.; Morgoev, I.D.; Zaseev, S.G. Analysis of non-sinusoidal voltage at metallurgical enterprises. *IOP Conf. Ser. Mater. Sci. Eng.* **2019**, *663*. [CrossRef]
3. Kabalci, Y.; Kockanat, S.; Kabalci, E. A modified ABC algorithm approach for power system harmonic estimation problems. *Electr. Power Syst. Res.* **2018**, *154*, 160–173. [CrossRef]
4. Hackl, C.M.; Landerer, M. Modified Second-Order Generalized Integrators with Modified Frequency Locked Loop for Fast Harmonics Estimation of Distorted Single-Phase Signals. *IEEE Trans. Power Electron.* **2020**, *35*, 3298–3309. [CrossRef]
5. Ashraf, M.M.; Ullah, M.O.; Malik, T.N.; Waqas, A.B.; Iqbal, M.; Siddiq, F. Robust Extraction of Harmonics using Heuristic Advanced Gravitational Search Algorithm-based Least Square Estimator. *Nucleus* **2017**, *54*, 219–231.
6. Ashraf, M.M.; Ullah, M.O.; Malik, T.N.; Waqas, A.B.; Iqbal, M.; Siddiq, F. A Hybrid Water Cycle Algorithm-Least Square based Framework for Robust Estimation of Harmonics. *Nucleus* **2018**, *55*, 47–60.
7. Labrador Rivas, A.E.; da Silva, N.; Abrão, T. Adaptive current harmonic estimation under fault conditions for smart grid systems. *Electr. Power Syst. Res.* **2020**, *183*. [CrossRef]
8. Enayati, J.; Moravej, Z. Real-time harmonic estimation using a novel hybrid technique for embedded system implementation. *Int. Trans. Electr. Energy Syst.* **2017**, *27*, 1–13. [CrossRef]
9. Yao, J.; Yu, H.; Dietz, M.; Xiao, R.; Chen, S.; Wang, T.; Niu, Q. Acceleration harmonic estimation for a hydraulic shaking table by using particle swarm optimization. *Trans. Inst. Meas. Control.* **2017**, *39*, 738–747. [CrossRef]
10. Chen, C.I.; Chang, G.W. Virtual instrumentation and educational platform for time-varying harmonic and interharmonic detection. *IEEE Trans. Ind. Electron.* **2010**, *57*, 3334–3342. [CrossRef]
11. Urbina-Salas, I.; Razo-Hernandez, J.R.; Granados-Lieberman, D.; Valtierra-Rodriguez, M.; Torres-Fernandez, J.E. Instantaneous Power Quality Indices Based on Single-Sideband Modulation and Wavelet Packet-Hilbert Transform. *IEEE Trans. Instrum. Meas.* **2017**, *66*, 1021–1031. [CrossRef]
12. Jain, S.K.; Singh, S.N. Harmonics estimation in emerging power system: Key issues and challenges. *Electr. Power Syst. Res.* **2011**, *81*, 1754–1766. [CrossRef]
13. Yao, J.; Wan, Z.; Fu, Y. Acceleration Harmonic Estimation in a Hydraulic Shaking Table Using Water Cycle Algorithm. *Shock Vib.* **2018**, *2018*. [CrossRef]
14. Yaghoobi, J.; Alduraibi, A.; Zare, F. Current harmonic estimation techniques based on voltage measurements in distribution networks. *Australas. Univ. Power Eng. Conf. AUPEC 2018* **2018**, 1–6. [CrossRef]
15. Garanayak, P.; Naayagi, R.T.; Panda, G. A High-Speed Master-Slave ADALINE for Accurate Power System Harmonic and Inter-Harmonic Estimation. *IEEE Access* **2020**, *8*, 51918–51932. [CrossRef]
16. Huang, C.L.; Yang, S.C. Sensorless Vibration Harmonic Estimation of Servo System Based on the Disturbance Torque Observer. *IEEE Trans. Ind. Electron.* **2020**, *67*, 2122–2132. [CrossRef]
17. Rodrigues, N.M.; Ramos, P.M.; Janeiro, F.M. Comparison of harmonic estimation methods for power quality assessment. *J. Phys. Conf. Ser.* **2018**, *1065*. [CrossRef]
18. Malagon-Carvajal, G.; Bello-Pena, J.; Ordonez-Plata, G.; Duarte, C. Time-to-frequency domain SMPS model for harmonic estimation: Methodology. *Renew. Energy Power Qual. J.* **2017**, *1*, 825–830. [CrossRef]
19. Sahu, B.; Dhar, S.; Dash, P.K. Frequency-scaled optimized time-frequency transform for harmonic estimation in photovoltaic-based microgrid. *Int. Trans. Electr. Energy Syst.* **2020**, *30*, 1–23. [CrossRef]
20. Kiani, A.T.; Nadeem, M.F.; Ahmed, A.; Khan, I.; Elavarasan, R.M.; Das, N. Exponential Function-Based Dynamic Inertia Weight Particle Swarm Optimization. *Energies* **2020**, *13*, 4037. [CrossRef]
21. Ahmed, A.; Faisal, M.; Ali, I.; Bo, R.; Khan, I.A. Probabilistic generation model for optimal allocation of wind DG in distribution systems with time varying load models. *Sustain. Energy Grids Networks* **2020**, *22*, 100358. [CrossRef]
22. Ray, P.K.; Subudhi, B. Ensemble-Kalman-filter-based power system harmonic estimation. *IEEE Trans. Instrum. Meas.* **2012**, *61*, 3216–3224. [CrossRef]
23. Singh, S.K.; Sinha, N.; Goswami, A.K.; Sinha, N. Several variants of Kalman Filter algorithm for power system harmonic estimation. *Int. J. Electr. Power Energy Syst.* **2016**, *78*, 793–800. [CrossRef]
24. Karimi-Ghartemani, M.; Ooi, B.T.; Bakhshai, A. Application of enhanced phase-locked loop system to the computation of synchrophasors. *IEEE Trans. Power Deliv.* **2011**, *26*, 22–32. [CrossRef]
25. Xi, Y.; Tang, X.; Li, Z.; Cui, Y.; Zhao, T.; Zeng, X.; Guo, J.; Duan, W. Harmonic estimation in power systems using an optimised adaptive Kalman filter based on PSO-GA. *IET Gener. Transm. Distrib.* **2019**, *13*, 3968–3979. [CrossRef]
26. Durdhavale, S.R.; Ahire, D.D. A Review of Harmonics Detection and Measurement in Power System. *Int. J. Comput. Appl.* **2016**, *143*, 975–8887.
27. Lu, Z.; Ji, T.Y.; Tang, W.H.; Wu, Q.H. Optimal harmonic estimation using a particle swarm optimizer. *IEEE Trans. Power Deliv.* **2008**, *23*, 1166–1174. [CrossRef]

28. Kiani, A.T.; Nadeem, M.F.; Ahmed, A.; Sajjad, I.A.; Haris, M.S.; Martirano, L. Optimal Parameter Estimation of Solar Cell using Simulated Annealing Inertia Weight Particle Swarm Optimization (SAIW-PSO). In *Proceedings of the 2020 IEEE International Conference on Environment and Electrical Engineering and 2020 IEEE Industrial and Commercial Power Systems Europe (EEEIC/I&CPS Europe), Madrid, Spain, 9–12 June 2020*; IEEE: Piscataway, NJ, USA, 2020; pp. 1–6. [CrossRef]

29. Bettayeb, M.; Qidwai, U. A hybrid least squares-GA-based algorithm for harmonic estimation. *IEEE Trans. Power Deliv.* **2003**, *18*, 377–382. [CrossRef]

30. Mishra, S. Hybrid least-square adaptive bacterial foraging strategy for harmonic estimation. *IEE Proc. Gener. Transm. Distrib.* **2005**, *152*, 379–389. [CrossRef]

31. Subudhi, B.; Ray, P.K. A hybrid Adaline and Bacterial Foraging approach to power system harmonics estimation. In *Proceedings of the 2010 International Conference on Industrial Electronics, Control and Robotics, Rourkela, India, 27–29 December 2010*; IEEE: Piscataway, NJ, USA, 2010; pp. 236–242. [CrossRef]

32. Ji, T.Y.; Li, M.S.; Wu, Q.H.; Jiang, L. Optimal estimation of harmonics in a dynamic environment using an adaptive bacterial swarming algorithm. *IET Gener. Transm. Distrib.* **2011**, *5*, 609–620. [CrossRef]

33. Singh, S.K.; Sinha, N.; Goswami, A.K.; Sinha, N. Optimal estimation of power system harmonics using a hybrid Firefly algorithm-based least square method. *Soft Comput.* **2017**, *21*, 1721–1734. [CrossRef]

34. Biswas, S.; Chatterjee, A.; Goswami, S.K. An artificial bee colony-least square algorithm for solving harmonic estimation problems. *Appl. Soft Comput. J.* **2013**, *13*, 2343–2355. [CrossRef]

35. Singh, S.K.; Sinha, N.; Goswami, A.K.; Sinha, N. Power system harmonic estimation using biogeography hybridized recursive least square algorithm. *Int. J. Electr. Power Energy Syst.* **2016**, *83*, 219–228. [CrossRef]

36. Kabalci, Y.; Kockanat, S.; Kabalci, E. A differential search algorithm application for solving harmonic estimation problems. In Proceedings of the 2017 9th International Conference on Electronics, Computers and Artificial Intelligence (ECAI), Targoviste, Romania, 29 June–1 July 2017; pp. 1–5. [CrossRef]

37. Singh, S.K.; Kumari, D.; Sinha, N.; Goswami, A.K.; Sinha, N. Gravity Search Algorithm hybridized Recursive Least Square method for power system harmonic estimation. *Eng. Sci. Technol. Int. J.* **2017**, *20*, 874–884. [CrossRef]

38. Xu, Y.; Liu, Y.; Li, Z.; Li, Z.; Wang, Q. Harmonic estimation base on center frequency shift algorithm. *Mapan J. Metrol. Soc. India* **2017**, *32*, 43–50. [CrossRef]

39. Saxena, A.; Kumar, R.; Mirjalili, S. A harmonic estimator design with evolutionary operators equipped grey wolf optimizer. *Expert Syst. Appl.* **2020**, *145*, 113125. [CrossRef]

40. Carbajal, F.B.; Olvera, R.T. An adaptive neural online estimation approach of harmonic components. *Electr. Power Syst. Res.* **2020**, *186*, 106406. [CrossRef]

41. Cupertino, F.; Lavopa, E.; Zanchetta, P.; Sumner, M.; Salvatore, L. Running DFT-based PLL algorithm for frequency, phase, and amplitude tracking in aircraft electrical systems. *IEEE Trans. Ind. Electron.* **2011**, *58*, 1027–1035. [CrossRef]

42. Mirjalili, S.; Mirjalili, S.M.; Lewis, A. Grey Wolf Optimizer 1 1. *Adv. Eng. Softw.* **2014**, *69*, 46–61. [CrossRef]

43. Li, Y.; Deng, Z.; Wang, T.; Zhao, G.; Zhou, S. Coupled harmonic admittance identification based on least square estimation. *Energies* **2018**, *11*, 2600. [CrossRef]

Article

Analysis and Control of Battery Energy Storage System Based on Hybrid Active Third-Harmonic Current Injection Converter

Yibin Tao [1,2,*]**, Jiaxing Lei** [3,*]**, Xinzhen Feng** [2]**, Tianzhi Cao** [4]**, Qinran Hu** [3] **and Wu Chen** [3]

1   School of Electric Power Engineering, South China University of Technology, Guangzhou 510640, China
2   China Electric Power Research Institute, Nanjing 210003, China; fengxinzhen@epri.sgcc.com.cn
3   School of Electrical Engineering, Southeast University, Nanjing 210096, China; qhu@seu.edu.cn (Q.H.);
    chenwu@seu.edu.cn (W.C.)
4   State Grid Jibei Electric Power Company Limited, Beijing 100045, China; cao.tianzhi@jibei.sgcc.com.cn
*   Correspondence: taoyibin@epri.sgcc.com.cn (Y.T.); jxlei@seu.edu.cn (J.L.)

Abstract: This paper applies the emerging hybrid active third-harmonic current injection converter (H3C) to the battery energy storage system (BESS), forming a novel H3C-BESS structure. Compared with the commonly used two-stage VSC-BESS, the proposed H3C-BESS has the capability to reduce the passive components and switching losses. The operation principle of the H3C-BESS is analyzed and the mathematical model is derived. The closed-loop control strategy and controller design are proposed for different operation modes of the system, which include the battery current/voltage control and the injected harmonic current control. In particular, active damping control is realized through the grid current control, which could suppress the LC-filter resonance without the need of passive damping resistors. Simulation results show that the proposed topology and its control strategy have fast dynamic response, with a setup time of less than 4 ms. In addition, the total harmonic distortions of battery current and grid currents are only 2.54% and 3.15%, respectively. The amplitude of the injected harmonic current is only half of the grid current, indicating that the current injection circuit generates low losses. Experimental results are also provided to verify the validity of the proposed solution.

Keywords: battery energy storage system; third-harmonic current injection; high efficiency; active damping

Citation: Tao, Y.; Lei, J.; Feng, X.; Cao, T.; Hu, Q.; Chen, W. Analysis and Control of Battery Energy Storage System Based on Hybrid Active Third-Harmonic Current Injection Converter. *Energies* **2021**, *14*, 3140. https://doi.org/10.3390/en14113140

Academic Editors: Irfan Ahman Khan and Teuvo Suntio

Received: 21 March 2021
Accepted: 25 May 2021
Published: 27 May 2021

Publisher's Note: MDPI stays neutral with regard to jurisdictional claims in published maps and institutional affiliations.

1. Introduction

1.1. Background

With the massive penetration of renewable energy resources such as the photovoltaic and wind power, the grid voltage and frequency stability are being threatened because of the high volatility of the renewable energy [1,2]. The battery energy storage system (BESS) has been suggested as a promising solution to suppress the power fluctuations and thus enables more integration of renewable energy [3,4]. In addition, the BESS could also be used to protect important loads from grid fault conditions. Due to these merits, the BESS has attracted tremendous attention worldwide [5,6].

1.2. Motivation

The power conversion system (PCS) is one of the key components of the BESS. The classic PCS solution is the two-stage topology composed of a bidirectional DC–DC converter and a voltage source converter (VSC) [7], as shown in Figure 1. The DC–DC converter enables a flexible combination of low-voltage batteries and the VSC is able to control the active and reactive power at the grid side. In applications where galvanic isolation is required, the DC–DC converter can be implemented by dual active-bridge (DAB) or resonant converters with high-frequency transformers [8,9]. Due to its good controllability at both the grid side and the battery side, the two-stage VSC-BESS is widely adopted in practical applications. However, it suffers from relatively high power loss since both the DC–DC converter and VSC work in high-frequency chopping mode. The single-stage VSC-BESS can be applied to reduce the loss, which directly connects batteries to the DC-link of VSC and eliminates the DC–DC converter [10]. Nevertheless, due to the high voltage at the DC-link of VSC, the single-stage VSC-BESS lacks the flexibility to match different battery voltage level and requires a large number of battery cells in series. The VSC shown in Figure 1 is a two-level topology, which can be replaced with the three-level VSC (3L-VSC) so as to reduce the loss to some extent [11]. For the medium- and high-voltage applications, the cascaded H-bridge converter (CHB) [12–14] and modular multilevel converter (MMC) [15–18] are possible solutions. Batteries can be installed at the DC-link of the sub-modules of CHB or MMC. Benefiting from the modularity and low-switching operation, the reliability and efficiency of CHB and MMC are high. However, they are not cost-efficient solutions for low-voltage applications.

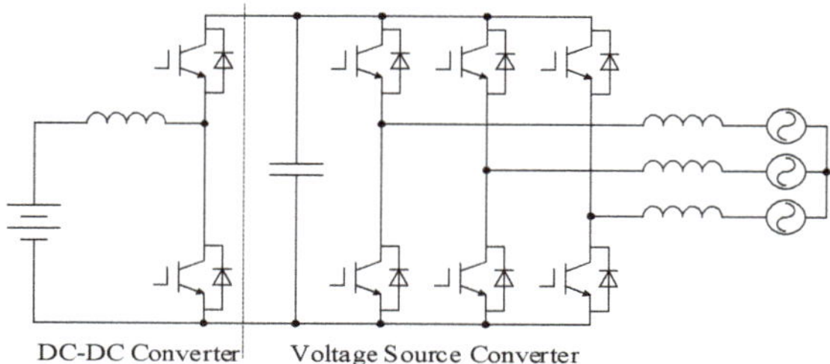

Figure 1. System topology of the conventional two-stage VSC-BESS.

Current source converter (CSC), which is also known as AC–DC matrix converter (MC), has been suggested as a novel solution for BESS. CSC is a buck-type converter from the grid side to the DC-link with a wide DC voltage control range [19]. Similar to the two-stage VSC-BESS, CSC-BESS could also adapt to batteries with different voltages but avoid the need of an additional DC–DC converter. CSC eliminates the large DC-link capacitors and bulky grid side filter inductors, which helps to improve the power density of BESS. In particular, the switching loss of CSC-BESS is much lower than VSC-BESS, because of the single-stage conversion and low voltage stress under commutation, making it a highly efficient solution. When galvanic isolation is required, CSC is also able to output both positive and negative voltages at the primary side of high-frequency transformers, eliminating the need of an additional primary DC–AC converter [20]. However, the extendibility of CSC is relatively low when multi-battery modules need to be integrated.

1.3. Related Work

In recent years, the hybrid active third-harmonic current injection converter (H3C) [21], which is also a buck-type AC–DC converter, has attracted attention. Like CSC, H3C also requires much smaller passive components. Moreover, only a few semiconductor devices work in high-frequency chopping mode while the rest are commutated at very low frequency. In addition, the semiconductor devices on the current path of H3C are less than CSC. Therefore, the achievable efficiency of H3C is much higher than CSC and VSC [22]. To date, H3C is mainly used in unidirectional rectifier applications or the AC–AC conversion system [21–26], while its application in BESS has seldom been explored. The existing studies mainly focus on the control performance at the load side, and the active power is mainly flowing through the grid to the load. However, the BESS application requires sufficient bidirectional power flow control capability. In particular, satisfactory control performance in the dynamic process and at the steady state is desired both at the battery side and the grid side. In addition, like other current source converters with a grid-side LC filter [27], the H3C could suffer from filter resonance. Active damping control is preferable to achieve high performance at the grid side, especially for the BESS application. However, the grid side converter of H3C is not fully controllable. As a result, the active damping control has not been practically realized in the literature.

1.4. Contribution of This Paper

Therefore, this paper utilizes the merits of H3C to construct a highly compact and efficient solution for BESS applications. This paper focuses on the operating principle and control strategy of the H3C-BESS, considering the bidirectional power flow and the requirement of control performance at both the battery side and the grid side. In addition, a novel control structure is proposed to achieve the active damping control at the grid side, which is realized both in simulation and experimental results. The proposed active damping control is able to suppress the filter resonance without the need of a passive damping resistor.

The rest of this paper is organized as follows. The system operation principle is analyzed in detail and the mathematical model is derived, which are the main contents of Section 2. The system control strategy regarding the operation mode is proposed and formulated in Section 3. Simulation and experimental results are provided in Section 4 for verifying the effectiveness of this new system associated with the control strategy. Conclusions are drawn in Section 5.

2. Operation Principle and Mathematical Model of H3C-BESS

2.1. System Description

Topology of the H3C-BESS is shown in Figure 2, which is composed of a bidirectional DC–DC converter at the battery side, a third harmonic injection circuit, a grid voltage selector (GVS), and the LC filter at the grid side. The harmonic injection circuit includes a half-bridge and an inductor. The GVS is composed of three-phase bridges and three bidirectional switches. Like the conventional two-stage VSC-BESS, several battery modules can be easily connected to the DC-link of GVS by extending multi-DC–DC converters.

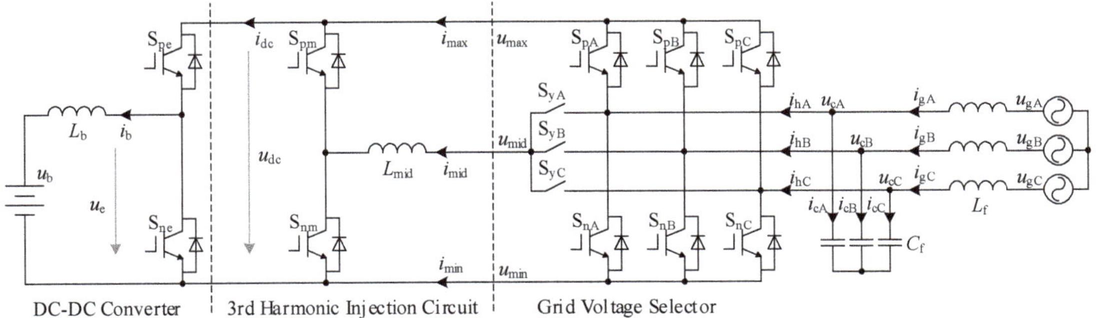

Figure 2. System topology of the proposed H3C-BESS.

In Figure 2, S_{px} and S_{nx} denote the upper switch and lower switch of the bridge x (x = e, m, A, B, C) separately. S_{yA}, S_{yB}, S_{yC} represent the bidirectional switch. L_b is the filter inductor at the battery side, L_{mid} is the filter inductor for the injected harmonic current, L_f is the filter inductor at the grid side, while C_f is the filter capacitor at the GVS side. u_b and i_b are the battery voltage and current, respectively, while u_e is the output voltage of the DC–DC converter. u_{dc} is the DC-link voltage and i_{dc} is the current flowing into the DC–DC converter. u_{cA}, u_{cB}, and u_{cC} are the capacitor voltages while i_{cA}, i_{cB} and i_{cC} are the currents flowing into the filter capacitor C_f. u_{gA}, u_{gB}, and u_{gC} are the three-phase grid voltages, while i_{gA}, i_{gB}, and i_{gC} represent the grid currents. i_{hA}, i_{hB}, and i_{hC} are the three-phase currents at the GVS side. i_{max}, i_{mid}, and i_{min} correspond to the currents flowing out of the GVS.

2.2. Operation Principles

As the input of GVS is a voltage source imposed by the filter capacitor C_f, the three-phase bridges in the GVS actually work as a three-phase bidirectional synchronous rectifier. This means that the maximum value (u_{max}) and minimum value (u_{min}) of capacitor voltages (u_{cA}, u_{cB}, u_{cC}) should always be imposed on the positive and negative DC-link, respectively. Accordingly, only the bidirectional switch corresponding to the middle value u_{mid} of capacitor voltages can be switched on. The three-phase currents are also determined by the switching states of the GVS, as shown in Table 1, where the sector division is determined by the relationships among the capacitor voltages shown in Figure 3.

Table 1. Switching states of the GVS.

Sector	On Switches			u_{max}	u_{mid}	u_{min}	i_{hA}	i_{hB}	i_{hC}
I	S_{pA}	S_{yB}	S_{nC}	u_{cA}	u_{cB}	u_{cC}	i_{max}	i_{mid}	i_{min}
II	S_{yA}	S_{pB}	S_{nC}	u_{cB}	u_{cA}	u_{cC}	i_{mid}	i_{max}	i_{min}
III	S_{nA}	S_{pB}	S_{yC}	u_{cB}	u_{cC}	u_{cA}	i_{min}	i_{max}	i_{mid}
IV	S_{nA}	S_{yB}	S_{pC}	u_{cC}	u_{cB}	u_{cA}	i_{min}	i_{mid}	i_{max}
V	S_{yA}	S_{nB}	S_{pC}	u_{cC}	u_{cA}	u_{cB}	i_{mid}	i_{min}	i_{max}
VI	S_{pA}	S_{nB}	S_{yC}	u_{cA}	u_{cC}	u_{cB}	i_{max}	i_{min}	i_{mid}

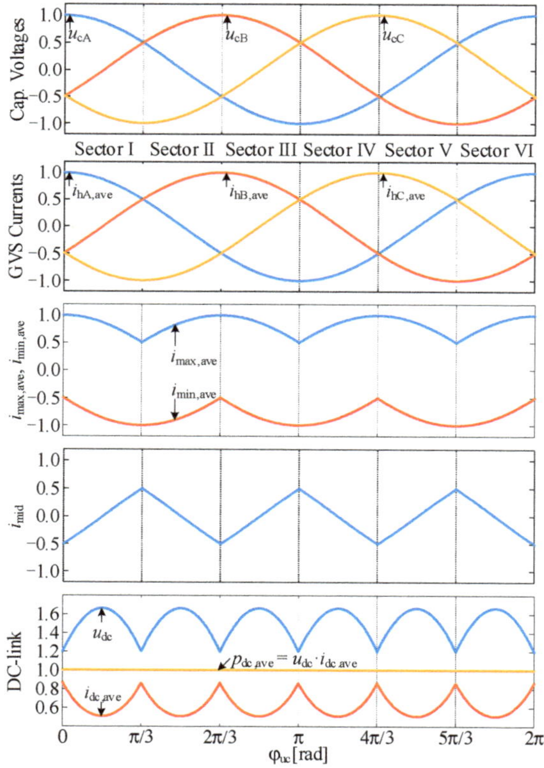

Figure 3. Key waveforms (in p.u.) of the GVS and the third harmonic injection circuit. The word "ave" in the subscript denotes the low-frequency component of the corresponding variable in each switching period. φ_{uc} is the phase angle of the capacitor voltage vector.

As can be found from Figure 1, Figure 3, and Table 1, only the phase current corresponding to the middle phase voltage can be directly controlled through the third harmonic injection circuit, and the remaining two-phase currents are passively determined by the positive and negative DC-link currents of GVS. In addition, it can also be known that the three-phase bridges in GVS are line-commuted, which means their semiconductor devices are switched at the fundamental grid frequency. Similarly, the bidirectional switches in GVS are switched at twice the grid frequency. The half-bridge in the third harmonic injection circuit is switched on/off at the middle current i_{mid}, while i_{mid} has the minimum absolute value. Therefore, the switching loss of the H3C-BESS is much lower than the VSC-BESS.

The DC-link voltage u_{dc} of H3C-BESS is equal to $u_{max}-u_{min}$, which has large fluctuations even under ideal grid voltages. However, through the real-time correction of the duty cycle of the DC–DC converter, the battery current i_b can be regulated at the desired value. Therefore, the DC-link current i_{dc} at the buck-converter side also fluctuates while the DC-link active power p_{dc} maintains constant, as shown in Figure 3.

Due to the absence of DC-link passive components, grid currents have to be passively determined by the battery current and the injected harmonic current. However, the battery current is also determined by the control of itself. As a result, the key point of H3C is how to generate sinusoidal grid currents by controlling the third harmonic injection circuit, considering the fact that the transferred instantaneous active power is constant. Under balanced and sinusoidal grid conditions, the three-phase sinusoidal capacitor voltages can be expressed as

$$\begin{cases} u_{cA} = u_{cm} \cos\left(\omega_g t\right) \\ u_{cB} = u_{cm} \cos\left(\omega_g t - 2\pi/3\right) \\ u_{cC} = u_{cm} \cos\left(\omega_g t + 2\pi/3\right) \end{cases} \tag{1}$$

where u_{cm} is the capacitor voltage amplitude. The desired low-frequency components of the GVS currents are expressed as

$$\begin{cases} i_{hA,ave} = i_{hd} \cos \omega_g t + i_{hq} \sin \omega_g t \\ i_{hB,ave} = i_{hd} \cos\left(\omega_g t - 2\pi/3\right) + i_{hq} \sin\left(\omega_g t - 2\pi/3\right) \\ i_{hC,ave} = i_{hd} \cos\left(\omega_g t + 2\pi/3\right) + i_{hq} \sin\left(\omega_g t + 2\pi/3\right) \end{cases} \tag{2}$$

where i_{hd} and i_{hq} are the active and reactive components of GVS currents, respectively.

Take Sector I as an example to illustrate how H3C can generate sinusoidal grid currents, namely $\omega_g t$ in (1) is within $[0, \pi/3]$. In this case, u_{max}, u_{mid} and u_{min} satisfy:

$$u_{max} = u_{cA}, \ u_{mid} = u_{cB}, \ u_{min} = u_{cC} \tag{3}$$

Therefore, the DC-link voltage and current are:

$$\begin{cases} u_{dc} = u_{max} - u_{min} = u_{cA} - u_{cC} \\ i_{dc,ave} = \dfrac{P_{dc,ave}}{u_{dc}} = \dfrac{P_{dc,ave}}{u_{cA}-u_{cC}} \end{cases} \tag{4}$$

where $P_{dc,ave}$ is the average instantaneous active power transferred from the DC–DC converter.

Since $u_{max} \geq u_{mid} \geq u_{min}$, the half-bridge in the third harmonic injection circuit is fully controllable and the actual current i_{mid} flowing through the inductor L_{mid} can be considered equal to its reference. In this case, it is:

$$i_{mid} = i_{hB,ave} \tag{5}$$

Supposing the duty cycle of the upper switch S_{pm} in the half-bridge is d_m, the following equation is obtained based on the voltage-second balance of L_{mid} at the steady state:

$$d_m u_{max} + (1 - d_m)u_{min} = u_{mid} \tag{6}$$

and thus:

$$d_{\mathrm{m}} = \frac{u_{\mathrm{mid}} - u_{\mathrm{min}}}{u_{\mathrm{max}} - u_{\mathrm{min}}} \tag{7}$$

According to Figure 2, the current i_{max} satisfies:

$$i_{\mathrm{max}} = i_{\mathrm{dc}} - d_{\mathrm{m}} i_{\mathrm{mid}} \tag{8}$$

To achieve sinusoidal grid currents, i_{max} in this case should be equal to its reference $i_{\mathrm{hA, ave}}$:

$$i_{\mathrm{max}} = i_{\mathrm{hA, ave}} = i_{\mathrm{hd}} \cos \omega_{\mathrm{g}} t + i_{\mathrm{hq}} \sin \omega_{\mathrm{g}} t \tag{9}$$

By substituting (1)–(8) into (9), the condition for (9) is resolved:

$$i_{\mathrm{hd}} = \frac{2 p_{\mathrm{dc, ave}}}{3 u_{\mathrm{cm}}} \tag{10}$$

Therefore, only if the active component i_{hd} of GVS currents is set based on (10) and the current i_{mid} is controlled to its reference value shown in (5), both the phase A and B currents can reach sinusoidal, and phase C is naturally sinusoidal because the sum of i_{hA}, i_{hB} and i_{hC} is always 0. Note that (10) is irrespective with i_{hq}, which means that the reactive power can be theoretically controlled to any desired value. When the capacitor voltages fall into other sectors, the operation principle can be analyzed in a similar way.

2.3. Mathematical Model

Equations presented in Part B of this section are sufficient to analyze the operation principle, but they do not include the dynamic behavior of the system. In order to formulate the system closed-loop control strategy, the mathematical model of H3C-BESS is developed.

According to Figure 2, the battery current i_{b} satisfies:

$$L_{\mathrm{b}} \frac{d i_{\mathrm{b}}}{d t} = d_{\mathrm{e}} (u_{\mathrm{max}} - u_{\mathrm{min}}) - R_{\mathrm{b}} i_{\mathrm{b}} - u_{\mathrm{b}} \tag{11}$$

where d_{e} is the duty cycle of the upper switch S_{pe} of the DC–DC converter; R_{b} is the parasitic resistance of the filter inductor L_{b}. The DC-link current i_{dc} at the DC–DC converter side is thus expressed as

$$i_{\mathrm{dc}} = d_{\mathrm{e}} i_{\mathrm{b}} \tag{12}$$

The current i_{mid} in the third harmonic injection circuit satisfies:

$$L_{\mathrm{mid}} \frac{d i_{\mathrm{mid}}}{d t} = (u_{\mathrm{mid}} - u_{\mathrm{min}}) - (u_{\mathrm{max}} - u_{\mathrm{min}}) d_{\mathrm{m}} - R_{\mathrm{mid}} i_{\mathrm{mid}} \tag{13}$$

where R_{mid} is the parasitic resistance of the filter inductor L_{mid}. The current i_{max} can thereby be expressed as

$$i_{\mathrm{max}} = i_{\mathrm{dc}} - d_{\mathrm{m}} i_{\mathrm{mid}} \tag{14}$$

The current i_{min} always satisfies:

$$i_{\mathrm{min}} = -(i_{\mathrm{max}} + i_{\mathrm{mid}}) \tag{15}$$

The GVS currents i_{hA}, i_{hB}, and i_{hC} are determined by i_{max}, i_{mid}, and i_{min} according to the switching states of GVS listed in Table 1. Grid currents (i_{gA}, i_{gB}, i_{gC}) and capacitor voltages (u_{cA}, u_{cB}, u_{cC}) satisfy:

$$\begin{cases} L_{\mathrm{f}} \frac{d i_{\mathrm{gX}}}{d t} = u_{\mathrm{cX}} - u_{\mathrm{gX}} - R_{\mathrm{f}} i_{\mathrm{gX}} \\ C_{\mathrm{f}} \frac{d u_{\mathrm{cX}}}{d t} = i_{\mathrm{hX}} - i_{\mathrm{gX}} \end{cases}, \ X = \{A, B, C\} \tag{16}$$

where R_{f} is the parasitic resistance of the filter inductor L_{f}. Together with Table 1, (11)–(16) construct the mathematical model of the proposed H3C-BESS, which can be used for designing the closed-loop control strategy. It should be noted that the relations shown in Table 1 make the H3C-BESS highly nonlinear. Therefore, only the piecewise linear model is available. Additionally, approximation needs to be applied when designing the controllers.

2.4. Comparison with Two-Stage VSC-BESS

The two-stage VSC-BESS has been widely adopted in practice. Therefore, it is necessary to compare the novel H3C-BESS with the VSC-BESS, so as to highlight its features. As it can be seen from Figures 1 and 2, the H3C-BESS has the following differences from the view of topology:

- The H3C-BESS does not need large capacitors at the DC-link of H3C-BESS [21], while a bulky capacitor is required at the DC-link of VSC-BESS;
- The grid side filter of H3C-BESS is a second-order LC filter, which is also much smaller than the filter inductor of VSC-BESS [28];
- An additional inductor L_{mid} is required in H3C-BESS. However, the maximum current flowing through it is only half of the grid current amplitude, as shown in Figure 3;
- The H3C-BESS requires 14 transistors (each bidirectional switch requires two transistors in practice). The VSC-BESS requires only eight transistors.

Therefore, the H3C-BESS requires more components and semiconductor chip area than the two-stage VSC-BESS. However, H3C-BESS could reduce the volume and weight of passive components. Moreover, the switching loss of H3C-BESS is much lower than VSC-BESS:

- Under the same input voltage, battery voltage and current, the DC–DC converter in H3C-BESS generates less switching loss than H3C-BESS, because the DC-link voltage of H3C-BESS is smaller than H3C-BESS [21];
- The GVS part in H3C-BESS transfers the main power. However, the switching frequency is quite low, only the fundamental or twice the grid frequency. Compared with the VSC operating at high switching frequency, the switching loss generated by GVS is ignorable [23];
- Although the half-bridge in the harmonic injection circuit works in chopping mode, the maximum chopped current is only half of the grid current amplitude. Therefore, the generated switching loss is also very small [21].

In summary, compared with the two-stage VSC-BESS, the H3C-BESS has the advantages of high power density and high efficiency if optimally designed, especially when the battery voltage is low. However, its drawback is the requirement of more components, which could result in higher cost in practice.

3. Proposed Control Strategy for H3C-BESS

3.1. Control Block Diagram

Subject to the power requirement of the grid and the state of charge (SOC) of batteries, the H3C-BESS could operate at three modes: constant current (CC), constant voltage (CV), and constant power (CP). In CP mode, the SOC of the battery is in the normal range (e.g., 20%–80%), and the battery can transfer active power to or absorb from the grid. CC and CV modes are activated to prevent battery damage due to excessive charge or discharge.

In contrast to the conventional two-stage VSC-BESS in which the large DC-link capacitor is able to isolate the battery and grid, the H3C-BESS directly couples the battery and grid. Therefore, the grid active power is always equal to that generated or absorbed by the battery, if the power loss could be ignored. The harmonic injection circuit has few influences on the transmitted active power but is able to make grid currents sinusoidal by changing the active power distribution among three phases.

Based on the operation principle analyzed in Section II, the system closed-loop control strategy is proposed for H3C-BESS, as shown in Figure 4. The proposed strategy measures grid voltages and further determines the switching states SGVS of GVS based on Table 1. The sorted grid voltages u_{max}, u_{mid}, and u_{min} are also selected based on Table 1. The battery current control is always the dominant loop of the system control strategy, irrespective of the operation mode, while the reference current i_b^* is set based on the operation mode. In CC mode, i_b^* is constant. In CV mode, i_b^* is generated from the closed-loop control of the battery voltage. In CP mode, i_b^* is calculated with the desired grid active power P_g^*. The closed-loop control of battery current generates the active current reference i_hd^*. The reactive current reference i_hq^* could be fixed at zero for simplicity. In particular, active damping control is realized by regulating the grid currents with proper controllers, which generates signals i_hd^* and i_hq^* to modify the references of battery current and injected harmonic current. The remaining parts of this section will describe the proposed control strategy in detail.

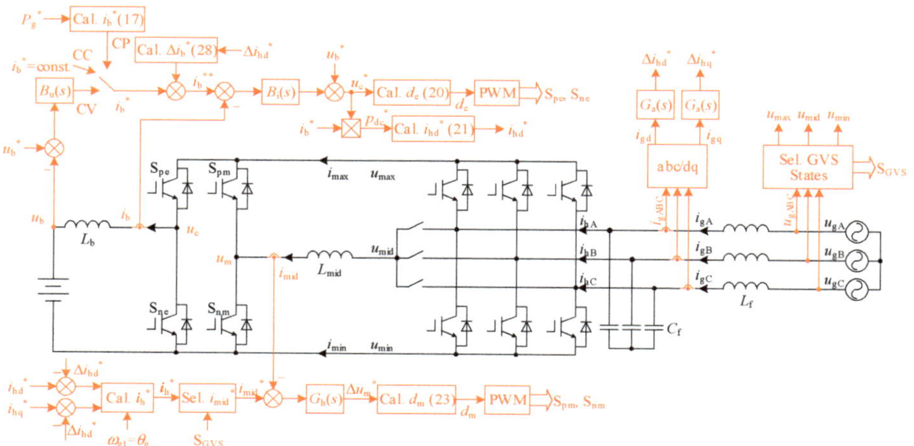

Figure 4. Control block diagram of the H3C-BESS.

3.2. Control of Battery Current and Voltage

The single-loop current control is applied to the battery in CC mode, where the reference i_b^* is set constant. In CV mode, the double-loop voltage and current control is applied, and thus i_b^* is generated from the voltage control. In CP mode, i_b^* is calculated with the desired grid active power P_g^* with the converter efficiency taken into consideration:

$$i_\text{b}^* = \begin{cases} \dfrac{\eta P_\text{g}^*}{u_\text{gm}}, & P_\text{g}^* \geq 0 \\[2mm] \dfrac{P_\text{g}^*}{\eta u_\text{gm}}, & P_\text{g}^* < 0 \end{cases} \tag{17}$$

where positive P_g^* means the battery absorbs active power from the grid while negative P_g^* indicates the inverse direction.

The control structure at the battery side is similar to the two-stage VSC-BESS. The typical proportional–integral (PI) controllers can be utilized to regulate the battery current and voltage, namely controllers $B_\text{i}(s)$ and $B_\text{u}(s)$ are expressed as

$$B_\text{i}(s) = K_\text{pbi} + K_\text{ibi}/s \tag{18}$$

$$B_\text{u}(s) = K_\text{pbu} + K_\text{ibu}/s \tag{19}$$

where K_pbi and K_ipi are the static gain and integral gain of the current controller; K_pbu and K_ipu are the static gain and integral gain of the voltage controller. The control parameters can be tuned based on the typical first-order system, which is mature for power converters.

However, the DC-link voltage in H3C-BESS is not constant as that in two-stage VSC-BESS, and thus feedforward control is indispensable when calculating the duty cycle of the DC–DC converter. The output of $B_\text{i}(s)$ represents the reference voltage drop on L_b.

Accordingly, adding the output of $B_i(s)$ and the battery voltage u_b generates the reference voltage u_e^* that the DC–DC converter should output. The duty cycle d_e of the switch S_{pe} is further calculated by

$$d_e = \frac{u_e^*}{u_{dc}} = \frac{u_e^*}{u_{max} - u_{min}} \tag{20}$$

The typical PWM technique for the DC–DC converter is then applied to generate the gate signals of S_{pe} and S_{ne}.

It should be noted that the feedforward control is able to suppress the influence of DC-link voltage fluctuation on the battery current. Therefore, any grid voltage disturbance (such as unbalance or distortions) can be prevented from transferring into the battery.

3.3. Control of Injected Harmonic Current

For the harmonic injection circuit, the reference current i_{mid}^* is selected from the reference GVS currents (i_{hA}^*, i_{hB}^*, i_{hC}^*) according to Table 1. The reference GVS currents can be obtained based on (2), where i_{hd}^* is calculated from the reference active power generated by the closed-loop control of battery current:

$$i_{hd}^* = \begin{cases} \dfrac{2i_b^* u_e^*}{3\eta u_{gm}}, & i_b^* \geq 0 \\ \dfrac{2\eta i_b^* u_e^*}{3u_{gm}}, & i_b^* < 0 \end{cases} \tag{21}$$

where the capacitor voltage amplitude u_{cm} is approximated by the grid voltage amplitude u_{gm}. As it is shown in Figure 3, the reference current i_{mid}^* is similar to a triangular wave with the frequency of $3\omega_g$, which means it contains harmonics with frequencies of $3\omega_g$, $9\omega_g$, $15\omega_g$, etc. To regulate such a kind of harmonic current, the controller $G_h(s)$ is designed as N vector-proportional–integral (VPI) controllers in parallel:

$$G_h(s) = \sum_{n=1}^{N} \frac{K_{phn}s^2 + K_{ihn}s}{s^2 + \left[3(2n-1)\omega_g\right]^2} \tag{22}$$

where K_{phn} and K_{ihn} are the static gain and resonant gain of the nth VPI controller. It can be known from Fourier series of the triangular wave that the content of the harmonic at the frequency of $21\omega_g$ or higher is less than 2% which has quite a limited effect on the grid current. Therefore, three VPI controllers (i.e., $n = 3$) are sufficient to regulate the injected harmonic current without generating observable steady-state errors.

The output of $G_h(s)$ represents the reference voltage drop Δu_m^* on the inductor L_{mid}, and thus the duty cycle of switch S_{pm} can be calculated based on (13):

$$d_m = \frac{u_{mid} - u_{min} - \Delta u_m^*}{u_{max} - u_{min}} \tag{23}$$

The typical PWM technique for the half-bridge is then applied to generate the gate signals of S_{pm} and S_{nm}.

The closed-loop control structure of the injected harmonic current is shown in Figure 5, where $G_{PWM}(s) = 1/(1.5T_s s + 1)$ and T_s denotes the sampling time. Similar to the design criteria of the typical PI controller which realizes zero-pole cancellation, parameters of the VPI are set as

$$K_{phn} = \frac{L_{mid}}{3T_s}, \quad K_{ihn} = \frac{R_{mid}}{3T_s} \tag{24}$$

Note that the expression of the VPI controller is similar to the common resonant controller (CRC). VPI and CRC have the same poles at the desired frequencies but have different numerators. Frequency responses of the closed-loop transfer function with VPI and CRC are shown in Figure 5b. It can be seen that, with CRC, undesired resonant peaks around the central frequencies are generated. In practice, the utility grid usually has a frequency variation up to 0.2 Hz. Therefore, such undesired peaks will result in the inaccurate control of the injected harmonic current and further undesired low-frequency harmonics in the grid currents. On the contrary, the frequency response with VPI is quite smooth around the central frequencies, without any undesired peaks. Therefore, VPI is more robust than CRC in practice considering the grid frequency variation.

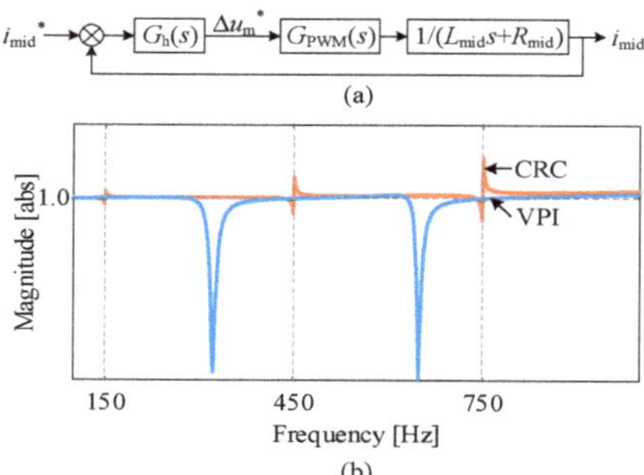

Figure 5. (**a**): Closed-loop control structure of the injected harmonic current; and (**b**) frequency response of the closed-loop transfer function.

It can be known from (21), generating the reference of the injected harmonic current is an open-loop method. In practice, precise parameters are unavailable, especially considering that the converter efficiency is variable with the working conditions. Under parameter uncertainties, the obtained reference of the injected harmonic current could be inaccurate. It can be inferred from Part B Section 2 that the inaccurate injected harmonic current directly leads to the deviation of the grid currents from the desired sinusoidal currents. As a result, the actual grid currents are distorted, reducing the power quality. Figure 6 shows the grid currents under different inaccuracies of injected harmonic current. It is clear that the inaccuracies directly result in grid current distortions. It should be noted that, the effect of input LC filter is ignored in Figure 6. The LC filter in practice could enlarge the harmonics in grid currents, leading to higher distortions.

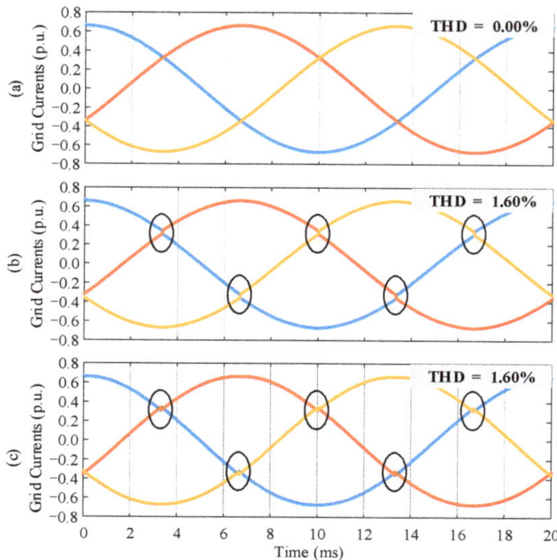

Figure 6. Grid currents considering the inaccurate reference of injected harmonic current: (**a**) 0% inaccuracy; (**b**) −0.5% inaccuracy; and (**c**) 0.5% inaccuracy. Distortion is indicated with circles.

3.4. Active Damping Control

Theoretically, according to the active power balance principle of H3C, grid currents could always be indirectly regulated without a closed-loop control. This helps to reduce the tuning effort of the controller. However, it is still beneficial to incorporate a closed-loop control of grid currents, so as to achieve an active damping function.

According to Figure 4, the transfer function from the GVS currents i_{hd} and i_{hq} to the grid currents i_{gd} and i_{gq} is determined by the grid-side LC filter, expressed as

$$G_{LC}(s) = \frac{1}{L_f C_f s^2 + R_f C_f s + 1} \tag{25}$$

where R_f is the parasitic resistance of the filter inductor. In practice, R_f is usually quite small, making $G_{LC}(s)$ a weakly damped second-order system. The frequency response of $G_{LC}(s)$ is shown in Figure 7. It is clear that without additional damping control, $G_{LC}(s)$ has a very high resonant peak. The PWM control of the converter could easily inspire the resonance of the grid-side filter, which deteriorates the input power quality.

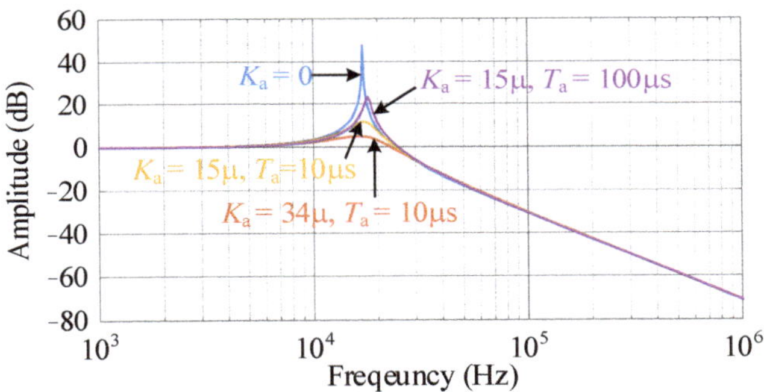

Figure 7. Frequency responses of the transfer function of the grid current control. The curve with $K_a = 0$ is for the $G_{LC}(s)$ without additional damping control. The other curves are for the $H_{LC}(s)$ with the proposed active damping control.

Therefore, additional damping control is indispensable for H3C. Usually, paralleling a damping resistor with the filter inductor L_f has good damping performance. However, the passive damping resistor generates power loss and increases the high-frequency harmonics in grid currents. As a result, it is preferable to adopt active damping control which suppresses the filter resonance through the control strategy.

In the proposed control strategy shown in Figure 4, the active damping function is realized by the closed-loop control of grid currents, where the controller is termed as $G_a(s)$. The closed-loop control structure of grid currents is shown in Figure 8. It can be noted that the controller $G_a(s)$ is located at the feedback path rather than the forward path. From Figure 8, the closed-loop transfer function is obtained:

$$H_{LC}(s) = \frac{G_{LC}(s)}{1 + G_{LC}(s)G_a(s)} \tag{26}$$

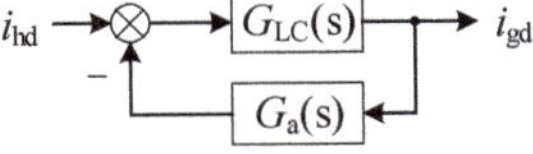

Figure 8. Closed-loop control structure of the d axis grid current. Q axis grid current has the same control structure. The dq axis coupling terms are ignored.

Clearly, by designing proper $G_a(s)$, it is possible to reshape the frequency response of $G_{LC}(s)$. In this paper, $G_a(s)$ is designed as

$$G_a(s) = \frac{K_a s}{T_a s + 1} \tag{27}$$

where the differential term with the coefficient K_a at the numerator contributes to the active damping function. As the differential operation could enlarge the high-frequency noise and harmonics in practice, it is necessary to include the low-pass filter in (27) with the time constant as T_a. Frequency responses of $H_{LC}(s)$ with different values of K_a and T_a are plotted in Figure 7. It can be concluded that the larger K_a and the smaller T_a are, the better damping performance is achieved. However, large K_a and small T_a mean large feedback signal, which could affect the normal operation of H3C-BESS. Therefore, trade-off should be made when selecting the values of K_a and T_a in practice.

As shown in Figure 4, controllers $G_a(s)$ at the dq axis generate the additional reference signals $\Delta i_{hd}{}^*$ and $\Delta i_{hq}{}^*$. Theoretically, $\Delta i_{hd}{}^*$ and $\Delta i_{hq}{}^*$ should be used to modify the GVS current references $i_{hd}{}^*$ and $i_{hq}{}^*$. However, as presented in Section II, the injected harmonic current can only influence parts of the GVS current. For the complete realization of the active damping function, the reference battery current should also be modified with:

$$\Delta i_b^* = \frac{1.5 \Delta i_{hd}^* u_{gm}}{u_b} \tag{28}$$

where the converter efficiency is assumed to be unity for simplicity.

4. Simulation and Experimental Verification

4.1. Prototype and Parameters

To verify the effectiveness of the H3C-BESS associated with the proposed system control strategy, a prototype of H3C is constructed. A photograph of the prototype is shown in Figure 9 and the experimental parameters are listed in Table 2. The GVS uses an intelligent power module (IPM) with part number PM25RLA120, while the switching frequency is 50 Hz. The DC–DC converter and the harmonic circuit shares one IPM with the same part number and their switching frequency is 16 kHz. The bidirectional switches are constructed using the discrete IGBT with part number IKW15N120H3 which are switched at 100 Hz. The digital signal processor (DSP) is TMS320F28379D with dual cores, which is powerful enough to handle all the computation. A large capacitor (5 mF) is used to emulate the behavior of the battery. The control parameters are optimally designed according to the filter parameters.

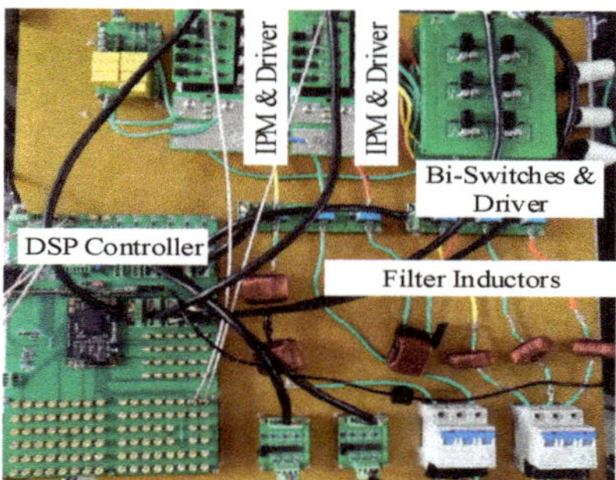

Figure 9. The H3C prototype used in the experiments.

Table 2. System parameters.

Variables	Description	Values
U_{gm}	Grid Voltage Amplitude	100 V
f_g	Grid Frequency	50 Hz
L_f	Grid Filter Inductor	0.5 mH
R_f	Resistance of L_f	35 mΩ
C_f	Grid Filter Capacitor	6.9 μF
L_{mid}	Filter Inductor of Third Harmonic Circuit	2.5 mH
R_{mid}	Resistance of L_{mid}	150 mΩ
L_b	Battery Filter Inductor	4.9 mH
R_b	Resistance of L_b	135 mΩ
U_b	Normal Battery Voltage	100 V
C_b	Equivalent Capacitance	5 mF
K_a	Gain of Active Damping Controller	15 μ
T_a	Time Constant of Active Damping Controller	10 μs
F_s	Switching Frequency of Chopping Switches	16 kHz

The corresponding simulation model was built in Matlab/Simulink. The simulation parameters are the same with those used in experiments, as listed in Table 2. A capacitor serves as the battery in CC and CV mode and is replaced with a constant voltage source in the CP mode, which is conducive to shorten the simulation time.

4.2. Simulation Results

The simulation results are shown in Figure 10. From 0.00 s to 0.06 s, the H3C-BESS works in CC mode with reference battery current i_b^* fixed at 4 A. From 0.06 s to 0.18 s, the system works in CV mode with the reference battery voltage fixed at 100 V. Between 0.18 s and 0.30 s, the system works in CP mode, with grid current reference stepping from 1.34 A to 2.68 A (the battery current stepping from 2 A to 4 A accordingly). During the remaining 0.06 s, the system is maintained in CP mode but the active damping function is removed.

As it is shown in Figure 10, the battery current i_b can always track its reference i_b^* during the steady state and the dynamic response is very fast with few overshoots. The setup time of the battery current is less than 4 ms. As for the battery voltage, it can also reach its reference with the expected trajectory. This indicates that the satisfactory performance of battery current and voltage control can be achieved by the H3C-BESS with the proposed system control structure. Note that the total harmonic distortion (THD) of the battery current ripple during the steady state is 2.54%, which can be further reduced by increasing the filter inductor L_b or switching frequency.

As is the case for the battery current, the injected harmonic current i_{mid} can also tract its reference i_{mid}^* without steady-state error and with a fast dynamic response. This shows that the adoption of multi-VPI controllers has achieved the desired control performance. Note that the amplitude of the low-frequency component of i_{mid} is only half of the grid current. As the switching loss of the power converter is proportional to the operating current, it can be expected that the third harmonic injection circuit generates only a few switching losses. In addition, increasing the filter inductor L_{mid} or the switching frequency is also able to reduce the high-frequency ripple of i_{mid}.

The grid currents automatically vary with the working operation modes and the current demands. Moreover, the grid current amplitude responds very fast and smoothly to its reference. The setup time is also less than 4 ms. The THD of grid currents is 3.15%. This demonstrates that the H3C-BESS exhibits satisfactory control performance at the grid side, without the need for PI controllers to regulate the grid current. This is an evidence that the H3C-BESS has lower control complexity than the two-stage VSC-BESS which must have multi-loop control for grid currents. It can be observed from the waveforms that the grid currents have some distortions at the sector boundary. This is a common issue of the H3C based converter system and could be addressed with rearranged filters and an additional control [29].

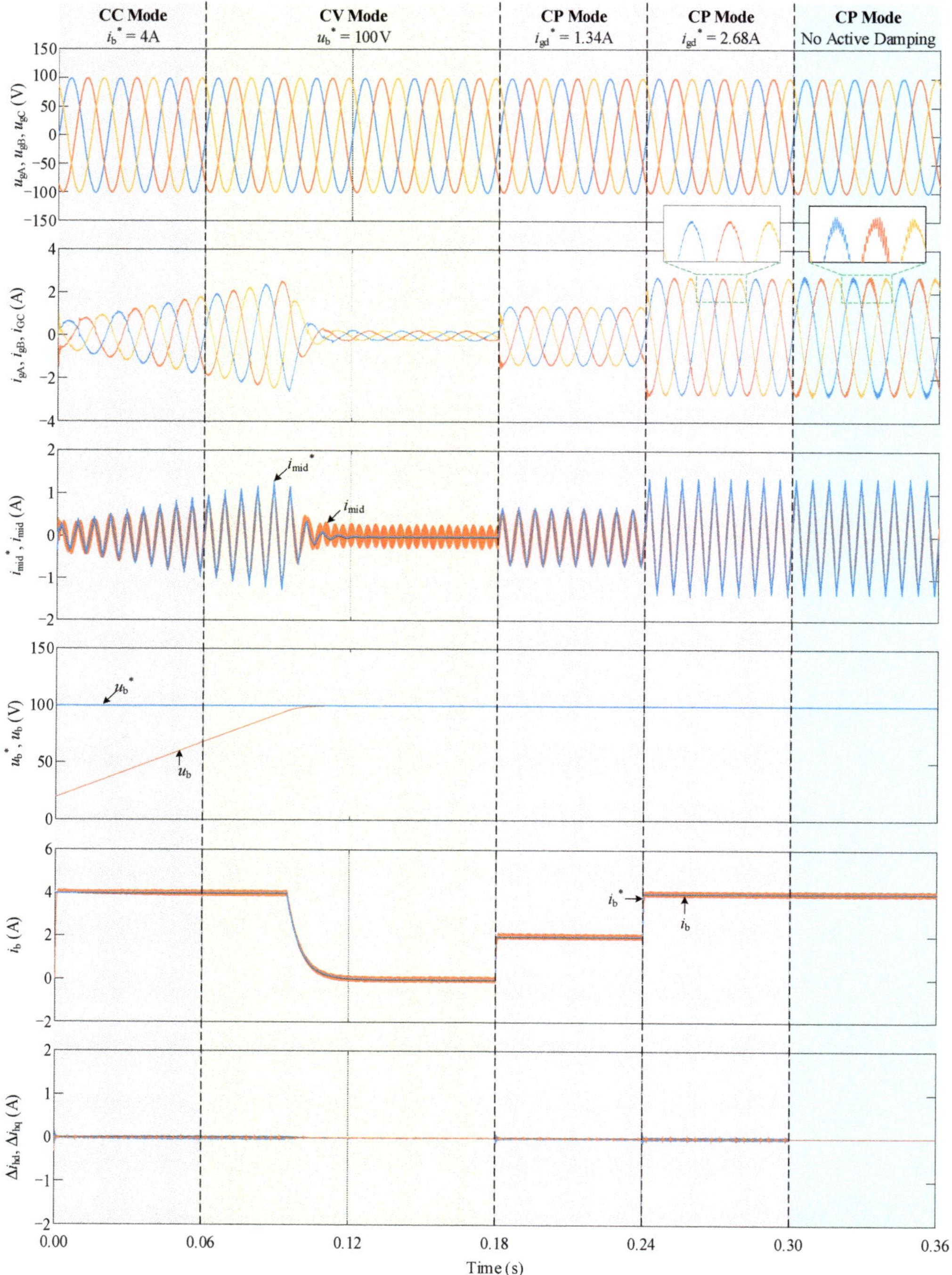

Figure 10. Simulation results of H3C-BESS under different working modes with and without active damping function.

In CC mode, the battery current is maintained as constant through the closed-loop control, while the battery voltage is increasing. Therefore, the active power flow to the battery is always increasing in CC mode, resulting in the increase in grid currents and the injected harmonic current. In practice, to avoid overcharging the batteries, CV control must be applied when the battery voltage reaches the predefined threshold. In CV mode, the battery voltage still increases and finally reaches the reference value. The battery current is maintained as constant firstly and then dropped gradually to zero until the battery voltage reaches the steady state. As a result, the transferred active power, grid currents and the injected harmonic current also increase firstly and then drop to zero. Therefore, all the waveforms in Figure 10 are consistent with the theoretical analysis.

With the proposed active damping control, the grid currents are very smooth and sinusoidal. The generated additional signals Δi_{hd}^* and Δi_{hq}^* are very small compared with i_b and i_{mid}, and thus have few influences on the normal operation of H3C-BESS. When the active damping control is disabled, significant oscillations are contained in the grid currents. This proves that additional damping control is necessary for the H3C-BESS and the proposed active damping control works correctly.

4.3. Experimental Results

In experiments, it was found that the overcurrent protection scheme of the prototype is easily triggered when the active damping function is disabled. Therefore, the active damping control is always included in the algorithm. The experimental results in CC mode and CV mode are shown in Figure 11. When the CC mode is activated, the reference battery current i_b^* is fixed at 4A. The actual battery current i_b reaches i_b^* very quickly without overshoot. In the CV mode, the voltage control loop is activated, which generates i_b^*. Due to the saturation of the PI controller, i_b^* is maintained at 4A for a short time and then decreases gradually to zero. Waveforms in the experimental results are almost the same with those obtained in simulation, which is another evidence that the H3C-BESS with the proposed control strategy works correctly in CC mode and CV mode.

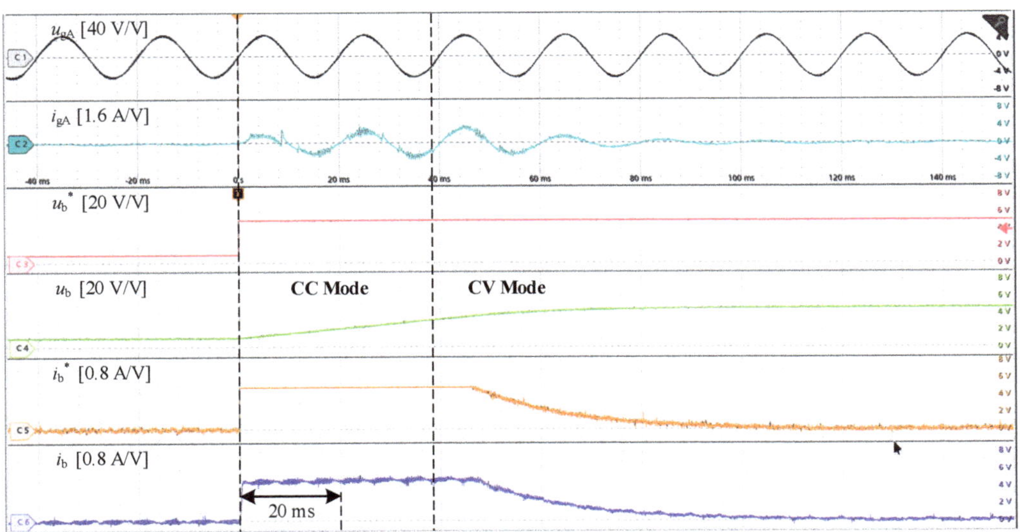

Figure 11. Experimental results of H3C-BESS in CC and CV modes.

Experimental results in the CP mode are shown in Figure 12. In this mode, the reference battery current i_b^* is calculated from the grid current reference i_{gd}^* in an open-loop way. i_b^* steps between 2 A and 4 A. It can be found that, in CP mode, both the battery current and the grid currents have a fast and relatively smooth dynamic response, proving that the proposed control strategy achieves good dynamic performance. Compared with the simulation results, more harmonics are contained in the grid currents. As presented in Part C Section 3, this is mainly because the reference value of the injected third harmonic current is calculated in an open-loop way. The inaccurate system parameters (especially the efficiency) result in the error of the reference harmonic current, which generates the distortion in grid currents. This is a common issue for this kind of converters. In previous publications [21,23], it can also be found that the grid currents are more or less distorted.

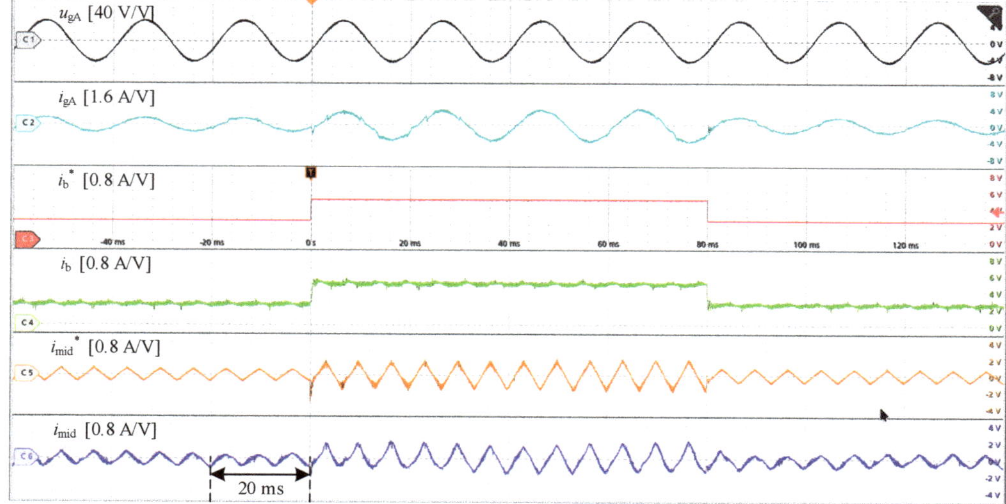

Figure 12. Experimental results of H3C-BESS in CP mode.

The measured efficiency of the prototype is about 92%, which can be improved by optimizing the selection of active and passive components. In summary, waveforms of the harmonic current are consistent with the theoretical analysis, demonstrating the operation principle of H3C-BESS.

5. Conclusions

This paper successfully applied the H3C to the BESS applications, forming a novel H3C-BESS. Compared with the commonly used two-stage VSC-BESS, the H3C-BESS has the capability to reduce the passive components and switching losses. The operating principle of H3C-BESS was analyzed in detail. The system control structure is proposed for the H3C-BESS. In particular, active damping control was realized through the grid current control, which could suppress the LC-filter resonance without the need of passive damping resistors. Simulation results proved that the proposed solution has a fast dynamic response with a setup time of less than 4 ms. In addition, the steady-state performance was also satisfactory with THDs of battery current and grid currents at 2.54% and 3.15%, respectively. The amplitude of the injected harmonic current is only half of the grid current, indicating that the corresponding circuit could generate low losses. Experimental results provided in this paper have also demonstrated the effectiveness of the H3C-BESS with the proposed system control strategy. It can be expected that the proposed H3C-BESS will be a promising solution in the BESS applications.

Closed-loop generation of the reference harmonic current is helpful to reduce the grid current distortions. In addition, the comprehensive comparison between H3C-BESS and the typical two-stage VSC-BESS is necessary to further demonstrate the advantages of H3C-BESS. These are all meaningful works in future studies.

Author Contributions: Conceptualization, Y.T. and J.L.; methodology, J.L. and Q.H.; software, X.F. and T.C.; validation, X.F. and J.L.; formal analysis, J.L.; investigation, Q.H.; resources, J.L. and W.C. data curation, J.L.; writing—original draft preparation, J.L.; writing—review and editing, Q.H.; visualization, X.F.; supervision, Y.T., W.C.; project administration, Y.T., J.L. and Q.H.; funding acquisition, J.L., Q.H., and W.C. All authors have read and agreed to the published version of the manuscript.

Funding: This work was supported in part by the National Key Research and Development Project of China under Grant 2020YFF0305800, in part by the Natural Science Foundation of Jiangsu Province of China under Grant BK20180396, and in part by the Zhishan Young Scholar Program of Southeast University.

Conflicts of Interest: The authors declare no conflict of interest.

References

1. Wei, S.; Zhou, Y.; Huang, Y. Synchronous Motor-Generator Pair to Enhance Small Signal and Transient Stability of Power System with High Penetration of Renewable Energy. *IEEE Access* **2017**, *5*, 11505–11512. [CrossRef]
2. Aliabadi, S.F.; Taher, S.A.; Shahidehpour, M. Smart Deregulated Grid Frequency Control in Presence of Renewable Energy Resources by EVs Charging Control. *IEEE Trans. Smart Grid* **2018**, *9*, 1073–1085. [CrossRef]
3. Azizivahed, A.; Arefi, A.; Ghavidel, S.; Shafie-Khah, M.; Li, L.; Zhang, J.; Catalao, J.P.S. Energy Management Strategy in Dynamic Distribution Network Reconfiguration Considering Renewable Energy Resources and Storage. *IEEE Trans. Sustain. Energy* **2020**, *11*, 662–673. [CrossRef]
4. Zhu, Y.; Liu, C.; Sun, K.; Shi, D.; Wang, Z. Optimization of Battery Energy Storage to Improve Power System Oscillation Damping. *IEEE Trans. Sustain. Energy* **2019**, *10*, 1015–1024. [CrossRef]
5. Faisal, M.; Hannan, M.A.; Ker, P.J.; Hussain, A.; Bin Mansor, M.; Blaabjerg, F. Review of Energy Storage System Technologies in Microgrid Applications: Issues and Challenges. *IEEE Access* **2018**, *6*, 35143–35164. [CrossRef]
6. Stecca, M.; Elizondo, L.R.; Soeiro, T.; Bauer, P.; Palensky, P. A Comprehensive Review of the Integration of Battery Energy Storage Systems into Distribution Networks. *IEEE Open J. Ind. Electron. Soc.* **2020**, *1*, 1. [CrossRef]
7. Vazquez, S.; Lukic, S.M.; Galvan, E.; Franquelo, L.G.; Carrasco, J.M. Energy Storage Systems for Transport and Grid Ap-plications. *IEEE Trans. Ind. Electron.* **2010**, *57*, 3881–3895. [CrossRef]
8. Tan, N.M.L.; Abe, T.; Akagi, H. Design and Performance of a Bidirectional Isolated DC–DC Converter for a Battery Energy Storage System. *IEEE Trans. Power Electron.* **2012**, *27*, 1237–1248. [CrossRef]
9. Chub, A.; Vinnikov, D.; Kosenko, R.; Liivik, E.; Galkin, I. Bidirectional DC–DC Converter for Modular Residential Battery Energy Storage Systems. *IEEE Trans. Ind. Electron.* **2020**, *67*, 1944–1955. [CrossRef]
10. Lo, K.-Y.; Chen, Y.-M.; Chang, Y.-R. Bidirectional Single-Stage Grid-Connected Inverter for a Battery Energy Storage System. *IEEE Trans. Ind. Electron.* **2017**, *64*, 4581–4590. [CrossRef]
11. Jayasinghe SD, G.; Vilathgamuwa, D.M.; Madawala, U.K. Diode-Clamped Three-Level Inverter-Based Bat-tery/Supercapacitor Direct Integration Scheme for Renewable Energy Systems. *IEEE Trans. Power Electron.* **2011**, *26*, 3720–3729. [CrossRef]
12. Maharjan, L.; Inoue, S.; Akagi, H.; Asakura, J. State-of-Charge (SOC)-Balancing Control of a Battery Energy Storage System Based on a Cascade PWM Converter. *IEEE Trans. Power Electron.* **2009**, *24*, 1628–1636. [CrossRef]
13. Maharjan, L.; Yamagishi, T.; Akagi, H. Active-Power Control of Individual Converter Cells for a Battery Energy Storage System Based on a Multilevel Cascade PWM Converter. *IEEE Trans. Power Electron.* **2012**, *27*, 1099–1107. [CrossRef]
14. Li, Z.; Lizana, R.; Lukic, S.M.; Peterchev, A.V.; Goetz, S.M. Current Injection Methods for Ripple-Current Suppression in Delta-Configured Split-Battery Energy Storage. *IEEE Trans. Power Electron.* **2018**, *34*, 7411–7421. [CrossRef]
15. Chen, Q.; Li, R.; Cai, X. Analysis and Fault Control of Hybrid Modular Multilevel Converter with Integrated Battery Energy Storage System. *IEEE J. Emerg. Sel. Top. Power Electron.* **2017**, *5*, 64–78. [CrossRef]
16. Li, N.; Gao, F.; Hao, T.; Ma, Z.; Zhang, C. SOH Balancing Control Method for the MMC Battery Energy Storage System. *IEEE Trans. Ind. Electron.* **2018**, *65*, 6581–6591. [CrossRef]
17. Zeng, W.; Li, R.; Cai, X. A New Hybrid Modular Multilevel Converter with Integrated Energy Storage. *IEEE Access* **2019**, *7*, 172981–172993. [CrossRef]
18. Judge, P.D.; Green, T.C. Modular Multilevel Converter with Partially Rated Integrated Energy Storage Suitable for Frequency Support and Ancillary Service Provision. *IEEE Trans. Power Deliv.* **2019**, *34*, 208–219. [CrossRef]
19. Feng, B.; Lin, H.; Wang, X. Modulation and control of ac/dc matrix converter for battery energy storage application. *IET Power Electron.* **2015**, *8*, 1583–1594. [CrossRef]
20. Varajao, D.; Araujo, R.E.; Miranda, L.M.; Lopes, J.A.P. Modulation Strategy for a Single-Stage Bidirectional and Isolated AC–DC Matrix Converter for Energy Storage Systems. *IEEE Trans. Ind. Electron.* **2018**, *65*, 3458–3468. [CrossRef]

21. Soeiro, T.B.; Vancu, F.; Kolar, J.W. Hybrid Active Third-Harmonic Current Injection Mains Interface Concept for DC Dis-tribution Systems. *IEEE Trans. Power Electron.* **2013**, *28*, 7–13. [CrossRef]
22. Schrittwieser, L.; Kolar, J.W.; Soeiro, T. 99% efficient three-phase buck-type SiC MOSFET PFC rectifier minimizing life cycle cost in DC data centers. *CPSS Trans. Power Electron. Appl.* **2016**, *2*, 1–8. [CrossRef]
23. Wang, H.; Su, M.; Sun, Y.; Yang, J.; Zhang, G.; Gui, W.; Feng, J. Two-Stage Matrix Converter Based on Third-Harmonic In-jection Technique. *IEEE Trans. Power Electron.* **2016**, *31*, 533–547. [CrossRef]
24. Wang, L.; Wang, H.; Su, M.; Sun, Y.; Yang, J.; Dong, M.; Li, X.; Gui, W.; Feng, J. A Three-Level T-Type Indirect Matrix Converter Based on the Third-Harmonic Injection Technique. *IEEE J. Emerg. Sel. Top. Power Electron.* **2017**, *5*, 841–853. [CrossRef]
25. Wang, H.; Su, M.; Sun, Y.; Zhang, G.; Yang, J.; Gui, W.; Feng, J. Topology and Modulation Scheme of a Three-Level Third-Harmonic Injection Indirect Matrix Converter. *IEEE Trans. Ind. Electron.* **2017**, *64*, 7612–7622. [CrossRef]
26. Zhang, G.; Wang, H.; Zhang, C. Modulated Model Predictive Control for 3TSMC. In Proceedings of the Energy Conversion Congress and Exposition (ECCE) 2020 IEEE, Detroit, MI, USA, 15–11 October 2020; pp. 4173–4177.
27. Lei, J.; Zhou, B.; Qin, X.; Wei, J.; Bian, J. Active damping control strategy of matrix converter via modifying input reference currents. *IEEE Trans. Power Electron.* **2015**, *30*, 5260–5271. [CrossRef]
28. Yagnik, U.P.; Solanki, M.D. Comparison of L, LC & LCL filter for grid connected converter. In Proceedings of the 2017 International Conference on Trends in Electronics and Informatics (ICEI), Tirunelveli, India, 11–12 May 2017; pp. 455–458.
29. Jankovic, M.; Darijevic, M.; Pejovic, P.; Kolar, J.W.; Nishida, Y. Hybrid three-phase rectifier with switched current injection device. In Proceedings of the 2012 15th International Power Electronics and Motion Control Conference (EPE/PEMC), Novi Sad, Serbia, 4–6 September 2012; pp. LS3e.2-1–LS3e.2-7.

MDPI
St. Alban-Anlage 66
4052 Basel
Switzerland
Tel. +41 61 683 77 34
Fax +41 61 302 89 18
www.mdpi.com

Energies Editorial Office
E-mail: energies@mdpi.com
www.mdpi.com/journal/energies